P9-AEZ-576

High Performance Polymers: Their Origin and Development

Proceedings of the Symposium on the History of High Performance Polymers at the American Chemical Society Meeting held in New York, April 15–18, 1986.

Editors:

Raymond B. Seymour
University of Southern Mississippi
Hattiesburg, Mississippi

Gerald S. Kirshenbaum
Celanese Engineering Resins
Chatham, New Jersey

Elsevier
New York • Amsterdam • London

7346-7546
CHEMISTRY

Elsevier Science Publishing Co., Inc.
52 Vanderbilt Avenue, New York, New York 10017

Distributors outside the United States and Canada:
Elsevier Applied Science Publishers, Ltd.
Crown House, Linton Road, Barking, Essex IG11 8J4, England

© 1986 by Elsevier Science Publishing Co., Inc.

This book has been registered with the Copyright Clearance Center, Inc. For further information please contact the Copyright Clearance Center, Inc., Salem, Massachusetts.

Library of Congress Cataloging in Publication Data

Symposium on the History of High Performance Polymers
(1986 : New York, N.Y.)
High performance polymers, their origin and development.

Includes indexes.
1. Polymers and polymerization – Congresses.
I. Seymour, Raymond Benedict, 1912– . II. Kirshenbaum, Gerald S. III. American Chemical Society Meeting (191st : 1986 : New York, N.Y.) IV. Title.
TP1081.S93 1986 668.9 86-24006
ISBN 0–444–01139–0

Current printing (last digit)
10 9 8 7 6 5 4 3 2 1

Manufactured in the United States of America

TP1081
S931
1986
Chem

High Performance Polymers

is dedicated

to our wives,

Frances Seymour

and

Edna Kirshenbaum

Contents

Indices

PREFACE

According to Johann Wolfgang Von Goethe's (1740-1832) Mineralogy and Geology, "The history of science is science."

A sesquicentennial later, one may state that the history of high performance polymers is the science of these important engineering polymers. Many of the inventors of these superior materials of construction have stood on the thresholds of the new and have recounted their experiences (trials, tribulations and satisfactions) in the symposium and in their chapters in this book.

Those who have not accepted the historical approach in the past, should now recognize the value of the historical viewpoint for studying new developments, such as general purpose polymers and, to a greater degree, the high performance polymers.

To put polymer science into its proper perspective, its worth recalling that historically, the ages of civilization have been named according to the materials that dominated that period. First there was the Stone Age eventually followed by the Tin, Bronze, Iron and Steel Ages. Today many historians consider us living in the Age of Synthetics: Polymers, Fibers, Plastics, Elastomers, Films, Coatings, Adhesives, etc. It is also interesting to note that in the early 1980's, Lord Todd, then President of the Royal Society of Chemistry was asked what has been chemistry's biggest contribution to society. He felt that despite all the marvelous medical advances, chemistry's biggest contribution was the development of polymerization.

Man's knowledge of polymer science is so new that Professor Herman F. Mark, the author of the first Chapter and a polymer science pioneer, has lived to see the first acceptance of the concept of the existence of chain-like macromolecules, the development of a multi-billion dollar industry based on general purpose polymers, and now a higher order of polymers which includes those which outperform all other materials of construction. Since synthetic polymers are so new it is important to record each historical breakthrough by the actual inventors and that is the purpose of this book and the ACS Symposium upon which the book is based.

Polyolefins, polyvinyl chloride, and polystyrene are good enough to justify annual sales of a score of millions of tons. Yet, they are not satisfactory for use at the temperature of boiling water. Thus, there was a need for polymers with greater strength and greater resistance to elevated temperatures. Fortunately, this need was fulfilled by the authors of the chapters in this book. Otherwise, our modern systems of transportation, construction, and communications would be nonexistant.

As stated in the first few chapters, much of the early developments in engineering plastics were associated with nylon-66. Nylon-66 fibers, which were invented by pioneer polymer scientist, Dr. W. H. Carothers, were one of the many polymers that were essential for the war effort by the Allied Forces during World War II.

The injection molding of these polymers provided a processable material of construction which could outperform classic materials, such as metals and ceramics. This important development, which has continued for a half a century, has led to other significant developments in other fields, which heretofore, would be considered impossible. As stated by Lawrence

Gillespie, "this concept launched an industry". In addition to spawning different types of engineering polymers, the development of Nylon-66 also led to the development of other nylons, such as Nylon-6, 11, and 12 which had properties that were somewhat different than those of the pioneer engineering polymer.

While less important in terms of the date of commercialization, polyphenylene sulfide, polycarbonates, the acetals, and polyphenlyene oxide are also important, both in their performance characteristics and in their unique chemistry. All of these materials, including nylon, are products of chemical reactions that are not found in classic textbooks.

Probably more important, unlike many classic reactions developed in university laboratories, most of these reactions which produce high performance polymers were developed in industrial research laboratories of firms such as DuPont, Phillips, G.E., Union Carbide, and Celanese.

The need for a thermally stable, solvent-resistant polymer was met by the use of a Wurtz-Fittig synthesis of polyphenylene sulfide. The need for an impact-resistant clear polymer was fulfilled by the reaction of phosgene and the sodium salt of bisphenol A. Admittedly, polymers of formaldehyde, which are found as a sediment in formalin bottles, was investigated by Nobel Laurate Hermann Staudinger in the 1920's but this polymer lacked integrity until DuPont and Celanese chemists produced thermally stable, injection moldable polyoxymethylene.

As described in later chapters, the polyethylene terephthalate fiber, like nylon fiber, was not readily moldable. However, moldability was accomplished by increasing the number of methylene groups present, as in polybutylene terephthalate, and by utilizing new knowledge of the crystallizability of thermoplastics. As a result of advances in molding technology, mechanical engineers can take well established polymers like PET, learn how to biaxially orient them to form a bottle that provides a carbonation barrier and opened a whole new industry. This development is described by N. Wyeth. Undoubtedly, the newly developed knowledge needed for the injection molding of less complex aromatic polyesters, paved the way for the injection molding of polyarylates which are discussed in a chapter bv Robeson and Tibbitt.

As discussed by Dr. A. S. Hay, polyphenylene oxide was extremely difficult to mold but a chance discovery of blends with polystyrene, made the commercialization of this important polymer feasible. The development of heat resistant polysulfones and polyaryletherketones are excellent examples of the application of good industrial research chemistry. Fortunately, the stories of these and other developments are told by the clever scientists who were responsible for these dramatic breakthroughs in polymer science.

The chapters on fluoropolymers are of equal interest and perhaps more intriguing. If Dr. Roy Plunkett had stored his tetrafluoroethylene in larger cylinders, he and his associates would probably have been injured by the explosion associated with the exothermic, instantaneous polymerization of this monomer. More important, if he had discarded the single cylinder which had no gas, he would not have discovered Teflon.

The latter, like nylon, and polyethylene, was essential during World War II for parachutes, radar cable and atomic energy. Today's engineers would probably find it extremely difficult to imagine a world without polyfluorocarbon plastics, elastomers, fibers and coatings.

Thermosets may be omitted in some lists of high performance polymers but they meet specifications and would not be used today if they were not high performance material. This criterion applies to carbon fibers, barrier resins, thermotropic polyesters, interpenetrating network polymers and many specialty elastomers. Likewise, the introduction of UV/EB curing techniques has brought about a new era in the polymer industry.

Because polymer science is a relatively new field, it continues to grow at an exciting pace. The thrust in application opportunities for polymers continues to center on the use of these materials in mechanical applications, improved membranes and medical applications. The currently used terms of 'engineering,' 'high performance' and 'advanced' are thus nebulous, yet, with evermore demanding meanings, the new thermoplastics and thermosets are synthesized and find applications. These developments are driven by the prospects of greater applications in aircraft, automobiles, appliances, and in general construction.

The approaches to the new polymer developments cover a multitude of areas such as new polymer systems such as the liquid crystalline polymers, new applications of old polymers such as carbon fibers which were used in Edison's light bulbs in the 1800's to their use today and in the future as advanced composites in aerospace structures and the exploding field of polymer blends and alloys.

These high performance polymers not only meet unusual requirements for long term durability at elevated temperatures, they may also have unique and diverse characteristics such as conductivity, insulating properties, chemical resistance, self extinquishing, selective gas barrier properties and are also used in normal applications as films, fibers, elastomers, adhesives, and coatings as well as injection moldable materials of construction.

Raymond B. Seymour
Gerald S. Kirshenbaum

High Performance Polymers: Their Origin and Development

INTRODUCTION

High Performance Polymers--- Natural and Synthetic

H. Mark
Polytechnic University
Brooklyn, New York

1. How Nature Does It

Nature has been designing and making high performance polymers - organic and ceramic - since several hundred million years as a crucial service for plants and animals. The base materials (all somewhat compatible with water) were proteins and carbohydrates (or copolymers thereof) together with a variety of clays.

Those properties which had to be brought up to exceptionally high levels were:

Modulus (hardness, rigidity) as for ivory, antlers, horns, wood, nuts and shells of all sizes

Tensile strength as for spider web, silk, flax, sisal, bast fibers

Toughness (work to rupture) as for the skin of shark, alligator, root of sequoia, grape of coconut stem

Elasticity (resilience) tendon (Kangaroo tail), fish-bladder, palm tree branches

Durability (eventually for several thousand years) wood (sequoia gigantua) skin and hair (mammoth), bones (dinosaurs) and large shells.

The construction principles were:

for linear systems (fibers, tendons, strands) very high molecular weight (up in the millions) flexible (proteins) or rigid (cellulosics) chains and for planar structures (skin, membranes, leaves) plyed and crosslinked layers (cellulosics) and beta pleated sheets (proteins).

There is also abundant application of intricate supermolecular (textural) designs, which was attainable by the slow growth processes of the different individual systems.

In the course of this impressive symposium it will become evident that our synthetic high performance composites are essentially using the same basic building principles; we even possess a much larger variety of chemical species to solve our problems but we must take into account one serious additional characteristic for them, namely serviceability at high temperatures - up to 1000°C for long periods; eventually for years.

A typical example of a natural composite is wood, of which there exist at present about two trillion tons on earth represented by many thousand different species. They all consist essentially of two major components. (1)

Copyright 1986 by Elsevier Science Publishing Co., Inc.
High Performance Polymers: Their Origin and Development
R.B. Seymour and G.S. Kirshenbaum, Editors

Cellulose, a fibrous, semicrystalline "filler" consisting of moderately rigid chains of very high (somewhere in the millions) molecular weight which is embedded in lignin, an amorphous, three-dimensional network of several aromatic compounds. Between these two essential components there exists a variety of resinous materials which provide for interlayers and eventually lead to true covalent bonds between cellulose and lignin. Such systems may be very light, resilient and strong such as the stem of a wheat plant or very heavy and hard such as the trunk of an ivory tree. The fatigue resistance of wood under ambient conditions is remarkable; the oldest living system on earth are trees, not one species but many hundreds of them.

This symposium is devoted to the description of the preparation and properties of synthetic high-performance polymers which are used in the form of fibers, films or manufacture of objects of many sizes and shapes. Their basic components are similar to those of their natural counterparts: long chain molecules, flexible or rigid, capable of aligning and crystallizing and of amorphous resinous networks in which the fibers are embedded. The fibrous and resinous species may be used alone or they may be assembled to multicomponent systems. Their chemical nature varies over the entire realm of organic and inorganic chemistry: from polyolefins to silicon carbide and from epoxides to bismaleiimides. The individual components may be homopolymers or copolymers and even more complex systems such as segmented and grafted macromolecules. Altogether four principles are used in the molecular design of high performance polymers: orientation, crystallization, crosslinking and the use of rigid chains. We have already encountered all of them in the natural protagonists.

Let us first take a closer look at the fibers:[2,3] two methods are being presently used to reach high performance in modulus (rigidity) and tensile strength; the elimination of all chain folding in flexible macromolecules and the spinning of rigid chains from liquid crystal (nematic) systems. In both cases,values around 1000g/dtex have been obtained for the tensile modulus and up to 35g/dtex for the tensile strength. These values are somewhat higher than those of the corresponding natural systems, the superextended content of the silk worm gland and the cellulosic fibers of Sisal, Heneguen, Flax, Hemp and Ramie.

Careful observations of the fiber forming process of silk have revealed that the unfolding and realignment of length of the fibroin molecule from the soluble random coil conformation to the insoluble solid beta-pleated sheet structure is fundamental for the fiber forming process. Kataoka (4) has found that it occurs in two stages, in the first stage the critical shear rate for the unfolding (beta-transition) of the protein chain is lowered by dehydration as the material moves from the posterior to the middle section of the gland. Upon reaching the anterior section "nuclei" for the beta- form are produced by the orientation of chain segments in the direction of the flow. In the second stage orientation and crystallization occur as a result of the large shear stress at the narrowest portion of the spinneret.

TABLE I

Average Tensile Moduli (rigidities) of Selected Fibers and Wires

	kg/cm^2	psi	N/dtex	g/den	g/dtex
silk normal		160,000		80	
silk superdrawn		920,000		460	
Sisal		500,000		250	
Lilienfeld silk		700,000		350	
Polyester		200,000		100	
Spectra 900		1.8 x 10^6		900	
Kevlar 49		2.2 x 10^6		1100	
Twaron		1.9 x 10^6		100	
Graphite	2.1 x 10^6	32 x 10^6			
S-glass	7.5 x 10^5	10 x 10^6			
Steel	2 x 10^6	30 x 10^6			

Approximate conversion factors to other units.

1N (Newton) = 100g
1g per den = 0.9g/dtex
1g per den = 2,000 psi
1kg/cm^2 = 15 psi
1kg/mm^2 = 100 kg/cm^2 = 1500 psi = 0.7 g/den

TABLE 2

Average Tensile Strengths and Elongation to Break of Selected Fibers and Wires

	kg/cm^2	psi	N/tex	g/den	g/tex	Elongation to break %
Silk normal		9,000		4.5		25
Silk superdrawn		130,000		15		3
Sisal		150,000				2
Lilienfeld silk		13,000		6.5		7
Polyester		9,000		4.5		25
Spectra - 900		60,000		30		5
Kevlar 47		48,000		24		5
Twaron		40,000		20		3
S-glass	18,000	280,000				12
Graphite	20,000	300,000				0.8
Steel	50,000	800,000				10

The principle of using rigid or semi-rigid chains in the formation of hard, strong and tough systems is extensively used in all plants. Obviously there is no distinct "spinning process" here but the macromolecules of the system slowly polymerize as the plant grows. Crystalline bundles of very long (MW up in the millions) relatively rigid chains of alpha cellulose are produced and embedded in an amorphous matrix of lignin with a variety of hemicelluloses serving as adhesive inter layers. During the formation of the structure the growing parts are highly swollen in water in order to permit the transportation of monomers for the chain growth. In most cases the ultimate textures of the resulting cell walls are extremely complex consisting of criss-crossing strands which are assemblies of cellulose materials and exist in right and left handed spirals providing for extra-ordinary mechanical properties. Orientation, crystallization and microfibrillar ply formation are the elements of these natural high performance systems.

It is interesting to remember that as early as 1926 Dr. Leon Lilienfeld prepared viscose rayon filaments with unusually high modulus and high tensile strength. He used highly xanthated cellulose, the chains of which were straightened by their polyelectrolytic character. The solution was - under tension - spun in concentrated sulfuric acid which produced instantaneous decomposition of the xanthate and maintained (at least partially) the anisotropy of the original spinning solution. The resulting filaments reflect this unidirectionality in tensile modulus and strength.

Thus the natural prototypes of high performance fibers provide a good recipe for the necessary steps of operation. In several instances they have been closely imitated with one important exception. Most natural components - protein and carbohydrates - have very high molecular weights (sometimes high as millions); all synthetic materials must be used in a much lower molecular weight range because of processing difficulties; because we do not yet know, on a broad front, how to dissolve, spin, cast and extrude economically and effectively macromolecules having molecular weights up in the millions.

2. How We Now Try To Do It

The program of this unique "macrosymposium" resembles a rich bunch of beautiful flowers each one having its own look, color and flavor. Forty three papers will be presented by lecturers who came from 15 different companies and academic institutions. Fittingly for the historic character of the show, the first two lecturers will give an account of two polymers which, almost 60 years ago - in the late 1920's - were prepared and studied on a laboratory scale by Wallace H. Carothers of DuPont and Paul Schlack of I.G. Farben: the two nylons - 6,6 and 6 - which inaugurated the age of the "Man-Made" synthetic fibers a decade later. Soon thereafter France joined the club with nylon 11 and 12. The main thrust of these early efforts was aiming at the replacement of silk, a high priced material of superior quality and performance. And, indeed, the first synthetic fiber formers, were costly newcomers. Polymer Science and Engineering was at its infancy: novel and unusual monomers had to be prepared with purity levels unheard of in earlier organic chemistry. Polycondensation reactions demanded the working with highly viscous melts for many hours at temperatures (up to 300°C) which were never used before at this level for such long times.

New concepts came up and had to be adhered to: linearity, regularity, crystallinity - penetration into the high molecular weight range (well above 10,000) with a minimum amount of smaller species (oligomers). All this had to be recognized and intergrated into the strategy for further progress to prepare and use other products of the same character and utility.

The lecturers on Polyphenlene Sulfide (Phillips) Aromatic Polycarbonate (GE), Acetal Homopolymers (DuPont), PBT (Celanese), PET (DuPont), Polysulfone and Polyarylate (UCC), PTFE (DuPont), PDFE (Pennwalt), Polyimides (Ciba-Geigy) and Polybenzimidazoles (Celanese) will discuss other linear homopolymers representing a wide variety of chemical compositions and uses.

The remaining reports (26) are going to deal with copolymers, block- and graft interpolymers, polymer blends, liquid crystalline systems, interpenetrating networks and other related compositions.

It will be a highly instructive and stimulating symposium for which we all are cheerfully expressing our thanks to its organizers, Professor Raymond B. Seymour of the University of Southern Mississippi and Dr. Gerald Kirshenbaum of Celanese.

REFERENCES

(1) Compare "Future Sources of Organic Raw Material" L. St. Pierre, C.R. Brown; Pergamon Press, Elmsford, N.Y. 1980.

R.H. Marchessault; Chem. Tech. September 1984; pages 542-552.

Specifically: Wood; "The Chemistry of Wood" B.L. Browning; J.Wiley N.Y. 1963,

Silk, Wool, Sisal and other natural fibers in "Fiber Chemistry"; M. Lewin and Eli M. Pearce Marcel Dekker, N.Y. 1985; pages 647 to 649 and 727 to 803.

(2) Compare H. Mark, "Physik und Chemie der Zellulose" in R.O. Herzog "Technologie der Textiefasern", Springer Berlin 1932 and Emil Ott; "Cellulose and Cellulose Derivatives" Interscience Publishers, New York 1943.
Specifically: Lilienfeld Rayon; in Mark, pages 27, 33 and in Ott pages 820 to 850.
Also DRP 448984 (1927)
USP 1,658 607 (1928)
1,722 929 (1929)
1,771 462 (1930)

(3) For Tables 1 and 2 compare: "Matthews Textile Fiber" Sixth Edition by H.B. Mauersberger, J. Wiley and Sons, 1954;
R.W. Moncrieff; "Man-Made Fibers" Fifth Edition J. Wiley and Sons, 1970.

(4) See, for instance: D. Coleman and F.D. Howit "Symposium on Fibrous Proteins", Soc. Dyers and Colorists, Bradford, UK (1946) pl 144.
Also, K. Kataoka: J.Soc.Fiber Science and Technology of Japan 34, 80 (1978) and J. Magoshi et al. IUPAC Fiber Symposium Oxford, UK (1982), page 614.

ENGINEERING PLASTICS: THE CONCEPT THAT LAUNCHED AN INDUSTRY

by
Lawrence H. Gillespie, Jr.
Engineering Polymers Division
E.I. du Pont de Nemours and Company, Inc.
Wilmington, Delaware

Engineering plastics: a family of materials based on technology that truly launched an industry.

Whether one regards the development of these remarkable materials as revolutionary or evolutionary, the fact remains that engineering plastics have changed the way we live. Indeed, continuing technological advances will have an even greater impact on our lives as engineering plastics further penetrate markets now dominated by metals.

Throughout history, we have witnessed trends where one material has evolved to replace another for efficiency, performance or economic reasons. Consider the fact that the ages of man to date have been broadly defined in terms of their key materials -- the Stone Age, the Bronze Age, the Iron Age and the Age of Steel.

These designations, of course, substantially simplify the total materials technology practiced during those periods. Man also made substantial use of nature's polymers in the form of wood, leather, bone, horn, shell, animal ligaments and tissues to make tools, clothing, houses, vehicles and weapons.

In the more recent past, chemistry has played a key role in the evolution of materials science, serving as a powerful engine of change through extensive research efforts into polymers and the way in which they can be combined to create even more sophisticated products.

Although concepts leading to the manufacture of plastics evolved during the 1860's, the official appearance is generally regarded as 1870 when John Wesley Hyatt patented celluloid. This material found many uses, such as stays for shirt collars and cuffs, but had a major deficiency in that it burned easily. In 1909, Leo Baekeland introduced a fire-resistant, all synthetic plastic called Bakelite which pointed the way to the invention of dozens of successor thermosets.

Over the next several years, chemists in Germany, Britain and the United States invented different varieties of thermoplastic materials, including acrylics, vinyls and cellulosics.

In the early 1930's, the pioneering work of Wallace Carothers and his associates at Du Pont gave us polymers such as neoprene synthetic rubber and nylon fibers. As defined at the time of its discovery, nylon is a generic term for any long-chain synthetic polymeric amide which has recurring amide groups as an integral part of the

Copyright 1986 by Elsevier Science Publishing Co., Inc.
High Performance Polymers: Their Origin and Development
R.B. Seymour and G.S. Kirshenbaum, Editors

main polymer chain, and which is capable of being formed into a filament in which the structural elements are oriented in the direction of the axis.

This is the foundation upon which engineering polymers were built.

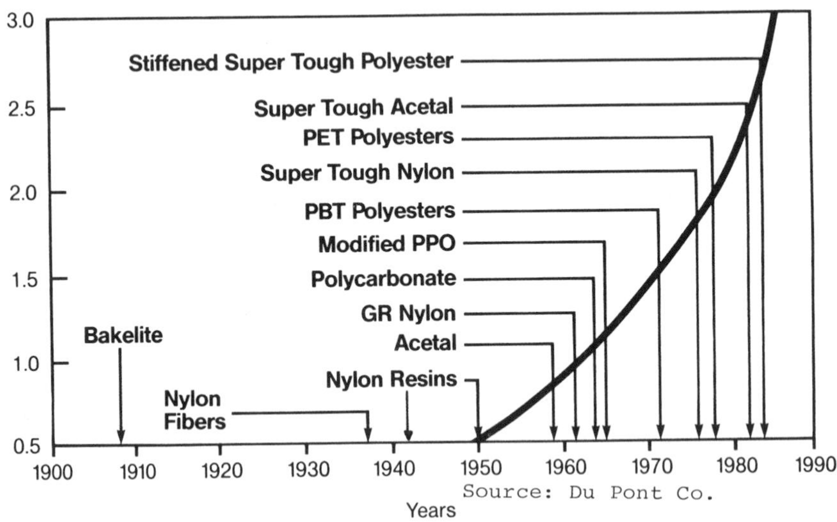

The first commercial use of nylon was in bristles for toothbrushes and hair brushes. Two years later, in 1940, nylon made its nationwide debut in women's hosiery. Its use in parachutes followed soon after.

Extending the processes developed to make nylon filaments and bristles, Du Pont scientists began investigating the potential use of nylon in a three-dimensional, rather than one dimensional form. In 1941, they found that it made an excellent molding powder and set about developing a feasible commercial means of extruding and molding it into small parts such as gears. After overcoming problems posed by the fact that nylon required special handling and special equipment when compared to conventional thermoplastics, Du Pont's Plastics Department began to develop a business in nylon resin. Commercial sales were interrupted from 1942 to 1945, when all nylon was dedicated to the war effort. Following the war, nylon was reintroduced to the public and substantial sales developed.

With the increasing importance of nylon molding resins, Du Pont developed the trademark "Zytel" in 1950 to distinguish it from the textile fiber product. Du Pont also developed a new concept -- that of engineering plastics -- to distinguish nylon from existing general purpose plastics, or what we now refer to as industrial or commodity plastics, which had major deficiencies. These deficiencies included the flammability of acetates, the brittleness of acrylics and the general structural weakness of polyethylene and vinyl, as well as the limited operating temperature range of all of these materials.

Du Pont's vision in the early 1950's was of a synthetic polymer which could be molded, shaped, cut, formed and used like metal; in fact, it would have properties as predictable as metal so that engineers could design new structures just as confidently with plastic as with metal. Resins were envisioned that could be described as "metal-like," as having "guts." To support this commitment, Du Pont and other resin producers made a heavy investment in resources to accelerate innovation and growth.

From the beginning, the primary force driving the adoption of engineering plastics was the replacement of metal, glass, wood and other natural materials. Among the reasons for their impressive inroads were an attractive balance of properties, favorable material and process costs, energy efficiency and less labor intensity in both assembly and finishing.

Within 10 years -- from 1959 to 1969 -- five new generics were introduced. In 1959, Du Pont introduced polyacetal, and Bayer and General Electric commercialized polycarbonate. Polyphenylene oxide was introduced by General Electric in 1962 and one of the first blends -- modified polyphenylene oxide -- debuted in 1965. Celanese commercialized polybutylene terephthalate (PBT) in 1968. The 1960's also saw companies such as ICI and Du Pont begin to produce nylon reinforced with small amounts of glass.

The definition of engineering plastics quickly evolved to mean the family of thermoplastic polymers to which standard metal engineering equations could be applied. They are capable of sustaining high load and stresses and can perform for prolonged periods over wide temperature ranges and in difficult chemical and physical environments. Their predictable engineering properties include electrical, thermal, chemical and physical properties -- especially strength, stiffness, elongation, wear, fatigue, impact resistance, flame retardancy and dimensional stability.

Until the 1970's, engineering resins were generally 90 percent polymer. During that decade, polymer scientists learned that the properties and performance of base polymers could be enhanced and extended by adding fillers and reinforcements in amounts up to 50 percent by weight. These fillers include particulate and fibrous forms of glass, carbon and a wide variety of chemicals and naturally occurring minerals which protected against deterioration caused by water, heat, solvents and other chemicals and radiation ... either alone or in combination.

Breakthroughs made possible by advances in alloying, blending and composites technology included supertough nylons, acetals and polyesters from Du Pont. Technologies such as these have enabled polymer producers to tailor properties to the specific requirements and preferences of customers in an increasingly competitive business environment.

In fact, technology developments emerging through alloys, blends and composites are likely to be the major factor driving engineering plastics' performance toward metals equivalency and a market opportunity delineated by the Fisher-Pry substitution theory to be as much as 18 billion pounds annually by the year 2000.[1]

ENGINEERING POLYMERS SUBSTITUTION FOR SELECTED METALS* (Volume Basis)

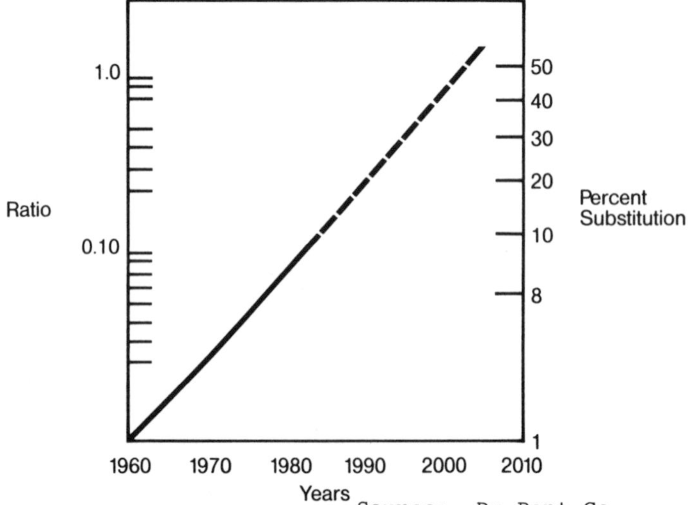

Source: Du Pont Co.

*Die Cast Metals; Rolled and Structural Steel

This view of advancing technology as a substitution process can be viewed as either evolutionary or revolutionary, depending on the time scale of the substitution. This model was developed by two General Electric scientists and is based on three assumptions:

- First, many technological advances can be considered as competitive with a current method of satisfying a need. Also, when a new technology is first introduced, it is less well developed than the older method it will replace. Therefore, it has greater potential for ongoing improvement and reduction in final part cost.

- Second, if a substitution has progressed as far as a few percent, it will proceed to completion.
- Third, the rate of substitution of new for old is proportional to the amount of old remaining to be substituted.

Based on industry data, substitution of the metals segment of die-cast metals, hot rolled steel and structural steel by engineering plastics is currently about 12.5 percent on a one-for-one volumetric basis. This segment currently accounts for 40 percent of all metals.

Using the Fisher-Pry theory to project the potential percent substitution beyond the year 2000, calculations place engineering plastics' substitution of selected metals by that time at about 50 percent, which translates to about 20 percent of all metal -- an 18-billion-pound potential opportunity for engineering plastics.

With estimates for 1986 volume of engineering polymers to be just under 4 billion pounds, a four-to-five-fold increase in plant and processing equipment will be necessary over the next 14 years to accommodate such market growth.

Because of further technology advances expected from this young industry, we believe engineering plastics will rapidly approach the property and performance parameters of metal and drive substitution further up this curve.

For example, in terms of stiffness in pounds per square inch, engineering plastics today are positioned between metals, such as magnesium, aluminum and carbon steel, and commodity plastics such as polyethylene and plasticized PVC. With advancing alloy and blend technology, plastics are approaching the stiffness properties of metals.

The toughness performance of all base resins is also advancing as we increase our understanding of the mechanisms involved. Rapid advancements in technology should bring the notched izod toughness of plastics close to the metals' value of 100 within the next 15 years or so. For example, notched izods of 75 with 50 percent volume glass loaded polyolefins have already been demonstrated in the laboratory. We believe that notched izods of 50 to 60 with only 20 percent reinforcement levels in nylon or acetal are achievable by the year 2000. These systems are well beyond the bounds of the alloy, blend and rubber-toughened polymer systems of today.

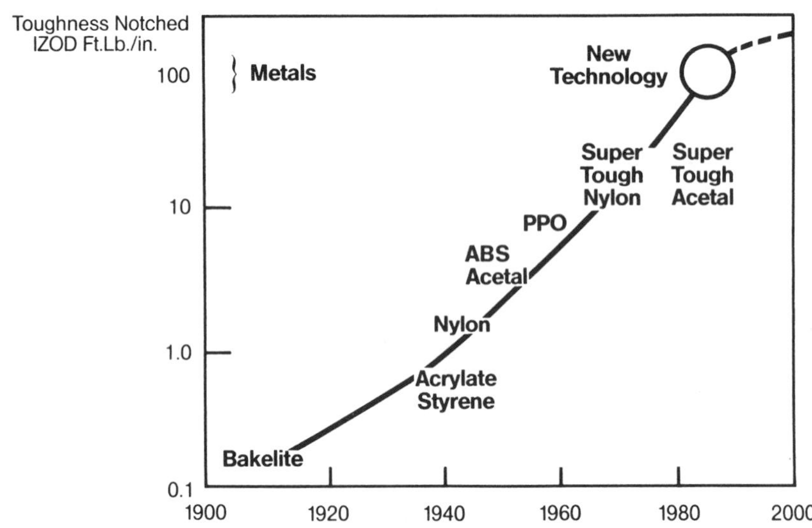

Source: Du Pont Company

Going beyond performance and properties, other major factors driving engineering plastics' replacement of metal include ease of processing and assembly, and parts consolidation. These factors all result in lower final part costs. Also, compared to the more mature metals industry, the plastics industry is young. As a result, we fully expect new technologies for producing resins, processing those resins into parts, and assembly/finishing operations will greatly improve productivity and economy.

Several markets hold large potential for engineering plastics and their extension into advanced composites. These include transportation, electrical, electronic, consumer, building/construction and medical. Growth will be accelerated by adoption in large structures in the building/construction industry and in automotive exterior body parts. Other markets that will maintain slow but steady growth are packaging, appliances, sporting goods, industrial machinery and business equipment.

What does the future hold for the producers of engineering plastics? What are the requirements for continued profitable participation in this highly competitive field?

The major players include companies such as Du Pont, General Electric, Bayer, Allied, Hoescht, Rhone Poulenc, BASF, Asahi Chemical, Celanese, ICI, Mitsubishi, Monsanto, Akzo, and Borg Warner.

The more successful companies will be those who are willing to make a major commitment of the resources necessary to pursue and capitalize on the total opportunity in this increasingly complex business. Success not only requires a fully integrated manufacturing position which starts with natural gas and oil ... world scale facilities for intermediates, polymer and compounded products ... demonstrated long term leadership in basic polymers and the technology of alloys, blends and composites ... but also an aggressive worldwide marketing organization with end-use expertise and strong technical support. For in handling engineering plastics, the quality of engineering is just as important to the marketplace as the quality of the plastic itself. Especially critical are the financial strength to invest the capital and resources required for worldwide leadership, as well as a strong management commitment to the future.

All of these factors have become even more important as engineering plastics have proliferated. Today, for example, there are significant overlapping regions in performance ranges -- a fact which intensifies competition in the marketplace.

The considerable upside potential for engineering plastics can be realized through unwavering dedication to all of the thrusts cited here -- technology development, property enhancement, process improvement and resource allocation. The revolution also depends to a large degree on conceptual changes in how we design, engineer and manufacture -- essentially, how we conceive and how we execute.

Great potential exists for the continuing evolution of engineering plastics and there will be, we expect, far reaching consequences. The opportunities are great and the times ahead promise to be both challenging and exciting.

[1]J. C. Fisher, R. H. Pry, "A Simple Substitution Model of Technological Change," American Elsevier Publishing Co. (1971)

Engineering Thermoplastics

THE HISTORY AND DEVELOPMENT OF NYLON-66

Melvin I. Kohan
E. I. Du Pont de Nemours & Company, Inc.
Wilmington, Delaware

ABSTRACT

In this review of the history and development of poly(hexamethylene adipamide), nylon-66, the following topics are briefly examined with emphasis on the plastic:

Discovery
Intermediates
Polymerization
Commercial Introduction and Growth
Nylon at War
Properties and Applications
The Business
Conclusion

Discovery

In December, 1926, Dr. C. M. A. Stine, Director of the Du Pont Chemical Department, explained to the Executive Committee his 1927 budget request for support of "fundamental research", the "principal raw material" of applied research characteristic of most industrial laboratories [1]. This led to the invitation to Dr. Wallace H. Carothers, an instructor at Harvard University, to head a research effort in organic chemistry.

Carothers, although relatively young and inexperienced (Ph.D., University of Illinois, 1924, at the age of 28, Ref. 2), had already authored eight papers and was given a free choice of area of study. He and his team of six chemists began their now classic studies on polymerization in 1928 [1]. They viewed a tiny polymer industry heavily dependent on modified natural products, and they saw a technical establishment still not fully convinced by the work of Staudinger, Svedberg, Meyer, and others that polymers were valence-bonded, high molecular weight entities and not special aggregates of small molecules (Table MK-1). A few trade journals in the paint, plastics, rubber and textile industries and five books (5) existed, but no scientific publications concerned with polymers (compared to over 100 periodicals and serial publications and innumerable books and handbooks of today). Even the definition of the term, polymer, was at issue [6].

Within a year Carothers had published his theory of polycondensation and compared it with polyaddition [7]. His initial efforts on polyesters were published at the same time [8], and this traditional and incontrovertible organic synthesis of high molecular weight species removed any lingering doubts as to the reality of such species [9]. Discussion of polycondensation, polymerization of ring and unsaturated compounds, and the

Copyright 1986 by Elsevier Science Publishing Co., Inc.
High Performance Polymers: Their Origin and Development
R.B. Seymour and G.S. Kirshenbaum, Editors

TABLE MK-1. Historical Perspective - 1928

1835	- Pelouze nitrated cellulose
1839	- Goodyear vulcanized rubber
1860's	- Parkes, Hyatt, others - development of celluloid
1870	- Hyatt, first practical process for celluloid
1894	- Cross and Bevan - cellulose triacetate
1899	- Kritsche and Spitteler - casein formaldehyde
1901	- Rohm studied polyacrylates
1905	- Miles developed moldable cellulose diacetate
1909	- Baekeland made the first synthetic polymer, phenol formaldehyde resin
1920	- Staudinger supports long chain structure for polymers. Begins studies, uses solution viscosity to promote concept in rubber, polyoxymethylene, and polystyrene.
1923	- Svedberg uses the ultracentrifuge to estimate a protein molecular weight of 50,000
1927	- Commercial production of poly(methyl acrylate)
1928	- Meyer, X-ray studies indicate celluosic fiber to consist of chains of glucose units
	- Processable PVC via copolymerization
	- Rossiter, urea formaldehyde resin

Source: Ref. 3,4

polymeric nature of cellulose and rubber, and precise definition of terminology still valid today were provided in a classic review only two years later [6]. Our subject is nylon-66, an invention of Carothers, a man whose broad talent and scientific contribution deserve more than this brief mention. He also was the discoverer of neoprene; his fertile mind yielded 54 papers and 52 patents in nine years at Du Pont before his untimely death in 1937 [10]. He also inspired an Ohio State alumnus to join him in 1934. This was the beginning of Paul Flory's theoretical contributions for which he received the Nobel Prize in 1974.

The polyesters made from hydroxy acids or dibasic acids and glycols were low melting solids (m.p. $\leq$ 108°C) of molecular weight in the 2500 to 5000 range. The only aromatic constituent used was phthalic acid. A hard material of moderate molecular weight was then obtained from epsilon-aminocaproic acid. A major advance resulted from use of the so-called molecular still in 1930. This was simply a device that facilitated removal of traces of water which Carothers recognized could be in equilibrium with the polymer and so be limiting conversion to higher molecular weight. Use with polyesters provided molecular weights of 10,000 to 25,000. Without this new-found ability to make so-called "superpolymers" the discovery of commercially useful polycondensates would not have been possible. The cold-drawability of these hard "superpolymers" to lustrous, tough fibers was an exciting observation, first made by co-worker Julian Hill in April, 1930, that provoked the search for a commercially feasible fiber. The low melting points of the polyesters directed attention to the superpolymer from poly(aminocaproic acid) which was subsequently judged to be intractable. Copolyester-amides were explored as a compromise but were found to be too low melting.

The polyamide from 9-aminononanoic acid was then made and found to have a too low melting point of 195°C but fiber properties comparable to silk. This lead initiated extensive studies on polymers from aminoacids or diamines and dibasic acids. On February 28, 1935, the superpolymer from hexamethylenediamine and adipic acid was first made. Its properties and melting point over 260°C, safely higher than ironing temperature, labeled it as the logical fiber candidate. Many a talent would have to be called on, however, for the difficult, unchartered course from laboratory to commercial manufacture of the first synthetic fiber [11,12]. Some of the problems are shown in Table MK-2 together with the time and effort to overcome them.

INTERMEDIATES

Ingredients, product, and names, both old and new, are shown in Fig. MK-1 together with their commercial status in 1935. As noted, new syntheses for both hexamethylenediamine (HMD) and adipic acid (ADA) were necessary. While these were in development Carothers and his team explored many alternatives but could find none superior to nylon-66. The numbers identify the number of carbon atoms first in the diamine and secondly in the diacid.

The initial synthetic route to this new polymer converted benzene to cyclohexanone via phenol and cyclohexanol. Air oxidation of the ketone gave adipic acid which was the source of HMD (Fig. MK-2). The air oxidation of cyclohexane to alcohol and ketone followed by air oxidation of pure cyclohexanone was developed soon thereafter (Fig. MK-3). The post-war years saw the introduction of nitric acid oxidation of the alcohol/ketone mixture. It also saw the development of HMD syntheses independent of adipic acid manufacture as well as use of hexanediol from adipic acid to make ADN (Fig. MK-3). The search for improved synthetic procedures is unending. The literature (e.g., 13, 14, 15) cites many techniques, but few become commercial. Significant developments are shown in Fig. MK-4. The value of a process depends not only on its chemistry but also on the cost and availability of raw materials. Increasing cost of phenol in the late 1930's prompted use of cyclohexane. The increase in price of CH by a factor of 4-5 from 1972 to 1974 together with the higher cost of energy has raised questions about preferred processes and yields. Better routes to phenol and changing cost ratios of cyclohexane to propylene or butadiene

TABLE MK-2. Some Problems in Development of Nylon-66.

- New, commercially practical, large scale synthesis of adipic acid and hexamethylenediamine required.
- No precedent for commercial polymerization of a crystallizable, high melting material of low thermal conductivity.
- No precedent for melting or pumping an organic material at carefully controlled, high temperatures where sensitivity to oxidation becomes acute.
- No precedent for filtering and spinning a fiber at high temperatures rapidly (>2500 ft/min) and effectively.
- Special problems in fiber development had to be overcome, e.g. drawing, twisting, sizing, and winding at very high speeds. The unforeseen wrinkling of hose during dyeing had to be eliminated.

Accomplishment

- Concept to pilot plant - less than two years.
- Concept to successful, commercial, fiber plant - less than five years.
- Laboratory to plant design - 230 chemists and engineers.
- First thirteen years - nearly 500 patents, over 200 for novel devices for manufacture of fiber.

FIG. MK-1. Nylon-66 and Intermediates.

$n\ H_2N(CH_2)_6\ NH_2$ = Hexamethylenediamine (HMD)
1,6-Hexanediamine, mol. wt. = 116
1935 = laboratory curiosity

+

$n\ HOOC(CH_2)_4COOH$ = Adipic Acid (ADA)
Hexanedioic Acid, mol. wt. = 146
1935 = commercial in Germany (from fats or phenol)

$H\text{-}[NH(CH_2)_6NH\overset{O}{\overset{\|}{C}}(CH_2)_4\overset{O}{\overset{\|}{C}}]_n\text{-}OH + (2n-1)H_2O$ = Poly(hexamethylene adipamide)
Poly(iminohexamethyleneiminoadipoyl)
Nylon-66, mol. wt./CONH = 113

$n = 50 - 150$ $\bar{DP}_n = 100 - 300$ $\bar{M}_n = 11{,}000 - 34{,}000$

FIG. MK-2. Monomer Synthesis, 1939

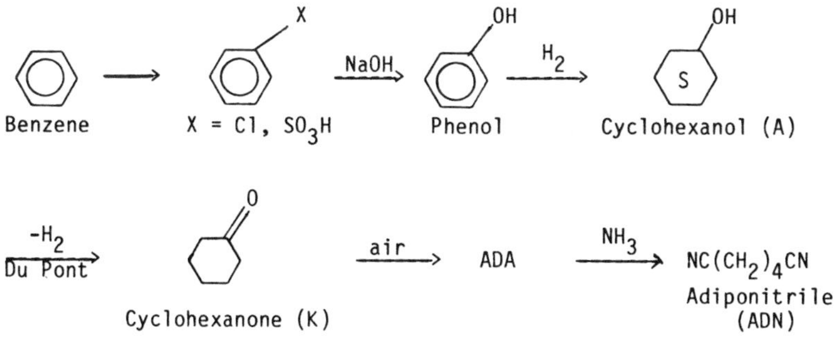

Benzene → X = Cl, SO_3H $\xrightarrow{NaOH}$ Phenol $\xrightarrow{H_2}$ Cyclohexanol (A)

$\xrightarrow[\text{Du Pont}]{-H_2}$ Cyclohexanone (K) $\xrightarrow{air}$ ADA $\xrightarrow{NH_3}$ $NC(CH_2)_4CN$ Adiponitrile (ADN)

$\xrightarrow{H_2}$ HMD

Also, A $\xrightarrow[\text{IGF}]{HNO_3}$ ADA

FIG. MK-3. Synthesis of Intermediates in the 1940's and 1950's

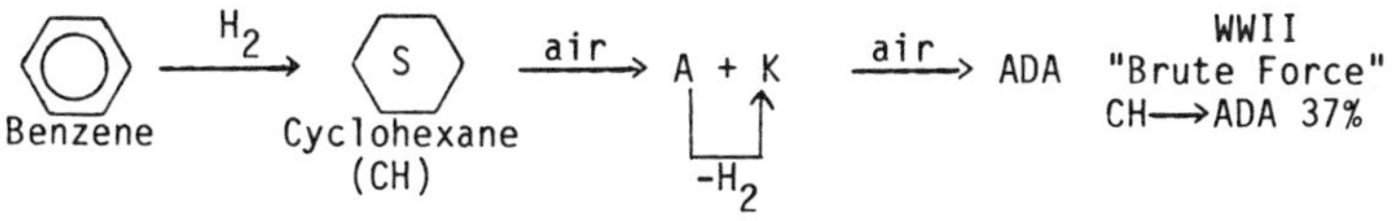

Benzene $\xrightarrow{H_2}$ Cyclohexane (CH) $\xrightarrow{air}$ A + K ($-H_2$: A → K) $\xrightarrow{air}$ ADA

WWII "Brute Force" CH⟶ADA 37%

A + K $\xrightarrow{HNO_3}$ ADA (Du Pont)

Furfural $\xrightarrow{-CO}$ $\xrightarrow{H_2}$ Tetrahydrofuran $\xrightarrow{HCl}$ $\xrightarrow{NaCN}$ ADN (Du Pont)

Butadiene $\xrightarrow{Cl_2}$ $\xrightarrow{NaCN}$ $\xrightarrow{H_2}$ ADN (Du Pont)

ADA $\xrightarrow{H_2}$ Hexanediol $\xrightarrow{NH_3}$ ADN (Celanese)

FIG. MK-4. Noteworthy Developments in Synthesis

CH $\xrightarrow{\text{air}}$ K + A Many patents for yield + conversion improvement, use of boric acid

CH $\xrightarrow[\text{low } P_{H_2O}]{\text{air, boric acid}}$ high A/K ratio (Scientific Design)

High A/K $\xrightarrow[\text{Cu + Mn catalysts}]{\text{air}}$ ADA

$CH_2{=}CHCH_3$ (Propylene) $\xrightarrow{NH_3}$ $CH_2{=}CHCN$ (Acrylonitrile (AN)) $\xrightarrow{\text{Electrohydro-dimerization}}$ ADN (Monsanto)

$CH_2{=}CHCH{=}CH_2$ (Butadiene) $\xrightarrow{2HCN}$ ADN (Du Pont)

---

could make for radical changes in intermediate syntheses. Excellent reviews of the synthesis, properties, and economics of adipic acid, adiponitrile, and hexamethylenediamine have been provided by Luedeke [16,17,18].

POLYMERIZATION

Obtaining high molecular weight polymer was recognized to require conversions in excess of ninety-nine percent, high for any organic reaction. High purity and precise stochiometry of the monomers were obviously necessary. Flory's theoretical contribution [19] relieved concern with respect to the effect of molecular weight on reactivity. An easy and simple way to assess stochiometry was rapidly recognized to be the pH corresponding to the inflection point in a titration curve of an aqueous solution of the salt, hexamethylenediammonium adipate [20]. Synthesis of pure salt could be accomplished by adding adipic acid to an alcohol solution of HMD. The salt rapidly precipitates with impurities remaining in solution. Salt could also be made by adding HMD to a stirred aqueous dispersion of ADA, but charcoal decolorization was necessary for product of good color until it was discovered that careful HMD refining, particularly freedom from by-product cis-1,2-diaminocyclohexane, would obviate that need [18]. Maintaining salt balance through polymerization was another problem to be overcome because of the higher volatility of HMD (b.p. 200°C at 760 mm Hg) relative to that of adipic acid (b.p. 205.5°C at 10 mm Hg). Equilibrium of salt with its components would supply more free HMD as volatilization occurred.

Initially polymerization was a batch process in which a salt solution was concentrated from about 50 to 75% solids before transfer to a vessel purged of oxygen. A 50% solution is convenient for storage because of moderate heating demand to avoid freeze-up; it corresponds to saturation at 20°C (68°F). Heating to about 210°C generates a pressure of about 17 atm. Heating is continued and pressure maintained at this level while bleeding steam until a temperature of 275-280°C is attained. This procedure permits formation of low molecular weight polymer with minimal loss of HMD and provides a high enough temperature to preserve the liquid state as water is removed and the freezing point of the polymer is approached. Finally, the pressure is reduced to atmospheric and the vessel is maintained at 275-280°C under an atmosphere of steam until the desired molecular weight is achieved. Application of vacuum to facilitate water removal and raise the molecular weight is feasible.

The economies of continuous polymerization were sought and realized but not easily. Atmospheric polymerization with excess diamine to compensate for volatilization and polymerization in solution have been explored as well as high-pressure melt polymerization [21]. Problems of heat transfer, continuous pressure let-down to achieve desired molecular weight without freezing the melt, avoidance of gel from excessive hold-up even of a very small fraction of the polymer, output rate versus entrainment and so forth have been resolved by various companies with different designs [21]. Batch as well as continuous processes remain useful, however, not only because of improved batch technique but also because of their relatively facile adaptability. The polyamidation process allows for the use of a variety of diamines, dibasic acids, and lactams in copolymerization. It also permits addition to the polymerizer of modifiers and additives such as antioxidants, carbon black, colorants, delusterants, nucleating agents, plasticizers, and UV and thermal stabilizers [22]. Some such modifications are better accomplished during polymerization rather than by subsequent dry or melt blending. It is indeed this flexibility that has allowed nylon to adapt to market needs and retain its preeminent position among the engineering thermoplastics.

Solid phase polymerization was considered by Flory [23] and has been variously pursued through the years [21]. Direct conversion of dry salt to solid polymer has not become important although increasing the molecular weight of polymer by solid state polymerization is a viable technique [24].

COMMERCIAL INTRODUCTION AND GROWTH

The first successful stocking was made in 1937 from the first pilot plant in Arlington, New Jersey. The first product sold, however, was a brush bristle, now normally classified as a plastic product, in 1938. Public announcement of nylon took place on September 21, 1938. In the following year the first public sale of hosiery occurred in Wilmington, Delaware. Also in 1939 the first non-filamentary product was made - textile machinery gears. Nylon hosiery was featured at the World's Fair in New York in 1939 and 1940. The first fiber plant went into full production at Seaford, Delaware, in early 1940 using intermediates from Belle, West Virginia. Initially designed for three million annual pounds, it went on stream at eight million [11]. The first commercial offering of molding powder was in 1941 and found initial use in electrical coil forms.

Nylon-66 was made in a 1939 pilot plant by the Badische division of IGF in Germany. They made nylon-66 and 66/6 copolymers commercially in 1941 [25]. Production of 1.7 million pounds of 66 and 2.1 million pounds of copolymer was reported for 1942 [25, 26]. ICI in Great Britain began manufacture of fiber in 1940 but did not enter the plastics business until 1955 [27]. Monsanto first made 66-fiber in 1954; its plastics business began in 1966 [28]. Akzo in Holland produced 66-yarn in 1954 and plastics in 1955 [29]. Other producers, exclusive of Communist nations, of nylon-66 besides Celanese in the U.S. now exist in Belgium, France, West Germany, Holland, Italy, Spain, the United Kingdom, and Japan [30]. This does not include the number of organizations that purchase polymer from primary suppliers and convert it into forms such as rod and slab for prototyping or machining or compound it into special lubricated, filled, or reinforced resins. An early leader in this field was the Polymer Corporation founded in 1946 to make stock shapes of nylon-66 for industry [31]). I will not cite the several Du Pont expansions beyond the one mentioned below except to note the plant at Parkersburg, West Virginia, built in 1948 and dedicated to the plastics business. The first facility devoted to technical service was established outside of Wilmington in 1955. The list of companies and countries now involved with nylon one way or another runs into the hundreds.

NYLON AT WAR

In 1942 all production went on allocation to the war effort and did not again become available to the public until 1945 [1]. In response to immediate market acceptance, a second plant comparable to the Seaford facility had gone into production at Martinsville, West Virginia in 1942 [11]. Belts, body armor, boot uppers, brushes, clothing, gloves, hammocks, parachutes, rafts, ropes, straps, tents and tire cord comprise an incomplete list of the military service of nylon, regarded not as a substitute for traditional materials but as a high quality product of exceptional durability [1, 12]. Nylon exhibited a previously unknown combination of strength, toughness, and resistance to heat, rot and wear. Utility in warm climates of high humidity solved otherwise difficult to impossible problems. Allocation occurred again during the Korean conflict with a big demand for coating of assault wire [32].

PROPERTIES AND APPLICATIONS

The versatility of nylon is attested to by its range of applications. I quote a comment on nylon from Modern Plastics, "Where can you find a material that runs from combs to ship propellers to ladies' stockings?" [33]. It is used in appliances found in the bathroom, bedroom, kitchen or laundry; in cars and trucks in the doors, interiors, and under the hood; in business equipment; in consumer products, e.g. brushes, film, fishing line, kitchen tools, sporting goods and toys; in electrical and electronic devices, e.g. connectors, controls, and switches; in hardware, e.g. door, furniture and window parts, lawn and garden implements, marine equipment, and tools; in machinery in the agricultural, food, handling, mining, motor, oil, printing and textile industries [34].

Like its plastics predecessors nylon-66 had low thermal and electrical conductivity, light weight, and excellent corrosion resistance. Unlike its predecessors it was semi-crystalline with a melting point of 269°C. Utility as a fiber rested upon properties that accrue to nylon because of its crystallinity which also has great significance in the plastics area. In injection molding [35] careful control of temperature, a special nozzle, and assured dryness were required to cope with the sudden transition from solid to a melt lower in viscosity than traditionally experienced by about an order of magnitude. Extrusion requirements including barrel and screw design were also cited [36]. With the advent of reciprocating screws in injection molding, a multiplicity of single screw and twin screw options for extrusion and compounding, and increasing need to optimize processes for the myriad of generic and non-generic products because of extensive competition, virtually every process has become special. This has eliminated the idea that processing nylon-66 requires a unique level of sophistication.

Stiffness, lubrication-free low friction, fatigue and wear resistance, non-toxicity, and absence of taste have resulted in the use of nylon-66 in bearings, bushings, and gears in a wide variety of equipment with moving parts including food handling devices. Washing machine applications rely on its hydrolysis resistance as well as its mechanical properties. Its resistance to fatigue and repeated impact is outstanding and has suited it for oscillating machinery parts, textile shuttles, auto door striker plates and many other such applications. Elevator gibs, cable ties, wire jackets and timing sprockets are among the applications that take advantage of its strength and abrasion resistance. Good resistance to creep is another property plus. Many of these properties reflect the crystalline character of nylon-66. Chemical resistance results from both its structure and its crystallinity. Oil and solvent resistance is outstanding and is responsible for its wide use, among other things, in automobiles as gas emission canisters. Crystallinity is the glue that holds nylon-66 together right up to the melting point which is about 190°C above its glass transition temperature. This was a radical change from earlier thermoplastics and contributed to its superior high temperature stability.

Many applications require still higher levels of resistance to thermal and photooxidation, stiffer or less stiff variants, improved toughness and/or reduced notch sensitivity, lower user cost via resin modification or improved processability, better lubricity, higher or lower melt viscosity for specific processing needs, colorants, enhanced fire retardation, and so forth. As often noted a major strength of nylon-66 is its capacity for modification. Many of these goals are achieved by changing molecular weight, making copolymers, or using additives such as antioxidants, nucleating agents, mold release agents and UV stabilizers. These techniques were developed by several companies prior to 1960. This is also true of enhanced flexibility via plasticizers. Glass reinforcement with short fibers, a major development, appeared in the 60's. Largely dependent on perceived need, fire retardant compositions were developed by various companies from the late 60's through the early 80's. Mineral filled compositions became available in the early 70's and were soon followed by hybrids with glass particularly to reduce the warpage of glass reinforced materials. The introduction by Du Pont of its "Super Tough

Nylon" in 1976 resulted in the toughest plastic yet known, one that showed no dependence of toughness on thickness, orientation direction, notch radius, or type of impact test. ICI has recently announced a composition with dispersible long fibers. These are the kinds of innovations that keep nylon-66 young and predict continued healthy growth.

THE BUSINESS

Growth curves are not readily available on an annual basis for the individual nylon plastics as it is for all nylons. Even in the years in the United States (1938-1954) preceding the appearance of non-Du Pont manufacturers some production involved copolymers of 66 as well as nylon-6 and -610. It is fair to say, however, that nylon-66 and those products consisting mostly of 66 (copolymers, filled, reinforced, toughened, fire retardant, etc.) have participated significantly in that overall growth. In the U.S. based on data for 1978 and 1983 [37], 61% of all nylon plastics is 66 and 26% is polycaprolactam (nylon-6). On a world-wide basis the numbers were about 35% 66 and 50% 6 for several years but they have changed in favor of 66 in the last decade. Reported for 1983 are 42.4% 66 and 49.2% 6 [37].

No doubt the versatility of nylon synthesis has been responsible for a performance approximating that of all plastics from 1950 to 1975 in spite of the introduction of many new resins in that interval (U.S. data, Fig. MK-5). Moreover, in the last decade nylon has done somewhat better than all plastics (Fig. MK-5). As seen in Fig. MK-6, nylon plastics in the U.S. roughly paralleled the growth of fiber and amounted to 7 to 8 percent of the fiber up to 1966. Fiber growth since then has been erratic and slower overall than plastics such that plastics reached the 10 per cent level in 1975 and has since increased to almost 17 per cent in 1984. A similar experience at higher levels has characterized Western Europe where plastic use was 22.7% of 66-fiber use in 1982 [41]. Of interest is the per capita consumption in various countries (Table MK-3) which suggests that many areas, particularly the U.S., have some catching up to do.

Following the pattern of consumption from 1962 to 1985 (Figs. MK-7A and -7B) more clearly reveals the uneven growth of plastics from year to year with peaks in 1966, 1973, 1979, and 1985 and lows in 1967, 1975, and 1982. In the U.S. electrical/ electronics, consumer products, and industrials approximated 40 million pounds each in 1985; transportation, the largest user at 125; all extrusion markets, 94; miscellany and export, 27 each; appliances, 12. Long term growth characterizes all areas with appliances lagging behind. The special class of filled/reinforced nylons, included in the above figures, has exhibited an overall growth of 19%/yr since 1966 with a possible slacking off in 1985 (Fig. MK-8). These applications add up to 402 million pounds in the U.S. in 1985 [38]. Another estimate of the total is 415 with world-wide use of 1332 million pounds [42]. This put U.S. use at about 45 per cent of the world total. This compares with a figure of about 33 per cent in the case of nylon fibers.

There is a lack of world data on applications but an interesting comparison with Western Europe can be made in spite of differences in categories. In 1983, contrary to the U.S., electrical engineering exceeded

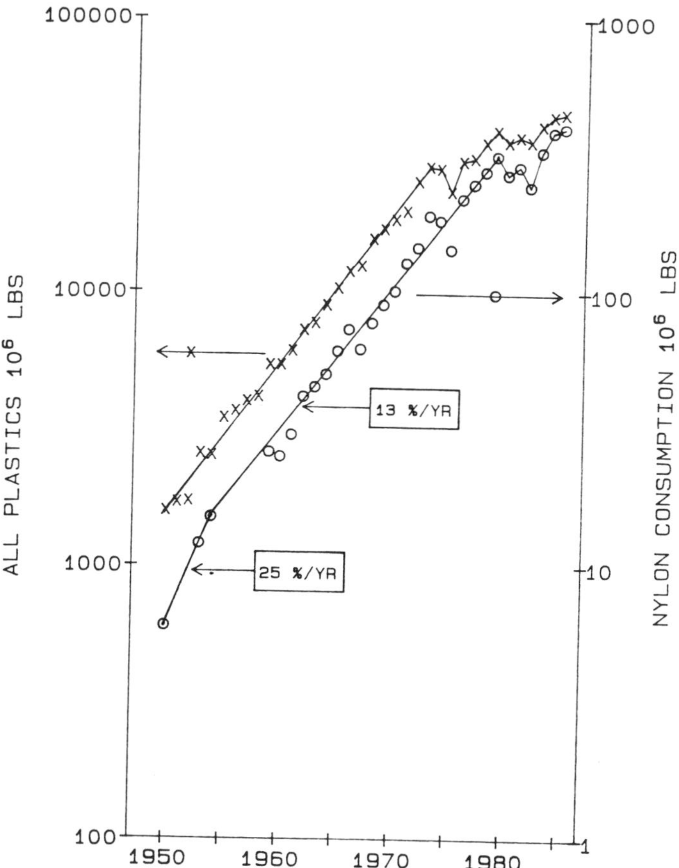
FIG. MK-5
GROWTH OF ALL PLASTICS AND NYLONS, 1950 - 1985
UNITED STATES ONLY
Source: Ref. 38
100000
10000
1000
100
1000
100
10
1
1950
1960
1970
1980
ALL PLASTICS 10^6 LBS
NYLON CONSUMPTION 10^6 LBS
13 %/YR
25 %/YR

FIG. MK-6
GROWTH OF NYLON PLASTICS AND FIBER, 1950 - 1985
UNITED STATES ONLY
Sources: Ref. 38, 39, 40

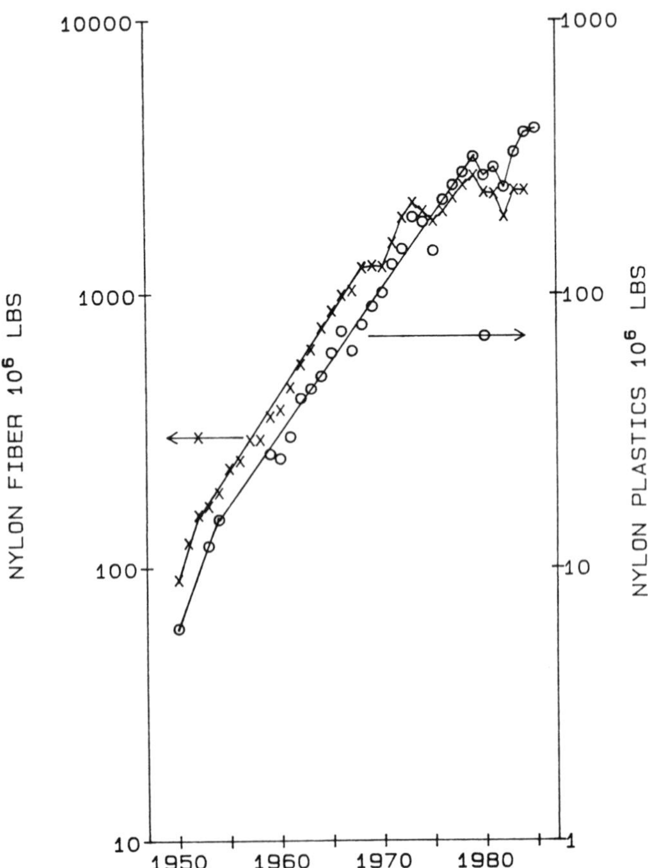

FIG. MK-7A
PATTERN OF CONSUMPTION FOR NYLONS, 1962 - 1985
UNITED STATES ONLY
Source Ref. 38

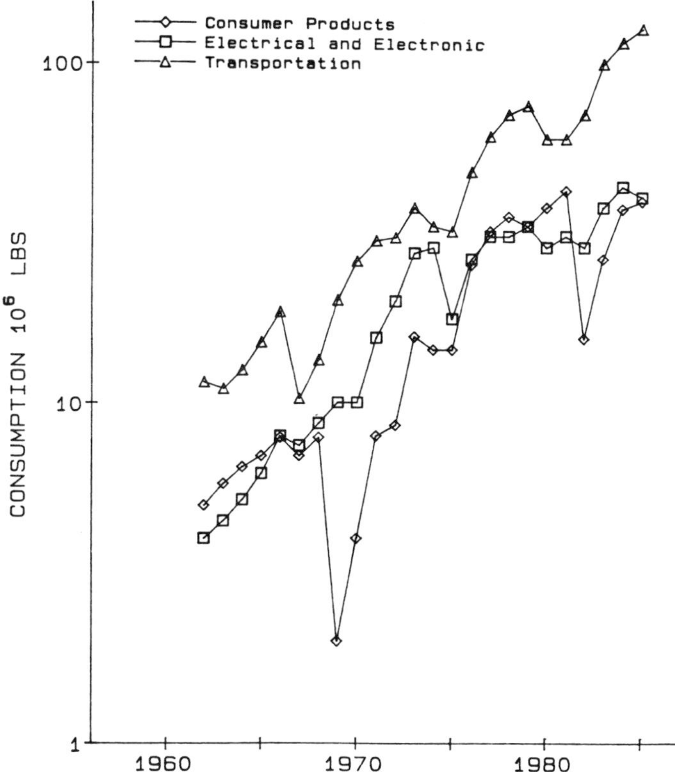

FIG. MK-7B
PATTERN OF CONSUMPTION FOR NYLONS, 1962 - 1985
UNITED STATES ONLY
Source: Ref. 38

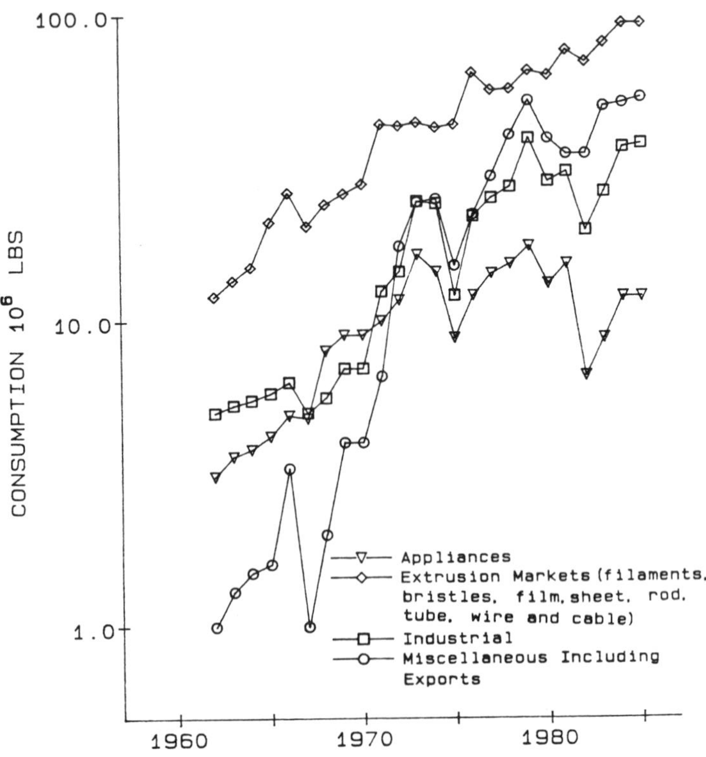

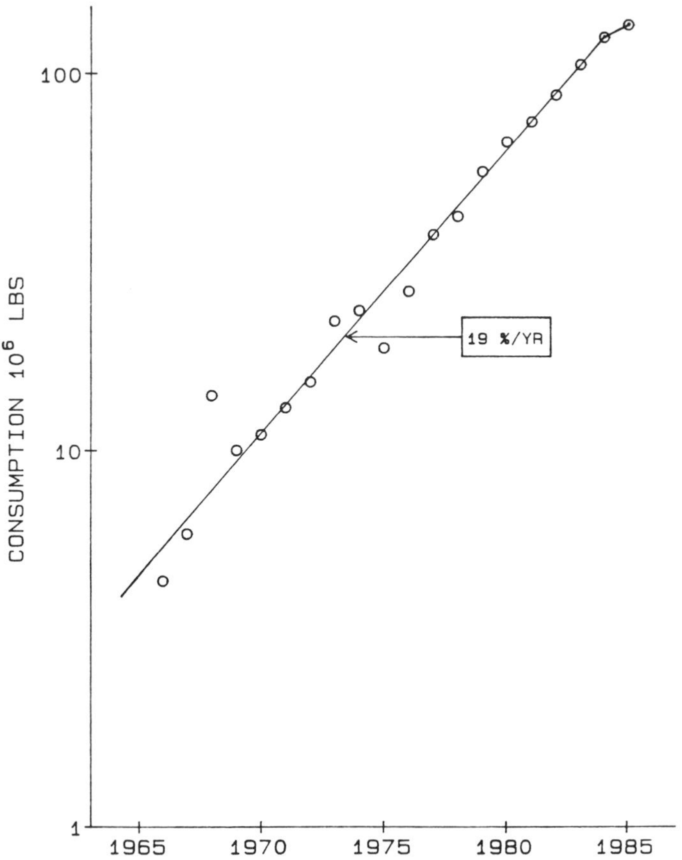
FIG. MK-8
CONSUMPTION OF FILLED/REINFORCED NYLONS, 1966 - 1985
UNITED STATES ONLY
Source: Ref. 38
CONSUMPTION 10^6 LBS
100
10
1
1965
1970
1975
1980
1985
19 %/YR

FIG. MK-9
STANDARD NYLON-66 AND CPI HISTORY, 1945 - 1985
UNITED STATES
Sources: Ref. 38, 43, 44

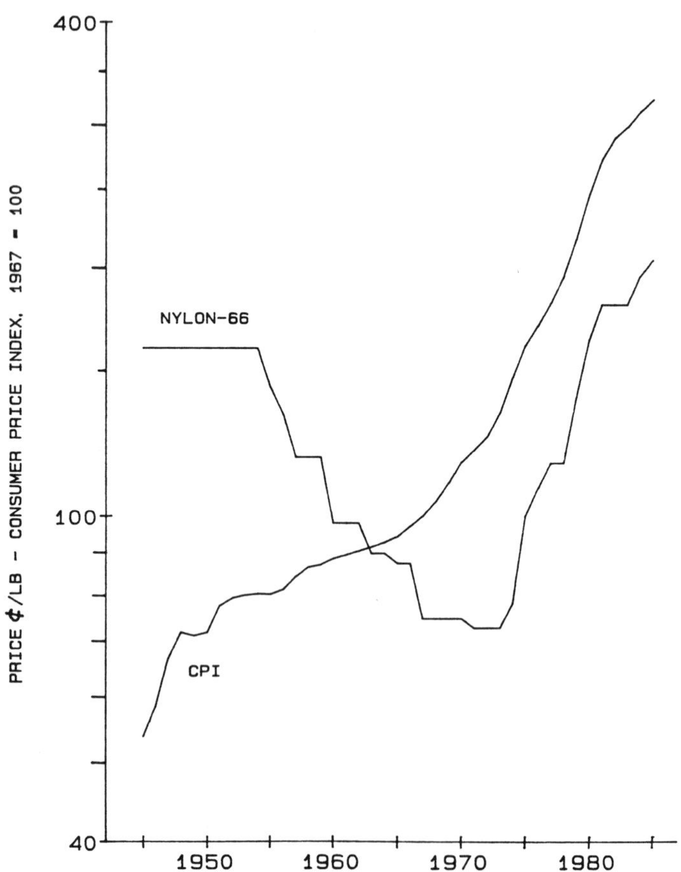

TABLE MK-3. Annual Per Capita Consumption in Pounds in 1978 [37].

West Germany	2.42
France	1.25
Sweden	1.21
United States	0.99
Japan	0.88

the vehicle industry, 103 to 95 million pounds [37]. The per cent of total that is injection molded and extruded has varied somewhat from year to year but has been about 65 and 23% in Western Europe and an approximately similar 62 and 25% in the U.S. [37, 38].

The U.S. price history of standard grade nylon-66 for its first three decades reflects the usual reduction as volume and technology advance (Fig. MK-9). It also shows the profound effect of oil prices from 1973 on and the attempt to keep up with the increase in the consumer price index.

CONCLUSION

In this review we have briefly examined the technical and business events that mark the history and development of nylon-66. The central message is one of continuing adaptation, the clue to a growth profile that persists even as its fiftieth anniversary in 1988 nears.

REFERENCES

1. Anon., "Nylon, The First 25 Years," Brochure of E. I. Du Pont de Nemours and Co., 1963.
2. R. Adams in Ref. 10, p. XV.
3. M. Kaufman, "The First Century of Plastics," Iliffe, London, 1963.
4. H. Mark, Amer. Scientist, 72, 156 (1984).
5. G. M. Kline, Mod. Plas. 57 (2), 66 (Feb., 1980).
6. W. H. Carothers, Chem. Rev. 8 (3), 353 (1931).
7. W. H. Carothers, J. Am. Chem. Soc. 51, 2548, (1929).
8. W. H. Carothers and G. A. Garvin, J. Am. Chem. Soc. 51, 2560 (1929).
9. P. J. Flory, "Principles of Polymer Chemistry," Cornell University Press, Ithaca, 1953, p. 23.

10. High Polymers, Vol. I, "Collected Papers of Wallace H. Carothers on Polymerization," H. Mark and G. S. Whitby, Eds., Interscience, NY, 1940.

11. E. K. Bolton, Ind. Eng. Chem. 34, 53 (1942).

12. W. Chambless, "Nylon is 40," article in Du Pont magazine, "Context", 1978, No. 2, pp 24-29.

13. E. H. Pryde and J. C. Cowan in "Condensation Monomers," J. K. Stille and T. W. Campbell, Eds., Vol. 27 in High Polymers Series, Wiley-Interscience, NY, 1972, pp. 39-55 (ADA).

14. P. T. Kan, ibid., pp. 185-194 (HMD).

15. Anon., Chem. and Eng. News 62 (18), 28 (April 30, 1984).

16. V. D. Luedeke in "Encyclopedia of Chemical Processing and Design," V2, J. J. McKetta, Ed., Dekker, NY, p. 122 (1977). (ADA)

17. V. D. Luedeke, ibid, p. 146. (ADN)

18. V. D. Luedeke, personal communication, article accepted for future volume of "Enclyclopedia of Chemical Processing and Design". (HMD)

19. P. J. Flory, loc. cit. in Ref. 9, p. 102.

20. W. H. Carothers, U.S. Patent 2,130,947 (September 12, 1939 to E. I. Du Pont).

21. D. B. Jacobs and J. Zimmerman, Chap. 12 in "Polymerization Processes," C. E. Schildknecht and I. Skeist, Eds., Vol XXIX in "High Polymers," Wiley-Interscience, NY, 1977.

22. M. I. Kohan, "Nylon Plastics," Wiley-Interscience, 1973, pp. 20-29 and 413-416.

23. P. J. Flory, U.S. Patent 2,172,374 (September 12, 1939 to E. I. Du Pont).

24. D. A. Beaton, U.S. Patent 3,821,171 (June 28, 1974 to E. I. Du Pont).

25. W. F. Seydl of BASF, Personal Communication, September, 1985.

26. G. M. Kline, Mod. Plast 23 (2), 152A (October, 1945).

27. M. T. Whitfield and R. Watson of ICI, Personal Communications, September, 1985.

28. R. D. Chapman of Monsanto, Personal Communication, September 1985.

29. W. J. Mijs of Akzo, Personal Communication, September, 1985.

30. Chemical Economics Handbook 580.0822Q (December 1983).

31. J. A. Rusnock of Polymer Corporation, Personal Communication, October 1985.

32. Anon., Mod. Plast. 28 (5), 63 (Jan. 1951) and 29 (5), 86 (Jan. 1952).

33. Anon., Mod. Plast 37 (5), 218 (Jan., 1960).

34. E. T. Darden, Chap. 20 in "Nylon Plastics," M. I. Kohan, Ed., Wiley-Interscience, 1973, pp. 619-641

35. Anon., Mod. Plas. 30 (5), 90 (Jan. 1953).

36. Ibid., 94 (Jan. 1953).

37. D. Michael, German Plastics 70 (10), 26 (Oct. 1980) and 74 (10), 18 (Oct. 1984).

38. Anon., Mod. Plast., January issues, 1950-1986.

39. Anon., Chem. & Eng. News. 49 (12), 126 (Mar. 20, 1961).

40. Anon., Tex. Organon LI (2), 13 (Feb., 1980); ibid LVI (6), 116 (June 1985).

41. Chemical Economics Handbook 664.1000P (March 1984).

42. Anon., Chem. and Eng. News 64 (3), 10 (Jan. 20, 1986).

43. Anon., Plast. Tech., Monthly issues, Jan. 1980 - Jan. 1986.

44. Anon., U.S. Dept. of Commerce.

HISTORY AND DEVELOPMENT OF NYLON 6

PAUL MATTHIES* AND WOLFGANG F. SEYDL**
*BASF Aktiengesellschaft, Ludwigshafen, Germany; ** BASF Corporation, Engineering Plastics, Bridgeport, New Jersey

INTRODUCTION

Nylons were introduced to the market in the late 1930's. Of the various types, nylon 66 was the first. As is well-known, this product was developed by the DuPont Company as the first truly synthetic fiber with a broad range of applications. Soon after, about 1939/1940 nylon 6 fibers were introduced by the I.G. Farbenindustrie in Germany. Simultaneously, however, the I.G. also developed nylon 6, nylon 66, and nylon copolymers for plastics use.

Whereas the main outlet for nylons was, and still is, the field of fibers, nylon polymers are also firmly established as engineering resins. This is due to a combination of such valuable properties as strength, stiffness, toughness, abrasion resistance and high service temperature as compared to previously available thermoplastics, and to their easy modification making them suitable for a wide spectrum of applications.

In the present paper, the review is confined to the history of nylon 6 emphasizing the plastics aspect. Development of caprolactam monomer and polymer processes is described as well as the principal modifications of the basic resin.

EARLY RESEARCH

The period about 1900

Nylon 6 is commercially produced by polymerization of ε-caprolactam. In the laboratory it can also be obtained by self-condensation of ε-aminocaproic acid. The preparation of both possible starting materials was first reported in 1899 by S. Gabriel and T.A. Maass at Berlin University [1]. In the course of fundamental research on seven-membered heterocyclic ring compounds they obtained aminocaproic acid by a lengthy synthesis starting from a dihalopropane. Then, by heating aminocaproic acid carefully above its melting point (202 - 203°C), water was set free and small amounts of caprolactam distilled off under vacuum.

The most important, present-day synthetic path to caprolactam was discovered as early as 1900 by O. Wallach at Göttingen University in Germany [2]. He converted cyclohexanone to the oxime, and by the Beckmann rearrangement of the oxime in sulfuric acid, he obtained caprolactam.

In 1907 J. von Braun, also at Göttingen University, confirmed that the dehydration of aminocaproic acid yielded some caprolactam (20 - 30 %) [3]. In addition, however, he noted that the larger part of the amino acid was converted into a

Copyright 1986 by Elsevier Science Publishing Co., Inc.
High Performance Polymers: Their Origin and Development
R.B. Seymour and G.S. Kirshenbaum, Editors

viscous mass. According to analytical data it was "isomeric" with caprolactam, "i.e. a polymeric product". He could not perform molecular weight determinations because suitable solvents were not known. It may be mentioned that A. Manasse [4], in 1902, and again von Braun [3] also reported the formation of polymeric products by thermal dehydration of ω-aminoheptanoic acid.

Thus, the formation of C 6 and C 7 lactam polymers from the corresponding amino acids was well known at the turn of the century. The work leading to this knowledge, however, was purely scientific; furthermore, the concept of high polymers was not well established at the time; and the valuable properties of synthetic polymers could not be anticipated. So this early work remained without practical consequences.

The period about 1930

Nearly 30 years later, W.H. Carothers commenced his famous studies [5] on polycondensation at the DuPont Company. He and his coworkers prepared fiber-forming polyesters and polyamides and discovered the cold-drawing phenomenon. In 1935, nylon 66 was prepared for the first time. Carothers' work finally resulted in the commercial development of nylon 66 at DuPont [6].

During the course of these studies, W.H. Carothers and G.J. Berchet in 1930, extending the early work of von Braun, investigated the self-condensation of ε-aminocaproic acid [7]. They obtained a low-molecular polyamide. From cryoscopic measurements in phenol they inferred a degree of polymerization of at least 10. In the same paper they stated that caprolactam "does not polymerize under the conditions of formation of the polyamide either in the presence or absence of catalysts". As practically no experimental details were given, one can only speculate about the reason of this failure. Perhaps they were influenced by the fact that the corresponding 5- and 6-membered lactams did not polymerize either - an explanation offered later by P. Schlack [8].

In 1931 the work on aminocaproic acid was carried on by W.H. Carothers and J.W. Hill [9]. The low molecular-weight polyamide already mentioned was heated in a molecular still for 2 days at 200°C in order to increase the molecular weight. A considerable change in physical properties was observed which indicated also a large increase in molecular weight. They also reported that the polymer softened at 210°C "with considerable decomposition", an observation which is puzzling today. From this they concluded that "the polyamide described above is too infusible and insoluble to allow a ready test of its ability to furnish fibers". Carothers also "suspected that the viscosity of amides was so high that the reaction stopped prematurely" [10]. Later on, Schlack conjectured about these findings that the aminocaproic acid used was impure and that the polymer was cross-linked [11].

Carothers' outstanding contributions to polymer science and the technological importance of his discoveries need not be further emphasized. On the other hand, the fact that he rejected caprolactam as seemingly unsuitable for polymerization turned out to be a piece of luck for the competitors.

Paul Schlack, mostly referred to as the "inventor of Perlon", the nylon 6 fiber, was born in Stuttgart in 1897 [12]. Early scientific activity was on pyrrole and polypeptides. In 1924 he joined the AGFA Company at Wolfen (near Leipzig) where he worked on the development of a cellulose acetate fiber. When the I.G. Farbenindustrie was established in 1925 by a merger of BASF, Bayer, Hoechst and others, AGFA also was taken over. In 1926, the I.G. and Glanzstoff (Vereinigte Glanzstoff-Fabriken) formed a joint venture, the Aceta GmbH in Berlin-Lichtenberg, for the manufacture of acetate fibers. Schlack was appointed head of its scientific laboratory. In 1930 Glanzstoff withdrew from the enterprise, and Aceta became wholly owned by I.G. Farben.

As Schlack recalls [13, 14] he began work on condensation polymers about the year 1930. He did the work without orders, more out of scientific interest and without putting much effort into it. He chose aminocaproic acid as the starting material, but the polymer obtained was not fiber-forming. He then began to explore other routes to polyamides and contacted colleagues to arouse their interest for preparing samples of other starting materials, but in vain. During the world economic crisis this work was suspended.

THE DISCOVERY OF CAPROLACTAM POLYMERIZATION

Some years later, in 1937, Carothers' first patents describing the field of polyamides came to Schlack's knowledge. He realized his lost chances. Nearly all promising routes appeared to be blocked. Nevertheless,he renewed his work on condensation polymers.

At the time, Schlack knew the paper of Carothers and Berchet [7] only from an abstract, and he did not know their statement about their failure to polymerize caprolactam [8]. In his disappointment he first experimented with polyesters, but when he recognized their low dye affinity, he started again with polyamides. Schlack described how his invention came about with the following words [8] (Fig. 1):

"The real starting point was an observation made by us when we polycondensed the N-methoxy-carboxy-ε-aminocaproic acid prepared by acylation of the raw aminocaproic acid with methylchloroformate. This reaction led, probably with ε-isocyanatocaproic acid as an intermediate, directly to spinnable Nylon 6 together with a significant amount of caprolactam ... Since the yield of lactam ... varied with changing reaction conditions such as temperature and pressure we thought that an equilibrium had been set up ... Hence we concluded that caprolactam should be polymerizable at least in the presence of an appropriate initiator. The first such experiment in which ε-aminocaproic acid hydrochloride was used as a transamination catalyst ... yielded a Polyamide 6 which, without further purification, could be spun to a highly elastic filament ...". The crucial experiment, set up with the aid of his technician W. Ahrens, took place during the night of January 28 to 29, 1938. Subsequently Schlack continued his research in a feverish effort in order to safeguard the result and to develop it in more depth before he informed his management [15].

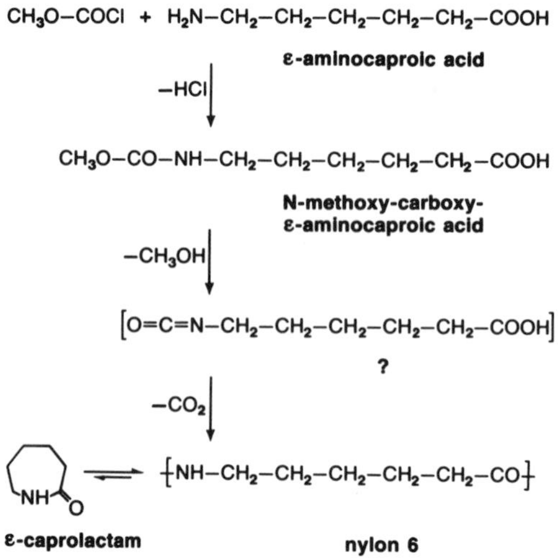

$$CH_3O{-}COCl + H_2N{-}CH_2{-}CH_2{-}CH_2{-}CH_2{-}CH_2{-}COOH$$

ε-aminocaproic acid

$$\downarrow -HCl$$

$$CH_3O{-}CO{-}NH{-}CH_2{-}CH_2{-}CH_2{-}CH_2{-}CH_2{-}COOH$$

N-methoxy-carboxy-ε-aminocaproic acid

$$\downarrow -CH_3OH$$

$$[O{=}C{=}N{-}CH_2{-}CH_2{-}CH_2{-}CH_2{-}CH_2{-}COOH]$$

?

$$\downarrow -CO_2$$

$$\text{ε-caprolactam} \rightleftharpoons {-}[NH{-}CH_2{-}CH_2{-}CH_2{-}CH_2{-}CH_2{-}CO]{-}$$

ε-caprolactam **nylon 6**

Fig. 1. First preparation of fiber-forming nylon 6 (P. Schlack 1937/1938)

Herbert Morawetz and Herman Mark, in a recent paper, arrive at the following appreciation of Schlack's achievement [16]: "Whereas Carothers' discovery of nylon 66 is an outstanding example of the vindication of a far-sighted policy of industrial research management, Schlack's achievement, arrived at as a hobby with the help of a single laboratory assistant, shows that success is possible even under adverse conditions for a determined researcher with a vision".

PROCESS DEVELOPMENT

Start-up

The basic patent application was submitted on June 10, 1938, in Germany [17]. The I.G. immediately recognized the potential of the new polymer as a counterpart to DuPont's nylon 66. They initiated an intense research and development program on polyamides. This included the synthesis of the necessary starting materials: caprolactam, and also hexamethylene diamine and nylon salt for 66 and copolyamides. (Adipic acid was already available). Work was taken up at several locations, mainly Ludwigshafen on Rhine, Leuna (near Leipzig) and Berlin. BASF at Ludwigshafen was chosen because they held a central position in plastics application development within I.G. Farben, and because some intermediates were already available there.

In 1939, I.G. Farben and DuPont concluded an agreement on licence exchange in the field of polyamides. This included a know-how exchange which came about just before war broke out.

Although the I.G. acquired the nylon 66 technology from DuPont, they favored nylon 6 for fibers. The reasons were [18]: production of monomer and polymer was considered simpler and, under the conditions in Germany, also cheaper; processing into fibers was easier because of the higher thermal stability of the melt; also, preference was given to their own development.

For plastics, as opposed to fibers, nylon 66 and several nylon copolymers were developed besides nylon 6. In 1939, highest priority was given to nylon 66. During the years 1940 to 1943, however, nylon 6/66 copolymers obtained the greatest importance due to military requirements.

Early caprolactam and polyamide development is closely connected with the name of Heinrich Hopff (1896 - 1977) [19]. Hopff studied at Munich University and took his Ph.D. under the direction of K.H. Meyer. He joined BASF at Ludwigshafen in 1921. His varied, productive activities covered a large area of organic chemistry including the plastics field. At the time of Schlack's discovery, he was head of the Intermediates and Plastics Laboratory. (In 1952, he left BASF for Zurich Technical University where he was elected Professor of Organic Chemical Technology.) His pioneering work on polyamides is documented by the first monograph in this field (in collaboration with Alfred Müller and Friedrich Wenger) [19a].

Caprolactam

The scope of this paper does not permit a detailed description of the caprolactam development. A competent, readable review, however, is available on this subject [20]. This deals also with later improvements by BASF and with alternative processes developed by other companies.

The synthetic route chosen for the first caprolactam process was the one already used by Wallach [2], taking phenol, available from coal-tar, as feedstock (Fig. 2). Cyclohexanone was at the time already a commercial product employed as a solvent and also for the manufacture of ketone resins (Kunstharz AW 2, BASF).

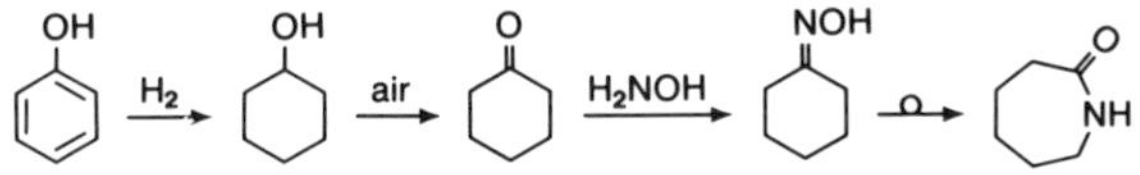

Fig. 2. First caprolactam process (H. Hopff, G. Wiest 1939)

In 1938, Schlack had carried caprolactam synthesis in the laboratory to the kilogram scale [21]. Subsequently a large-scale continuous process was developed by H. Hopff and G. Wiest at Ludwigshafen, and by J. Giesen and F. Korn at Leuna. Key to the successful operation was control of the exothermic Beckmann rearrangement. This was accomplished by separate, continuous addition of moist, molten oxime and concentrated sulfuric acid (later oleum) to the reaction mixture [22]. Production of caprolactam started at Ludwigshafen in 1939 in a pilot plant, reached about 130 tons/year in 1940 and peaked in 1943 with 1270 tons/year [23]. Capacity was 150 tons/month at Ludwigshafen. Another plant of similar capacity (150 - 200 tons/month) was put in operation at Leuna [18]. In the following years, production was low due to wartime and post-war difficulties.

With the advent of petrochemistry, BASF gradually replaced phenol by cyclohexane as the starting material. Development of a new continuous caprolactam process based on cyclohexane was started by BASF in 1950, and a full-scale plant went on stream in 1960 [20]. Similar processes based on cyclohexane were developed by DSM, Bayer and Inventa. A number of alternative processes were developed by other companies (Fig. 3). To name only some of them: the cumene/phenol based process of Allied (on stream in 1958), the photonitrosation process of Toyo Rayon (1962), the toluene based Snia process (1964), and DuPont's nitrocyclohexane process (in operation from 1961 to 1966).

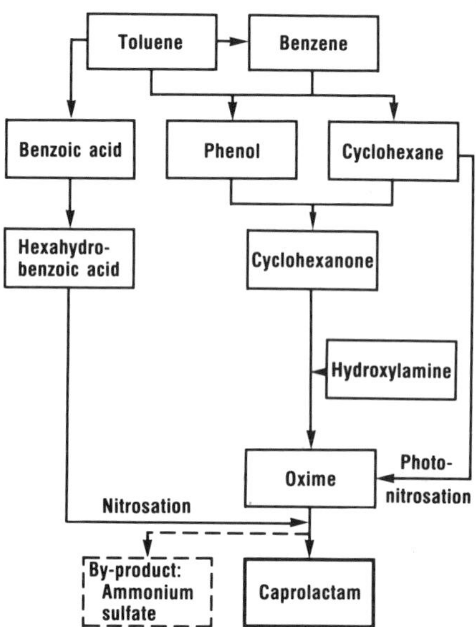

Fig. 3. Caprolactam processes

Nylon 6

In 1938, as already mentioned, laboratory investigation of caprolactam polymerization was carried out by Schlack in Berlin. H. Hopff and H. Ufer at Ludwigshafen started the development of a technical process for nylon 6 manufacture in 1939 [23].

Caprolactam polymerizes at temperatures in the range of

about 240 - 280 °C in the presence of water as initiator (so-called "hydrolytic" polymerization). At first, batch-wise operation in autoclaves was developed employing conditions similar to those of nylon 66 manufacture. Pilot plant facilities were set up in 1939, and about 40 tons/year of nylon 6 were produced at Ludwigshafen in 1940.

Other nylon types were manufactured in the same equipment, mainly nylon 66 and 6/66 copolymers. Total autoclave capacity for all nylon types at Ludwigshafen was increased to 400 tons/month in 1941. Batch size was 2 tons. High pressure steam of 60 - 100 atm. was used as a heat transfer agent. Production figures for 1943 are: about 790 tons of nylon 6, 60 tons of nylon 66 and 840 tons of nylon copolymers [23]. Additional quantities of nylon 6 were produced at other locations for fiber use.

Whereas nylon 66 manufacture requires a pressure polymerization step, nylon 6 can easily be polymerized at atmospheric pressure in the presence of initiators. An economic, continuous process in which these conditions were applied was developed by H. Ludewig [24] at Berlin in 1940, the so-called VK process (Fig. 4). The initials VK stand for "simplified continuous" from the German "vereinfacht kontinuierlich".

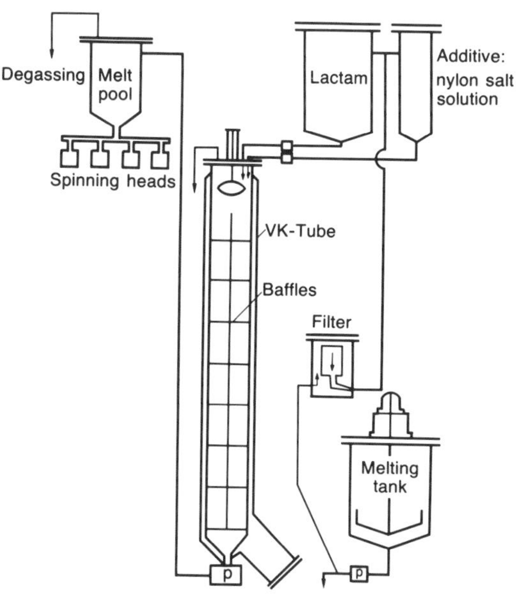

Fig. 4. First continuous caprolactam polymerization process (H. Ludewig, about 1940)

As a continuous process it lends itself to direct spinning. The first VK-tube reactors, 8 m long with 250 to 300 mm inner diameter, were installed in the early 1940's at Berlin-Lichtenberg and at Premnitz (near Berlin), some of them for the production of nylon 6 staple fiber by direct spinning [18]. A full-scale nylon 6 plant with a capacity of 450 - 600 tons/ month for staple manufacture at Premnitz, under construction at the end of the war, was not completed because higher priority was given to filament production for parachutes and tire yarns (at Landsberg/Warthe).

Molten polycaprolactam, at the end of equilibrium polymerization, contains 10 - 12 % low molecular-weight material consisting of unreacted monomer and of cyclic oligomers. For most end-uses, this low m.w. portion has to be removed. For this purpose, Hopff and Ufer developed a hot-water extraction process for nylon 6 chips [25]. As an alternative process, vacuum demonomerization of the polymer melt in the autoclave was practiced as early as 1939/1943 [26]. Hot-water leaching and vacuum extraction were further elaborated and incorporated in integrated, continuous processes for the manufacture of nylon 6 polymer (Fig. 5). In the case of vacuum extraction the integration may also include fiber manufacture by direct spinning [26].

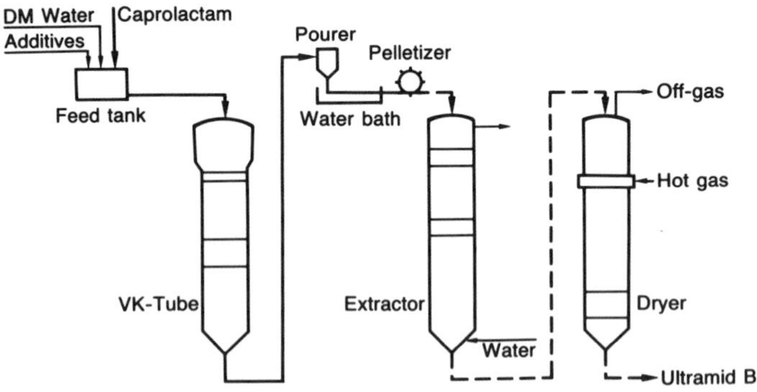

Fig. 5. Manufacture of ®ULTRAMID B (Nylon 6/BASF)

The original VK process was modified and refined in many aspects. Other continuous processes, comprising pressurized steps, were developed. By scale-up and rationalization the capacity of the single production lines was considerably increased (Fig. 6).

As regards continuous nylon 6 processes, reference is made to developments of Allied, BASF, Bayer, Enka und AGFA Wolfen as well as those of engineering companies, e.g. Inventa and Zimmer. Several reviews of this work have been published [26, 27, 28]. In the United States, commercial production of nylon 6 was taken up about 1955 by Allied. Discontinuous manufacture, being less economical, gradually passed out of use.

The reaction mechanism of the hydrolytic caprolactam polymerization is rather complex. Early kinetic investigations were carried out between 1940 and 1955 by A. Matthes at Wolfen. The basic reaction mechanism was elucidated during the years 1955 - 1960 by F. Wiloth at Glanzstoff and by two Dutch research groups led by P.H. Hermans at Utrecht (A.K.U.) and A.J. Staverman at Delft (T.N.O.). This extensive work has been reviewed [27, 29]. The investigations on kinetics, equilibria, and mechanism were later considerably extended by H.K. Reimschuessel [27] at Allied, by K. Tai et al. [30] at Unitika and by G. Bertalan et al. [31] at Budapest Technical University. This work now forms a basis for an optimization of the polymerization process [30].

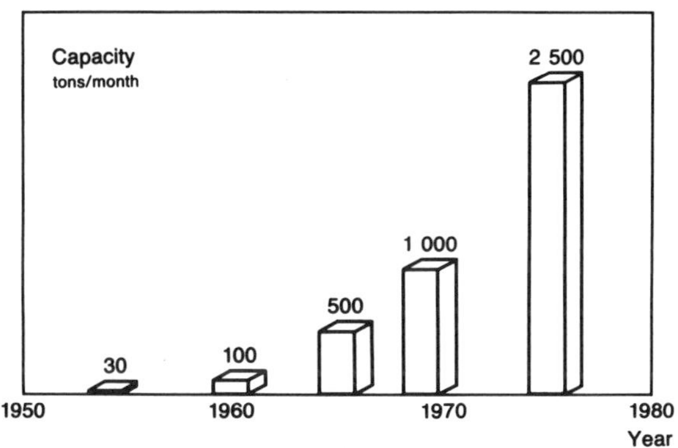

Fig. 6. Scale-up of the VK process

Caprolactam can also be polymerized under anhydrous conditions using strong bases as catalysts. This was already reported by Schlack in 1938 [17]. R.M. Joyce and D.M. Ritter [32] at DuPont discovered in 1939 that this anionic polymerization proceeded with a much higher rate than the hydrolytic process and that it could be carried out below the melting point of the polymer to yield products of low residual monomer content, so that they needed no extraction. In spite of these obvious advantages this type of polymerization at first did not enter into commercial practice because of poor reproducability, unstable melt viscosity and poor product color [29].

The situation changed in 1956/57 when E.H. Mottus and coworkers at Monsanto and independently H. Schnell and G. Fritz at Bayer as well as O. Wichterle and co-workers at Prague discovered that the process could be improved by using acylated caprolactam or acylating agents as activators [33]. During the following years the technology of activated anionic polymerization was developed at BASF, Bayer, and Monsanto [29, 34]. About 1960 it was introduced as a commercial process in the form of the so-called "monomer casting".

PROCESSING AND APPLICATIONS

Early developments [19a]

First processing trials with several nylon types and development of applications in the plastics field were taken up at Ludwigshafen in 1939. Nylon 66, nylon 6 and 66/6 copolymer (60 : 40) were introduced to the market in limited quantities during that year [23, 35]. The first technical information on injection molding of these types was issued in January 1940. It may be mentioned that nylon 6 fibers, which are actually outside the scope of this paper, were introduced by I.G. Farben in 1939 in the form of bristles and in 1940 as filaments [11].

In the plastics field the various polyamide types were first designated by code numbers, thus nylon 6 as "Polyamid 6000". In 1940, for an interim period, they received the name "Lupamid". This was changed in 1941 to the trade name "Igamid". Thus nylon 66 was designated "Igamid A", nylon 6 "Igamid B", and the 66/6 (60 : 40) copolymer "Igamid 6 A". In 1950, because of the liquidation of the I.G. Farbenindustrie, they were renamed "Ultramid[R]".

Nylon 6 and 66 were the first crystalline thermoplastics. The properties in which they differed most significantly from the previously existing thermoplastics are strength, toughness, abrasion resistance, high softening temperature, and solvent resistance [36]. To a large extent these properties are due to hydrogen bonding between carbonamide groups of neighboring chains and to crystallinity. R. Brill at Ludwigshafen studied the crystal structure of nylon 6 and 66 in 1940 as well as its thermal behavior, and he recognized the role of the hydrogen bonds [37].

High melting point, narrow melting range and low melt viscosity at these high temperatures at first presented some difficulties in processing [38]. In injection molding, heating of the cylinder and the nozzle had to be improved; shut-off nozzles and close-fitting dies were required. Nozzles with automatic needle-type shut-off were developed in 1942/43 by H. Beck [38] and by H. Gastrow [39]. Another problem was the absorption of moisture by the resin. Proper handling under dry conditions was needed to avoid adverse effects in melt processing. Early applications of Igamid A and B moldings during wartime were buttons, zip-fasteners and grenade fuses [18]. But it was also realized that nylons were excellently suited for applications as parts and components in mechanical engineering.

One of the most important applications of Igamid B during the 1940's was melt-cast tapes [23]. Polymerization and shaping were combined in one operation by casting the polymer melt directly from the autoclaves in tapes of various dimensions and profiles. The tapes were used, with or without stretching, with or without hot-water extraction, for instance as transmission belts [40].

Of the various Igamids, the 6 A type had the largest production volume during the years 1940 - 43. Films produced by casting 6 A from hot aqueous alcohol were used as liners for self-sealing aircraft gasoline tanks [18]. Large quantities of film were also kept ready to be used as protective "gas capes" in case of chemical warfare. Compression moldings from cross-linked, plasticized 6 A were used as leather substitute.

In the plastics field, nylon 6 was at first mostly used without prior extraction of the low molecular constituents. (Filament production, however, required extracted polymer for satisfactory spinning performance.) Monomer content in nylon 6 certainly increased impact strength, but it also caused plate-out on chill-rolls and in molds. Because of increasing quality requirements extracted molding-grades were largely introduced about 1955. Due to their excellent wear resistance and self-lubricating properties, nylons in general found wide use in bushings, bearings and gears. Also about 1955, high molecular-weight grades made by a process involving solid-state condensation of the polymer [41] came on the market for extrusion of films, sheets, rods, tubes, pipes and cable sheathing.

Development of injection molding

For one and a half decades, the injection-molding machines that were available for nylon 6 processing were of the plunger type. Plastication was limited, heating inhomogeneous. It was therefore a great improvement when the more efficient screw-type machines came on the market. The design of this new type originates in work by H. Beck at Ludwigshafen in 1943 [42] and was further developed by H. Goller of Ankerwerk Nürnberg, who introduced the first single-screw injection-molding machine in 1956 [43].

This development facilitated the technological break-through of injection molding in general and enabled the nylons in particular to be processed more economically and to higher quality products. Screw-type injection molding provided better temperature control and melt homogeneity; the mixing action led to shear-induced nucleation and thus produced a fine, uniform spherulitic structure which made for easier mold release [44]. Also larger shot volumes were possible and higher-viscosity grades could be processed.

Development of modified grades

During the period of about 1956 - 65 a number of modifications were developed and introduced which helped improve the processability and the performance of nylon 6 and broaden the range of applications [29]. The most important were lubrication, heat stabilization, nucleation, and glass-fiber reinforcement.

The incorporation of long-chain fatty acid derivatives as lubricating agents improved the melt flow properties and the mold-release behavior.

The long-term heat stability of nylon 6 (and other nylons) is somewhat deficient, so a demand for stabilized grades developed early. Heat stabilizers had in principle been known since the 1940's for nylon 6 [45]. It was, however, not before about 1960 that heat stabilized grades containing copper compounds and phenolic antioxidants entered the market to a large extent.

Nylon 6 crystallizes by nature somewhat less rapidly than nylon 66. The incorporation of nucleating agents finely dispersed in nylon 6 increased the rate of crystallization and the set-up in injection molding to the desired extent. By proper choice of nucleation level it was possible to attain economical molding cycles without losing impact strength. Nylon 6 grades containing e.g. colloidal silica or high-melting polyamides as nucleating agents were introduced in the late 1950's in Europe and subsequently also in the United States [46].

Glass-fiber reinforcement of unsaturated polyester resins, of epoxies, and polystyrene had been practiced for a number of years, resulting in a considerable improvement in strength, stiffness, and dimensional stability. Later, about 1960, glass-fiber reinforced nylon 66 was also developed, by Fiberfil Inc. in the United States [47] and by ICI in England [48]. This kind of modification was subsequently applied to nylon 6, in Germany particularly by BASF and Bayer.

More recently, additional modifications of nylon 6 (and of other nylons) were produced. Reference is made to grades containing mineral fillers, flame retardants, and toughening agents. Since these developments are still in full swing, they will not be treated here in detail.

One of the most important uses of nylon 6 is in electrical engineering [49]. Applications in this field, mostly based on modified grades, were mainly developed during the 1960's. The relevant properties are good volume and surface resistivity, high dielectric strength and high resistance to tracking. Resistance to heat and aging, and flame retardancy are additional factors contributing to the excellent electrical performance of nylon 6.

Film

Nylon 6 film has acquired great significance in the packaging sector because of its excellent processability, its good thermoforming properties, and low permeability to gases [50]. The main application is for foodstuff packaging, where composite films with low-density polyethylene are employed, which also feature good heat-sealability and low permeability to water vapor. Large quantities of nylon 6 have been used in this sector since the mid-sixties. As compared to nylon 66 (homopolymer), nylon 6 offers advantages with respect to its greater thermal stability in melt processing, absence of plate-out on the chill-rolls and superior thermoforming properties.

Monomer casting

The technology of "monomer casting" which is used to polymerize large parts at normal pressure directly in the mold was introduced about 1960 [26, 34]. The process is applied to manufacture heavy rods, tubes and plates which are subsequently machined into finished items. Furthermore, large containers such as fuel tanks are produced by rotational molding.

ECONOMIC ASPECTS

Production figures of caprolactam monomer covering the entire development are not available. In Fig. 7 world-wide capacities compiled from various sources are given. The bulk of caprolactam goes into fibers.

Production of nylon 6 plastics started in the early 1940's with, by today's standards, very small quantities, came to a standstill at the end of World War II, and modestly started again about 1950. Subsequently there was a steady increase, particularly pronounced in the early 1970's and temporarily interrupted by the oil crises. In 1985, production in the Western World attained about 285 000 metric tons. Development of nylon 6 production is shown in Fig. 8 (BASF estimate).

The proportion of nylon 6 on total nylon plastic production increased during the first decades, attained approximately 50 % about 1970, and remained in the range 45 - 50 % since then.

ACKNOWLEDGEMENT

We appreciate the assistance of many colleagues at BASF in preparing this paper. Particular thanks are due to Richard Pflüger for helpful suggestions and discussions.

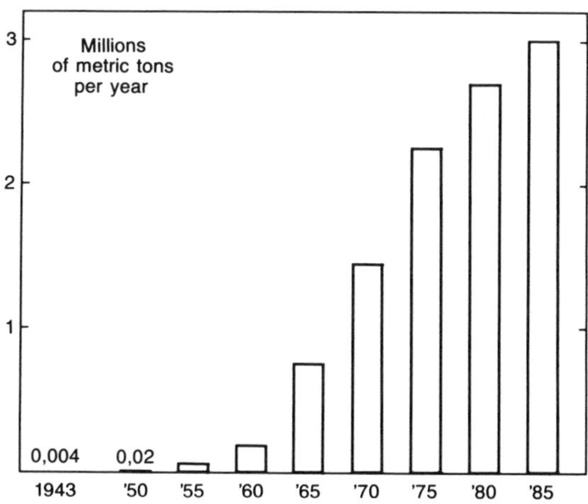

Fig. 7. World production capacity of caprolactam

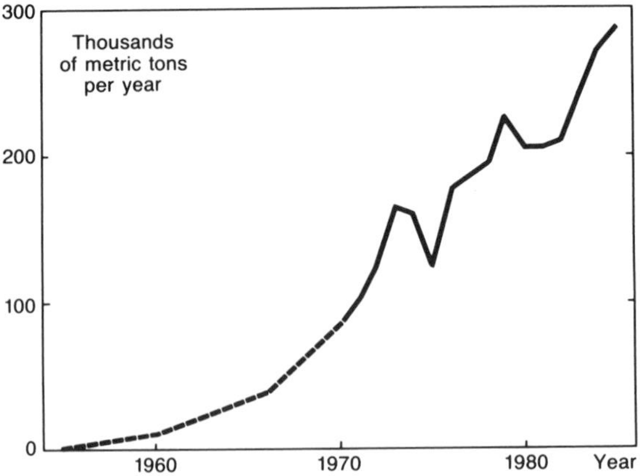

Fig. 8. Production of Nylon 6 Plastics, Western World

REFERENCES

1. S. Gabriel and T.A. Maass, Ber. 32, 1266 (1899)
2. O. Wallach, Ann. 312, 171 (1900)
3. J. von Braun, Ber. 40, 1834 (1907)
4. A. Manasse, Ber. 35, 1367 (1902)
5. H. Mark and G.S. Whitby (Eds.): High Polymers, Vol. I, Collected Papers of Wallace Hume Carothers, Interscience Publ., New York 1940
6. E.K. Bolton, Ind. Eng. Chem. 34, 53 (1942)
7. W.H. Carothers and G.J. Berchet, J. Am. Chem. Soc. 52, 5289 (1930)
8. P. Schlack, Pure & Appl. Chem. 15, 507 (1967); reproduced with permission of the publisher
9. W.H. Carothers and J.W. Hill, J. Am. Chem. Soc. 54, 1566 1932)
10. W.H. Carothers, "Early history of polyamide fibers", memorandum to A.P. Tanberg, Feb. 19, 1936; quoted by H. Morawetz in "Polymers: The Origins and Growth of a Science", John Wiley & Sons, New York 1985, p. 120
11. P. Schlack and K. Kunz, in R. Pummerer (Ed.): "Chemische Textilfasern, Filme und Folien", Ferdinand Enke Verlag, Stuttgart 1953
12. Biographical notes: Nachrichten Chem. u. Techn. 6,4 (1958); Textil-Praxis 18, 1 (1963)
13. P. Schlack, in R. Bauer and H.J. Koslowski (Eds.): "Chemiefaser-Lexikon", Deutscher Fachverlag, Frankfurt am Main, 9. Ed., 1983, p. 13
14. H. Klare, "Geschichte der Chemiefaserforschung", Akademie-Verlag, Berlin 1985
15. P. Schlack, quoted in Chemiefasern 17, 961 (1967)
16. Extracted with permission from: H. Morawetz and H. Mark, Chemtech 15, 345 (1985). Copyright 1985 American Chemical Society

17. DRP 748 253, 11.6.38/23.3.44; US 2 241 321, 20.7.38/6.5.41, P. Schlack, to I.G. Farbenindustrie
18. Mod. Plastics 25, March 1948, 125; April 1948, 148; J.M. DeBell, W.C. Goggin, and W.E. Gloor: "German Plastics Practice", DeBell and Richardson, Springfield Mass. 1946
19. Biographical note: Makromol. Chem. 177, 3473 (1976)
19a. H. Hopff, A. Müller, and F. Wenger, "Die Polyamide", Springer-Verlag 1954
20. B. Achilladelis, Chem. & Ind. 1970, 1549, 1584, 1608
21. E. Bäumler: "Ein Jahrhundert Chemie", Econ-Verlag, Düsseldorf 1963, p. 200
22. DRP 753 046, 6.5.39/20.7.44, G. Wiest, to I.G. Farbenind.
23. Unpublished documents, BASF archives
24. H. Ludewig, Faserforschg. Textiltechn. 2, 341 (1951); obituary: Acta Polymerica 35, 113 (1984)
25. DRP 766 120, 19.6.41/8.2.45, H. Hopff and H. Ufer, to I.G. Farbenindustrie
26. H. Klare, E. Fritzsche, and V. Gröbe: "Synthetische Fasern aus Polyamiden", Akademie-Verlag, Berlin 1963
27. H.K. Reimschuessel, J. Polymer Sci.: Macromol. Revs. 12, 65 (1977)
28. P. Matthies, in Ullmanns Encyklopädie der technischen Chemie, 4. Ed., Vol. 19, Verlag Chemie, Weinheim 1980, p. 39
29. R. Vieweg and A. Müller (Eds.): Kunststoff-Handbuch, Vol. 6, "Polyamide", Carl Hanser Verlag, München 1966
30. K. Tai and T. Tagawa, Ind. Eng. Chem. Prod. Res. Dev. 22, 192 (1983)
31. G. Bertalan et al., Makromol. Chem. 185, 1285 (1984)
32. US 2 251 519, 7.2.39/5.8.41, R.M. Joyce and D.M. Ritter, to DuPont
33. K. Schneider, Kunststoffe 55, 315 (1965)
34. Modern Plastics 39, Dec. 1961, p. 92
35. Kunststoff-Techn. u. Kunststoff-Anwendg. 9, 342, 424 (1939)
36. H. Hopff, Kunststoffe 31, 220 (1941)
37. R. Brill, Naturwiss. 29, 220 (1941); J. prakt. Chem. 161, 49 (1942)
38. H. Beck and F. Schaupp, Kunststoffe 32, 205 (1942)
39. H. Gastrow, Kunststoffe 33, 159 (1943)
40. A. Müller, Kunststoffe 40, 241 (1950)
41. DE 1 048 026, 30.9.55/31.12.58, H. Linge, to BASF
42. DE 858 310, 16.12.43/20.03.52, H. Beck, to BASF
43. R. Sonntag, Kunststoffe 75, p. V (1985)
44. H. Goller, Kunststoffe 49, 531 (1959)
45. DE 883 644, 10.1.43/16.10.52, M. Hagedorn and E. Schmitz-Hillebrecht, to Bobingen AG
46. L.L. Scheiner, Plastics Technology, April 1967, p. 37
47. Chem. Week, Jan. 25, 1964, p. 109
48. GB 950 656, 20.6.61/26.2.64, J. Maxwell and A. Rutherford, to ICI
49. K. Feser, M. Glück, and H.J. Mair, Kunststoffe 60, 155 (1970)
50. C.-D. Weiske, Kunststoffe 61, 518 (1971)

THE HISTORY OF DEVELOPMENT OF NYLONS 11 AND 12

G.B. APGAR AND M.J. KOSKOSKI, ATOCHEM INC., POLYMERS DIVISION,
1112 Lincoln Road, Birdsboro, PA

ABSTRACT

Nylons 11 and 12 are high performance polymers which combine the flexibility and toughness of the lower carbon number nylons with the property of low moisture absorption. These nylons have been very successful in tubing, hose and piping applications as well as cable insulation and jacketing. They are used as well for the injection molding of many items where dimensional stability, toughness and insensitivity to humidity are required. Both Nylons 11 and 12 are the base polymers of formulated coatings for metal items ranging from sea water heat exchangers to building siding panels. These very useful polyamides are the products of research based upon the earlier work of Wallace H. Carothers. The paper summarizes the development, both technical and commercial, of these resins from 1938 to the present.

NYLONS

Nylon is the common designation for aliphatic polyamide thermoplastic resins. They were first prepared by Wallace Hume Carothers in the 1930's at DuPont, which story is elsewhere recorded. The word "Nylon" started out as a trademark of DuPont, but as have the names of other commercial successes; it has evolved into a generic term.

Polyamides are polymer chains whose monomer units are joined by amide links:

$$R - \overset{\overset{\displaystyle O}{\|}}{C} - \underset{\underset{\displaystyle H}{|}}{N} - R'$$

The existence and frequency of these amide groups sets the polyamide resins apart from other polymers and from each other. The amide link is usually formed by condensation reaction between a carboxylic acid and a primary amine.

Monomers for Nylons

There are three commercial groups of monomers: equimolar mixtures or salts of diamines and diacids, lactams and aminoacids. Lactams are usually hydrolysed to the amino acid in the reaction vessel prior to polycondensation. Additional polymerization is also possible with unhydrolysed lactams.

A mixture of a diacid and a diamine was the starting material for the first commercial, high molecular weight, polyamide. Hexamethylene diamine or HMDA (1,6 diaminohexane) and adipic acid (1,6 hexanedioic acid) yield nylon 66, so named because each monomer contains six carbon

Copyright 1986 by Elsevier Science Publishing Co., Inc.
High Performance Polymers: Their Origin and Development
R.B. Seymour and G.S. Kirshenbaum, Editors

atoms.

HMDA has also been commercially reacted with the 9, 10 and 12 carbon diacids to prepare respectively nylons 69, 610 and 612. A long list of diamines and diacids of every conceivable length up to and including 13 carbon have been tried in various laboratories.

The second group of monomers, lactams, has been the raw materials for nylons 6, 8 and most of the production of nylon 12. Nylon 8 is no longer commercially available.

The third group, aminoacids, accounts for nylon 11 and a fraction of the world supply of nylon 12.

HIGH CARBON NUMBER NYLONS

Aliphatic polyamides are named by the number of carbon atoms in their monomer molecules. Naturally, the larger the number of carbons, the longer the monomer chain and consequently, the farther apart the amide groups will be from each other in the polymer chain.

Amide groups are polar and are capable of hydrogen bonding; to each other, to water, to salts and to a variety of other substances, many of which are deleterious to the well-being of the resin. The capability of the amide groups to hydrogen bond to each other leads, in a homopolymer resin where the amide groups are spaced at regular intervals along the chain, to crystallization. The frequency of the amide groups (amide density) affects the crystallinity, and through it, the melting point, modulus, SpG and other physical properties.

The capability of the amide group to bond water causes nylon to absorb moisture. The amide density then determines the extent of moisture absorption, and through it, the consistency of moisture dependent properties such as dimension, modulus and most electrical properties.

The ability to bond to other chemicals influences the retention of plasticizers and stabilizers, as well as chemical resistance.

High carbon number polyamides share with other nylons the characteristics of: toughness; high tensile elongation at failure; good general chemical resistance except to acids and phenols; flexibility; the ability to easily be colored, filled and reinforced; processability by extrusion, injection molding, blow molding and rotomolding; as well as many other desirable traits.

All of the differences between nylons 11 and 12 as a group and the lower carbon number nylons stem from the effects of amide density. Lower amide density in 11 and 12 lowers crystallinity. This in turn lowers melting points and moduli. This may be advantageous or disadvantageous depending on the application. Lower amide density means lower moisture absorption at any given temperature/time/humidity condition. This is always an advantage, as consistency of dimensions, physical and electrical properties is universally sought. Lower moisture absorption combined with a reduced tendency to bond to chemicals leads to improved chemical resistance. The tendency of 6 and 66 to stress crack in contact with calcium or zinc halides is almost absent in 11 and 12.

PROPERTIES OF NYLONS

To illustrate these principles the following table compares a very few selected properties of some common commercial nylons.

	6	66	11	12
Melting Point, °C	220	255	185	175
SpG	1.13	1.14	1.04	1.02
Moisture absorption				
Immersion, Saturation, %	10.5	9.5	1.8	1.7
Torsional Modulus, psi x 10^3				
DAM	500	500	180	200
50% R.H.	230	280	130	150
Dielectric constant				
DAM	6	5	3	2
50% R.H.	12	7	4	3
100% R.H.	>30	20	8	7

The effect of amide density is easily seen upon each of the foregoing properties.

HISTORY

The history of all nylon development begins with the discovery in 1935 of the first high molecular weight polyamide, Nylon 66. Wallace Hume Carothers, working for DuPont, synthesized the first resin from hexamethylene diamine and adipic acid. The first commercial announcement of this material in 1938 and publication of its properties had the effect of initiating parallel studies in Europe.

NYLON 11

In 1938, Joseph Zeltner, research director for Thann and Mulhouse, had an idea which was suggested by Carothers' work, that an omega aminoacid might be prepared from undecenoic acid and that it might be polymerized into a polyamide resin. His company was experienced in the cracking of the castor oil derivative ricinoleic acid to yield 10-undecenoic-acid plus a variety of seven carbon materials. Eventually he and his coworkers, notably Michel Genas and Marcel Kastner were to succeed in bringing to market a very useful commercial polymer.

There were numerous chemical problems to overcome. First, Zeltner wanted to prepare 11-bromoundecanoic acid by hydrobromination of the double bond in 10-undecenoic acid. This operation by a conventional hydrobromination reaction route would lead to the 10-bromo product instead, in accordance to Markownikoff's rule. Work by Karash suggested that hydrobromination in the presence of oxygen (they used peroxide) would proceed in an anti-Markownikoff direction. This proved to be true and commercially practical as well.

The second problem was to effect amination of the brominated carbon. At that time it was thought that amination of a bromocarboxylic acid wasn't possible if the bromine was far removed from the carboxyl group. Genas demonstrated the converse and produced the first few grams of 11-aminoundecanoic acid in 1940. The 10-bromo impurity is not as readily aminated however. This reluctance aids in later purification steps. Process development progressed only sporadically through the wartime years. The laboratory had to be moved twice due to the German occupation and the site of the first pilot plant was levelled by Allied bombing. Mr. Zeltner as well, was lost to the program because of the war. Furthermore, castor oil was almost unobtainable because of its strategic importance as a lubricant.

Despite the difficulties, the monomer process was eventually perfected at a Paris lab, by Kastner under the direction of Genas. In 1944 pilot production was begun in Normandy. 1950 saw the spinning of the first nylon 11 thread. Commercial production of monomer commenced in 1955 at Marseilles, where that plant continues today as the sole production unit of 11-aminoundecanoic acid. Production of polymer continued in the tiny Normandy town of Serquigny;this production constituted the total world capacity until 1971 when a U.S. subsidiary brought a 4000 T unit on-stream.

The commercial process of monomer production begins, of course, with castor oil:

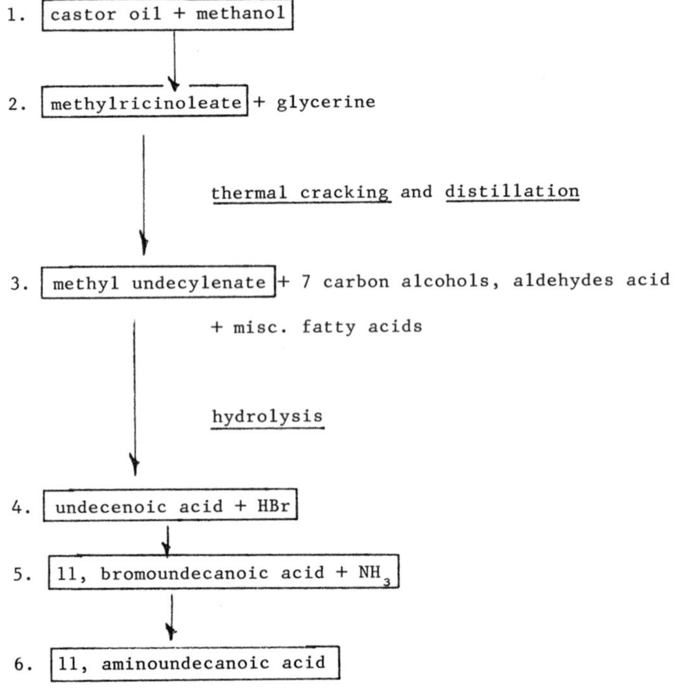

Polymerization is by straightforward polycondensation. Molecular weight is limited by the addition of an acid or base which cap one end of the growing chain. Carboxylic acids, phosphoric acid and potassium hydroxide are all used as chain limiters.

NYLON 12

In contrast to Nylon 11, Nylon 12 was developed simultaneously by several companies; Organico (the forerunner of ATOCHEM which had already pioneered Nylon 11), BASF, Emser and Chemische Werke Huels. Later on, Snia Viscosa and BP also did research work.

The feedstock for most commercial processes for dodecalactam, the usual monomer for nylon 12, is butadiene. In all the processes, butadiene is trimerized to form cyclododecatriene, but after that the similarity ends. Each of the major processes differs in its method of transforming the cyclododecatriene to dodecalactam. In general, research was begun in the late 1950's and led to commercial products from Huels and Emser Werke in 1966 and from Aquitaine Organico in 1971. A joint venture of Toray, Daicel, Huls, Toa Gosei and ATO CHEMIE was envisaged in 1974 for communal production of lactam in Japan, but this never materialized. Since 1979 UBE has been producing commercial Nylon 12 in Japan from 12-aminododecanoicacid monomer, made in turn from cyclohexane rather than butadiene.

Lets look at a brief history of each commercial process for Nylon 12.

Organico/Atochem

Picking up again with the same people who developed Nylon 11, Organico management was interested in the development of a high carbon number nylon that wouldn't be dependent on a vegetable oil and hence vulnerable to crop failures or excessive dependency on imported supply from a very few tropical nations. Genas noticed the publication by G. Wilke describing trimerization of butadiene to cyclododecatriene. This might provide a 12 carbon chain whose unsaturation could be exploited to attach reative groups. Cyclododecatriene proved indeed to be a good starting material not only for dodecalactam but also dodecanedioic acid, one of the monomers for Nylon 612.

The progression toward Organico's commercial production of lactam began in 1958. Initially, cyclododecatriene was epoxidized then hydrogenated to the alcohol. The alcohol was oxidized to the ketone, which was then transformed by reaction with hydroxylamine to the oxime which by Beckmann rearrangement yielded the lactam. They immediately began to work on a method to react nitrosyl chloride directly with dodecatriene and eliminate some intermediate steps. This program eventually led to the earliest version of the commercial process in 1962.

As with many of the commercial processes, butadiene is trimerized to cyclododecatriene or CDT (G. Wilke, 1957)

the CDT is then hydrogenated to cyclododecane or CDAN

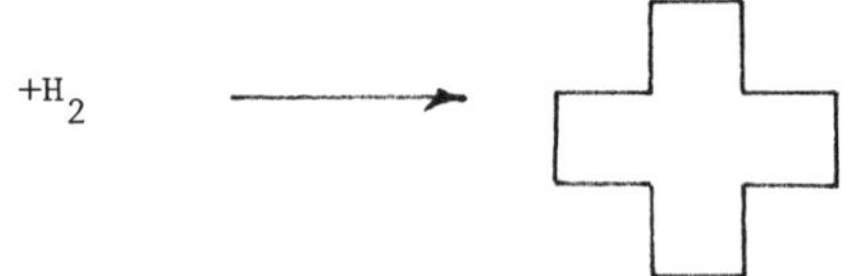

the CDAN is reacted with nitrosyl chloride to form cyclododecanone oxime or CDON oxime. The nitrosyl chloride is formed in situ from $HNOSO_4$, H_2SO_4 and HCl. The corrosion problems are incredible. The NOCl must be dissociated by UV light to make the reaction go.

The CDON oxime is heated in H_2SO_4 solution to effect Beckmann rearangement to dodecalactam or lauryllactam.

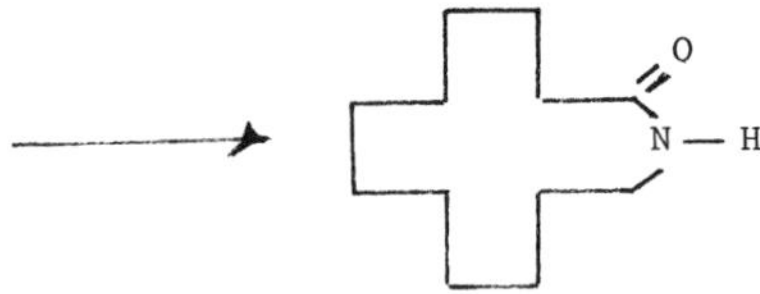

After purification by recrystallization, washing and often distillation, the lactam is ready for polymerization.

Polymerization is most often carried out in a two step polycondensation process. The lactam is first hydrolysed to the aminoacid at high pressure and temperature. Subsequently the pressure and temperature are reduced and polymerization proceeds to the desired viscosity.

Commercial lactam production was begun by Organico at Lacq in southwest France in 1970 and polymer production soon after at Serquigny, their existing site of Nylon 11 production.

Of no commercial importance but technically interesting was a process used by Genas early on in the program to prepare enough Nylon 12 to measure its properties, long before he was able to prepare lactam. Starting from 11-bromoundecanoic acid, an intermediate in the monomer process for Nylon 11, he reacted it with KCN to form the cyano derivative. He now had a 12-carbon chain acid terminating in a nitrogen which he hydrogenated to obtain the 12- aminododecanoic acid. A simple and direct way to see the properties of the intended product before investing in the monomer process development.

BASF

Badische was also involved in research at an early point. Their process, published in 1962, was described by K. Dachs and E. Schwartz.

The process was intended to produce either the 8 or 12 carbon lactam. They actually did produce the eight carbon capryllactam until about 1973 but never commerciallized Nylon 12. Their main monomer process points were:

CDT is reacted with acetaldehyde peracetate or hydrogen peroxide to form Cyclododecadiene epoxide.

This product is then hydrogenated to cyclododecane epoxide

The epoxide is then rearranged to the ketone, CDON, with the use of MgI_2 as a catalyst.

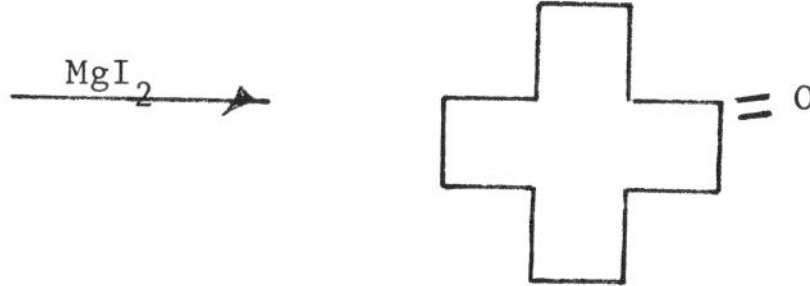

This is reacted with hydroxylamine to form the oxime, CDON oxime.

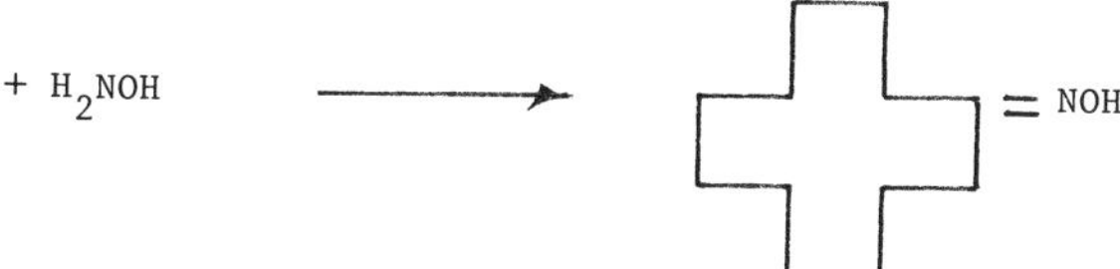

The balance of the reaction is as previously described, Beckmann rearrangement to the lactam in H_2SO_4 solution.

HUELS

Judging from the dates of their patents, Huels' lactam research began around 1959 and continued through 1964. Huels' early lactam work is reported to have been done in cooperation with Emser. Initial research was done by Muller and Weber. Later process improvements were chiefly by Strauss. They were the first to come to market with Nylon 12, starting up a 1200 T unit in 1966 at Marl, West Germany. Their monomer plant was set up in cooperation with Emser Werke. Published reports of plant capacities show volumes increasing to 3600 T in 1969, 4500 T in 1971, 6000 T in 1974 and 8000 T in 1976.

The process resembles, in the first and last steps, the Organico method. The oxime is created by a different route however.

CDT is hydrogenated to CDAN

CDAN is oxidized with air or peroxide to CDON

CDON is converted to the oxime, CDON oxime with hydroxylamine sulfate and sodium hydroxide.

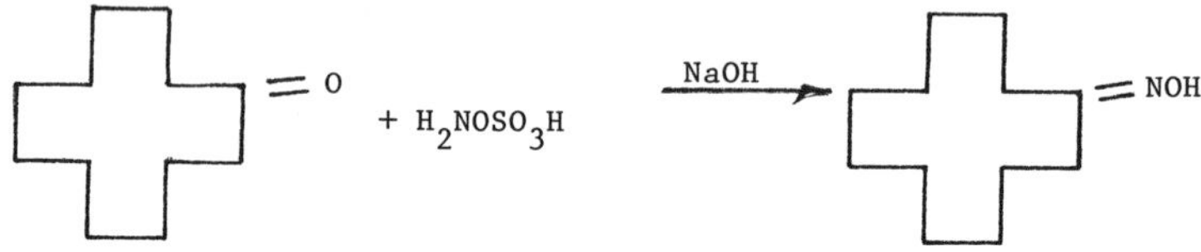

Finally as in previous reaction schemes, Beckmann rearrangement of the oxime yields the lactam.

This process, while more complicated than some of its competitors, yields monomer and polymer of excellent quality. Huels' products were for many years the standard of the nylon 12 industry.

Emser Werke

Emser's history is closely interwoven with that of Huels. At first, much of their monomer research was done in common with Huels. The first commercial lactam unit, although located at Marl, which is Huels' headquarters, was operated by Huels in cooperation with Emser. Emser backed out of the partnership upon starting up a unique plant which made caprolactam and lauryllactam simultaneously in approximatly 1:1 ratio. This process was described in 1970 and led to a production unit which started up in 1973. While Emser ceased to obtain lauryllactam from Marl at this time, they did continue to get their starting material CDON from the Marl process stream. This mixed lactam plant was shut down in 1975 when Emser began again to simply purchase lactam from Huels.

Emser's polymer production dates also from 1966. They constructed and commissioned a subsidiary plant at Sumter, SC in 1981 but soon ceased nylon 12 polymerization there, probably in 1982.

Emser's mixed lactam process was understandably similar to that of Huels. The difference was that the steps of the process from oximation on; oximation, rearrangement and initial purification, were done with a mixture of 6 and 12 carbon species. A mixture of cyclohexanone and CDON was fed to the oximation unit and allowed to remain mixed until caprolactam and lauryllactam were separated at the final step. This process had two main advantages; the low melting, more-soluble C6 species acted as process solvents for the high melting, less-soluble C12 rings, and the output of two monomers provided economy of scale to a caprolactam plant of relatively low volume. This monomer process would be ideal for a producer of 6/12 copolymers.

Ube

Ube Industries is the sole producer of nylon 12 monomer in Japan although an ambitious joint venture was envisaged years ago. In 1974 Toray, Daicel, Huels, Toa Gosei and ATO were going to put up a joint monomer plant in Japan but would continue to compete at the polymer end of business. This project never came to fruition, leaving Ube as Japan's sole domestic supplier. Ube's production came on-stream in 1979 with start-up of a 2500 T unit.

Ube's process was developed by British Petroleum in 1971 and is unusual for several reasons; first, it yields 12-aminododecanoicacid or 12 ADA instead of lauryllactam, and second, the starting material is cyclohexanone rather than butadiene. The advantage is that Ube starts from relatively available and inexpensive raw materials; cyclohexanone, ammonia and hydrogen peroxide. Reputed disadvantages are the danger of explosion during peroxidation and purity of product. The reaction scheme is the following:

Cyclohexanone is reacted with peroxide and ammonia to form 1,1'- peroxydicyclohexylamine or PXA.

$$\text{(cyclohexanone)}=O + HOOH + NH_3 \longrightarrow \text{PXA (O — O; N—H)}$$

PXA is pyrolysed over LiCl to form 11-cyanoundecanoic acid or 11 CUDA.

$$\xrightarrow{LiCl} \text{(cyclic imide, O H O / N)} \longrightarrow N \equiv C\text{-}(CH_2)_{10} - \overset{O}{\overset{\|}{C}}\text{-OH}$$

Hydrogenation of 11 CUDA yields 12 ADA.

$$+ H_2 \longrightarrow H_2N - (CH_2)_{11} - \overset{O}{\overset{\|}{C}} - OH$$

Polymerization of 12 ADA is identical to that for 11 AUA, the monomer for nylon 11. It doesn't require the high-pressure ring opening step needed for polycondensation when starting from the lactam.

Snia Viscosa

This 12 ADA route was never commerciallized, to the best of our knowledge. We are including it because it is unusual in several ways:

CDT is reacted with ozone and acetic acid,

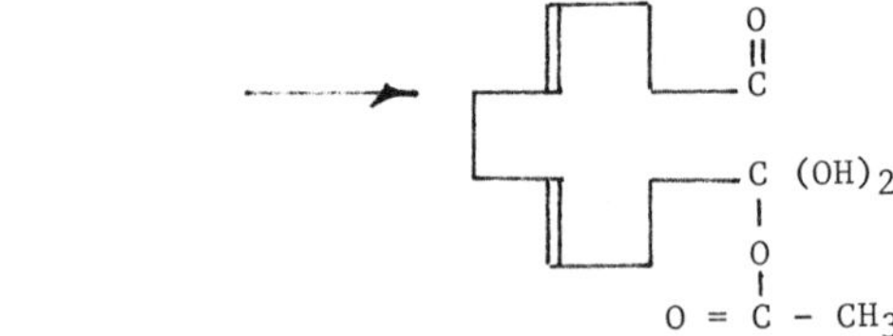

rearranged,

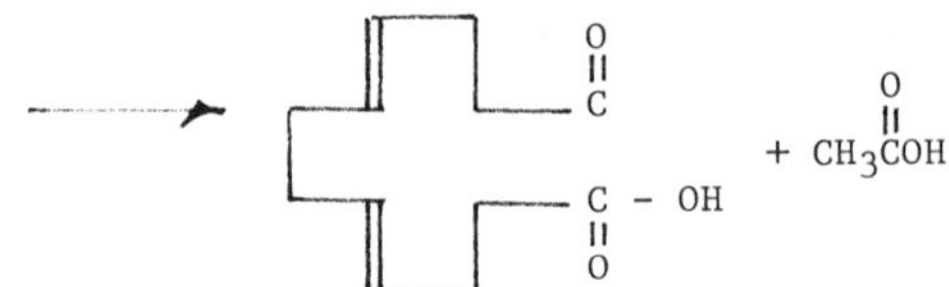

aminated

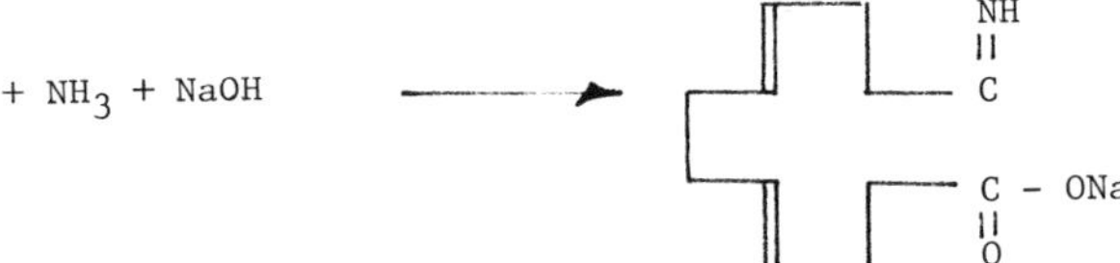

hydrogenated,

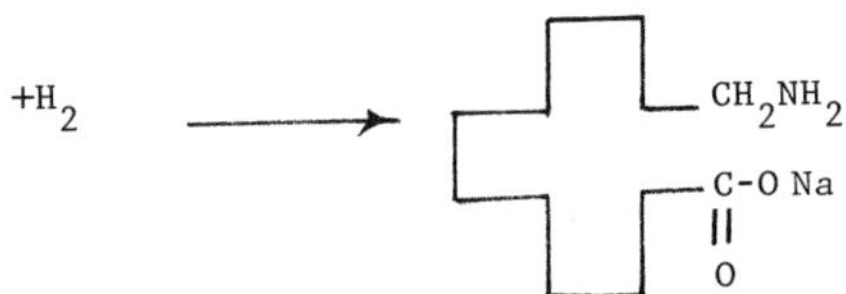

and acidified to 12 ADA

Applications

Nylons 11 and 12 are used extensively in the extrusion of tubing, hose and piping. End use situations are for example: automotive fuel lines, industrial and power transmission hose, petroleum production hose, and truck air brake tubing. Other extrusion applications include insulation and jacketing of wire and cable. Buffering of optical fiber for the communication and data processing industries is a relatively new use for these resins. Both resins are, in unmodified form, FDA listed; they are used for a variety of medical and food contact applications ranging from cardio-catheters to meat processing film. Injected molded items are made from high carbon number nylons when low moisture absorption and improved chemical resistance are required in addition to nylon's usual toughness.

Both 11 and 12 are used in the preparation of coating powders. Nylon powder coatings are tough, durable substitutes for paint which are applied in 100% solid form which eliminates the need for solvents. Both resins also are convertible by rotomolding and blow molding methods.

In summary, nylon 11 and 12 are the most moisture insensitive and chemically resistant examples of those tough and useful thermoplastics known as nylons.

References

1. C. Bernier, "Le Rilsan", unpublished, 1978
2. R.J. Hebert, private correspondence, 1985
3. W.K. Franke and K.A. Muller, Chimie - Ing. Techne 36, 960 (1964)
4. K. Dachs and E. Schwartz, Angew. Chem. 74, 540 (1962)
5. G. Wilke, Angew. Chem. 69, 397 (1957)
6. G. Strauss, Chemical Engineering, 7-28-69, 106
7. R. Feldmann, private correspondence, 1986
8. C. Berther and H.J. Schultze, Chimie et Industrie 103, 1235 (1970)
9. Chemical Week, 7-14-71, 27
10. European Chemical News, 5-14-71, 31
11. European Chemical News, 3-29-74, 18

HISTORY
Aromatic Polycarbonates

Daniel W. Fox
GENERAL ELECTRIC COMPANY
One Plastics Avenue
Pittsfield, MA

Aromatic polycarbonates were first disclosed by A. Einhorn in 1898. He reacted pyridine solutions of hydroquinone, resorcinol and catechol with phosgene to produce respectively (1) an insoluble and intractible hydroquinone polycarbonate, (2) a glassy resorcinol polycarbonate melting at about 200°C and (3) a simple cyclic catechol carbonate. Four years later C. Bischoff and A. von Hedenstrom produced and reported the same products employing transesterification. And then the subject of aromatic polycarbonates was apparently forgotten.

More than 50 years later Dr. Herman Schnell at the Uerdingen West Germany plant of Farbenfabriken Bayer A.G. and I at General Electric's Corporate Laboratory in Schenectady, N.Y. independently synthesized aromatic polycarbonates based on bisphenol-A. At the time of my initial work I was not aware of the 1898 and 1902 work because of neglect to check the literature. We were not aware of Dr. Schnell's polycarbonate activity until his publication in Argewante Chemie in 1956. When this publication, which was received as somewhat of a bombshell in Pittsfield, Massachusetts, occurred we were three years along on our commercial development program.

Inasmuch as the ultimate patent disposition could require years and place both Bayer's and General Electric's development in jeopardy, cross-licensing agreements were negotiated. The wisdom of the early decision to cross license was confirmed by the prolonged course of patents through the USA Patent Office. Bayer's basic patent claiming polycarbonate products made by either solution or interfacial processes was filed in 1954 and issued in 1962. General Electric's patent claiming products made by way of transesterification was filed in 1955 and issued in 1964.

Having established a basis for future cross-licensing General Electric and Bayer continued their very successful independent development and commercialization programs. Other companies have subsequently entered the polycarbonate business under license or on expiration of patents. Bisphenol-A polycarbonates have found broad application around the world as a premier engineering plastic.

Copyright 1986 by Elsevier Science Publishing Co., Inc.
High Performance Polymers: Their Origin and Development
R.B. Seymour and G.S. Kirshenbaum, Editors

Volume-wise, they are a close number two and gaining on polyamides for which the name "engineering plastics" was originally coined.

Features which qualify bisphenol-A polycarbonates as "Engineering Plastics" include outstanding toughness, dimensional stability and load bearing properties over a very broad temperature range, transparency, resistance to burning, excellent electrical properties, etc.

General Electric Background

One reported basis for the Bayer work which lead to bisphenol-A polycarbonates was their desire to expand and build on the properties of polyethylene terephthalate. This same polymer played a role in the General Electric development. A substantial program at General Electric's Corporate Research Center in Schenectady aimed at thermally upgrading magnet wire insulation had reached a successful conclusion in 1953. The basis for this new, high performance magnet wire insulation was a polyethylene terephthalate polymer modified by partial substitution of polyols for some of the ethylene glycol. This substitution enabled cross-linking to occur at very high application temperature. The product was commercialized and is still in use.

However, some last minute experimentation aimed at improving the hydrolytic stability of the polyethylene terephthalate backbone lead to bisphenol-A polycarbonate. The simple molecule, guaiacol carbonate, provided the key to this transition since it was known to be very resistant to hydrolysis. This lead to the concept of preparing bis-guaiacol polycarbonate as the basis for a more hydrolytically stable wire enamel. There was no bis-guaiacol in the stockroom so bisphenol-A was selected as a model.

Experimental

The first successful synthesis of bisphenol-A polycarbonate was effected by transesterification (a technique learned from our modified polyethylene terephthalate syntheses). Bisphenol-A and a slight molar excess of diphenyl carbonate were heated in a three neck flask fitted with thermometer, distillation head and a stainless steel shaft stirrer (no added catalyst). Phenol began to evolve at about 200°C. The temperature was gradually raised and the pressure was gradually reduced to maintain evolution of phenol. The viscosity of the polymer continued to increase indicating formation of the sought polycarbonate. After several hours the reactor temperature was approaching the thermometer limit of 360°C; the pressure was reduced to the limit of the aspirator; and the stirring motor stalled. The thermometer was salvaged and the experiment terminated.

On cooling, a combination of differential thermal coefficient of expansion and excellent adhesion of the polymer to glass caused delamination of the interior of the flask. The remnants of the flask were broken away to yield a hemispherical,glass fragment embedded, glob of plastic on the end of a steel stirrer shaft. The glob was pounded on the cement floor and struck with a hammer in abortive attempts to remove the remaining glass, and/or, shatter the plastic. The plastic pseudo mallet was even used to drive nails in wood. Eventually, the original glob of polycarbonate on the end of a steel rod was sawn and chiseled into small pieces for characterization and evaluation.

For General Electric the glob provided the seed of a new multibillion dollar industry.

Commercialization

Perhaps the first obstacle to commercialization was lack of a viable polymerization process. Scale-up of the glass flask reactor was technically feasible but commercially impractical. There was literally no equipment available at the time which could provide or promote the required effective heat and mass transfer under reduced pressure through an extremely viscous media at temperatures of 300°C and higher.

After a series of mechanical horror stories, attention was turned from transesterification to direct phosgenation. High molecular weight polymer was produced by passing phosgene into a stirred solution of bisphenol-A in a mixed methylene chloride/pyridine solvent. Excess pyridine and by-product pyridine hydrochloride was removed by water/acid washing. The polymer was recovered by addition of an anti-solvent such as alcohol or aliphatic hydrocarbon. This general process provided initial development quantities of polymer.

We learned later that Bayer was practicing a form of tertiary base catalyzed interfacial polymerization. This process consisted of metering phosgene and caustic solution into a stirred slurry of bisphenol-A and tertiary amine catalyst in a water/methylene chloride solvent mixture. The polymer forms at the aqueous/organic solvent interface and remains in the organic phase. The formed sodium chloride accumulates in the aqueous phase. The polymer may be recovered in a manner similar to that employed in solution polymerization.

The second obstacle to commercialization was the lack of polymer grade bisphenol-A. Available epoxy resin grade material was totally inadequate. Various purification processes were evaluated with the ultimate solution being backward integration of polycarbonate producers.

The third obstacle to commercialization was polymer conversion or fabrication of useful items from powdered polymer. Most available extruders and molding equipment could not provide the high temperature and pressure required. The quality of available polymer tended to aggravate the processing problems since occasional supposed linear polymer would rearrange and cross link at the high processing temperatures. Commercialization was a learning experience for our bisphenol-A monomer suppliers, for us and even more so for our fabricator customers.

The technical obstacles were overcome, the challenges met and today many hundreds of millions of pounds of polycarbonate polymers per year are serving us. They occur as shatter proof baby bottles, streetlights, bus/plane/train windows, astronaut/athletic/firemen helmets, protective or even bullet resistant windows in public buildings, sterilizable medical utensils, business machine housings, laser recording discs for symphonies and many other places where a combination of toughness, transparancy, fire retardance, dimensional stability and strength over a temperature range from -100° to + 150°C is required.

References

1. A. Einhorn, Liebig's Ann. Chem. 300, 135 (1898).

2. C. Bischoff and a. von Hedenstrom, Ber. 35, 3431 (1902).

3. H. Schnell, Ang. Chem. 68, 633 (1956).

4. D. Fox and W. Christopher, "Polycarbonates", Reinhold Publishing Corp., New York, 1962, p. 31.

5. H. Schnell, "Chemistry and Physics of Polycarbonates", Wiley-Interscience, New York, 1964, pp. 44-51.

6. D. Fox in Encyclopedia of Chemical Technology, Volume 18, Third Ed., 1982, John Wiley & Sons, Inc., p. 479-494.

THE HISTORY OF POLY(BUTYLENE TEREPHTHALATE) MOLDING RESINS

DONAL MCNALLY & JOHN S.GALL
Celanese Engineering Resins Inc.,
26 Main St. Chatham, N.J.

Poly(butylene terephthalate) is usually called PBT by analogy with PET, the well known fiber forming thermoplastic polyester. Like PET, PBT is a linear condensation polymer based on terephthalic acid, but instead of the ethylene glycol used in PET, PBT employs 1,4-butanediol. The more definitive name, poly (tetramethylene terephthalate) or PTMT, never did catch on, perhaps an indication of the power of three letter strings in the English alphabet.

The utility of PBT molding resins can be gauged from the extent to which their use has grown over the last fifteen years. As Fig. 1 shows, current world consumption is estimated to be in excess of 107,000 tonnes annually, or about 235 million lb., with a market value over a third of a billion dollars.

FIG. 1
ANNUAL CONSUMPTION OF PBT MOLDING RESINS
(1972 - 1985)

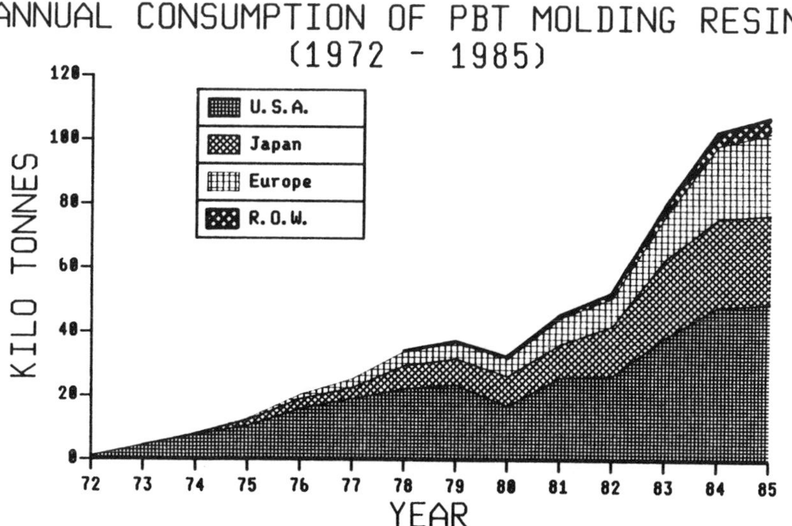

World-wide usage of engineering resins in general is growing at about 6%-7% per annum at the present time, with a somewhat higher growth rate in the U.S. [1], but PBT usage has been growing quite a bit faster, averaging close to 20% per annum over the last few years. Recent growth has been least in the U.S., at 15%, somewhat higher in Japan and Western Europe, at 18% and 26% respectively, and higher again in the rest of the world, at 46%. These increasing growth rates of course reflect smaller bases in each case.

Copyright 1986 by Elsevier Science Publishing Co., Inc.
High Performance Polymers: Their Origin and Development
R.B. Seymour and G.S. Kirshenbaum, Editors

The long term growth of PBT-based molding resins was driven in the beginning by the combination of good properties and excellent moldability. As the original products became more mature, further growth was fueled by the materials's remarkable ability to incorporate fillers and additives and to blend or alloy with other polymeric materials. Indeed, the relatively recent surge in marketplace interest in polymer blends and alloys is due in part to the flexibility and utility of PBT as a base material for formulations of this type.

Returning to the beginning (as described by Korshak and Vinogradova [2], we see that polyesters of polybasic acids and polyhydric alcohols were first synthesized by none other than Berzelius, and that the line from him runs forward through many distinguished investigators until it reaches W. H. Carothers, justly famous for his far-ranging work on linear fiber-forming polymers which provided the foundation for a major twentieth century industry. It's only a short step onward from him to J. R. Whinfield and J. T. Dickson, who first prepared the terephthalates in the 1940's [3], but on the other side of the Atlantic from Carothers.

This early work by Whinfield and Dickson and others such as Flory [4], Batzer & Fritz [5], Hill & Walker [6], Izard [7], et al. showed the utility of poly(methylene terephthalates) as compared to terephthalate polyesters wherein diols are not straight chain. As Table I shows, use of diols with side groups generally leads to amorphous low melting polymers, whereas the straight chain diols give crystalline (crystallizable) materials. However, as the aliphatic segment of the diol becomes longer, the product becomes more like polyethylene, and without nylon's hydrogen bonding the melting point drops off as shown in Fig. 2.

FIG. 2
MELTING POINTS OF
POLY(METHYLENE TEREPHTHALATES)
C2 - C10

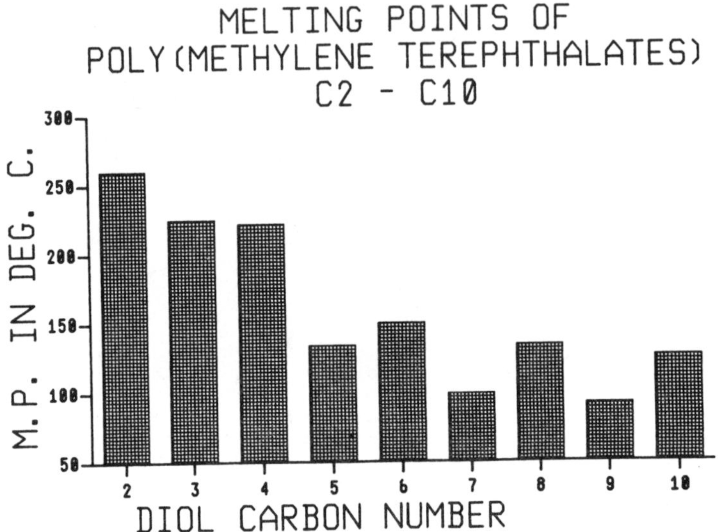

TABLE I. Terephthalate Polyesters with straight chain and branched diols.

Glycol	M.P. in deg. C.	Crystallinity
1,3-propane diol	225	Yes
propylene glycol	122	No
1,4-butanediol	222	Yes
1,3-butandiol	81	No
1,5-pentanediol	134	Yes
1,6-hexanediol	150	Yes

The initial success of PET as a fiber and as a film, combined with the low cost of ethylene glycol, made inevitable the extensive investigation of its properties as a molding material. It quickly became clear that the "cold crystallization" phenomenon was a major obstacle to successful use of PET in injection molding applications, as was its sensitivity to moisture during molding. The moisture problem comes about because the effect of water on PET around the melting point is to give a virtually instantaneous hydrolysis [8]. Adequate drying is enough to handle this problem, but the cold crystallization presents greater difficulty.

The temperature at which PET crystallizes most rapidly is about 190 degrees C. (Fig.3) To get fully crystallized parts (about 50% crystallinity), plain PET must be molded in a hot mold on a relatively long cycle. For economic reasons, most molders run cold molds and short cycles, and, in PET, this conduces to the production of amorphous or partially amorphous molded parts.

FIG. 3
CRYSTALLIZATION HALF-TIMES AND INDUCTION PERIODS OF PET

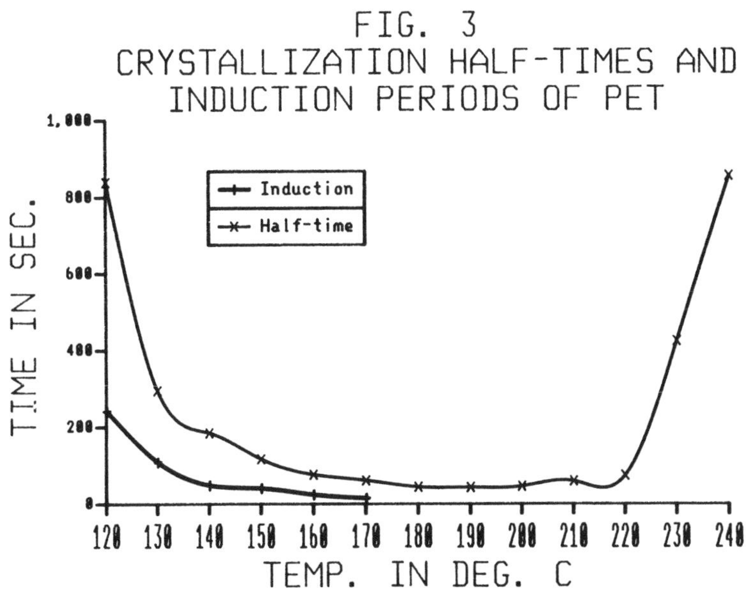

As a result final parts were subject to post-molding crystallization over time, were variable from cavity to cavity and sometimes from part to part and showed significant dimensional instability. The earliest attempts to develop PET molding resins did not fully address these problems, so that the use of hot (oil heated) molds was necessary. Mold heating systems of this sort were not commonly found in the U.S., with the result that in the late 60's, PET enjoyed modest acceptance only in Japan and Western Europe, where oil heated molds were not uncommon. Teijin and Akzo respectively were the main suppliers in these two areas. In the U.S. PET molding resins were virtually unknown, setting the stage for the emergence of PBT materials which crystallize very rapidly, even at room temperature. In fact, one must take considerable pains to get amorphous PBT in any shape other than the thinnest. A thick part cannot be cooled quickly enough to avoid crystallization in the interior.

Though Whinfield and Dickson decribed linear terephthalate polyesters from glycols containing 2 through 10 methylene groups, commerical interest did not develop in the tetramethylene polymer until the 60's. However, there is earlier evidence of some interest in the material, in the form of patents issued to ICI over the period 1957 through 1962, which deal with formation of PBT from both dimethyl terephthalate/1,4-butanediol and terephthalic acid/1,4-butanediol. This work at ICI led to the evaluation of PBT in fiber end uses from which it was seen that PBT fibers exhibited superior recovery characteristics, making them suitable for use in carpets and stretch fabrics. At present, PBT is still used in stretch fabrics, but the earlier carpet use has fallen by the wayside, primarily because of cost and poorer performance in the "pill" test, a method of comparing flame spread characteristics of carpeting. Nevertheless, it was this late 60's carpet application for PBT fiber which provided the volume manufacturing base from which the injection molding PBT materials were developed.

Fiber Industries Inc., a manufacturing concern at that time jointly owned by ICI and Celanese Corporation, was making PBT polymer for fibers. In common with other major PET manufacturers, Celanese was then looking at ways to improve on the performance of PET in injection molding applications, and the ultra-rapid crystallization of PBT suggested it as a candidate for molding evaluation. The initial assessment by Celanese Plastics of fiber grade PBT confirmed its excellent moldability, surface gloss and natural color. The mechanical properties of this lower molecular weight material were moderate and its impact resistance limited, so fiber glass reinforcement was evaluated, resulting in greatly enhanced physical properties [9]. Consequently, in 1969, PBT was introduced to the injection molding marketplace by Celanese Plastics as a glass reinforced material under the tradename X917, a designation later changed to Celanex® Thermoplastic polyester.

As often happens in technology, use of PBT as a solution to the PET molding problems occurred to other investigators at about the same time as it did at Celanese. Based on the studies of poly(methylene terephthalates) carried out in the mid 60's by Smith, Kibler and Sublett [10], glass reinforced PBT was independently developed at Tennessee Eastman [11]. However, the product was taken to the market by Celanese, immediately gaining recognition in the form of the IR100 award, an industry honor bestowed on the 100 most significant new products introduced in that year [12].

The significance of PBT molding resins was succinctly stated by a consultant - "It is the first commercial thermoplastic polyester to provide molding properties that make it a formidable competitor of other engineering thermoplastics and thermosetting resins". [13]. This judgement was based on the fact that glass reinforced PBT combines many of the desirable properties of the ideal thermoplastic engineering resin: high heat deflection temperatures, high rigidity, broad chemical resistance, low creep, low moisture absorption, excellent electrical properties and superior dimensional stability at elevated temperatures.

The initial favorable reception given the material resulted in an explosion of grades and suppliers. The first major addition to the range of products was a flame retardant grade with a limiting oxygen index of about 29% [14]. This grade was given the SE-O rating (as it was then known) in the U.L. 94 test. An important aspect of the new product was that it attained a high level of flame retardation without loss of mechanical or electrical properties. Then, as now, the flame retardant package used bromine with an antimony synergist [15].

The early end-uses of PBT materials were concentrated in the electrical and automotive areas, primarily in replacement of thermosets and to some extent, metals. Initial electrical/ electronic applications were in trimmer potentiometer housings, printed circuit connectors, terminal strips, coil forms and bobbins, motor commutators and integrated circuit carriers [16]. In the automotive area the major applications were in distributor caps and other ignition system components such as ignition coil forms and caps. The beginning of integrated ignition systems can be traced to the GM High Energy Ignition developed in the mid 70's, and the success of this HEI system was due in significant measure to the excellent electrical properties of PBT and to its ability to resist the under-hood environment as well as the improved economics of production associated with a thermoplastic material [17].

Automotive body parts got their start in PBT at about the same time as the HEI system. The first significant part molded in PBT was a window louver at Fisher Body division of GM. This metal replacement end use was for a while the largest volume single commercial application of an engineering resin [18].

As the PBT business was getting up a good head of steam, a number of other companies entered the market, chief among them being General Electric's Plastics Dept., so that by 1973 the U. S. market was served by Celanese, G.E., Eastman and Goodyear, with the water also being tested by LNP, Goodrich and Hooker. In Europe, entrants amounted to ICI, BASF, Hoechst, Bayer and Ciba-Geigy. Japan also had a number of suppliers - Toray, Toyobo, Teijin and Mitsubishi Rayon as well as Polyplastics, an affiliate of Celanese. A Celanese report from that time lists 51 grades available from the 17 named suppliers [19], and there were undoubtedly quite a few more. By 1976, a study by Kossoff & Associates showed, in addition to the U.S.

companies listed above, DuPont, GAF, Allied Chemical, Plenco, Mobay, Amoco, Rohm & Haas, FMC, Hercules, Diamond Shamrock and Cyanamid as having an interest in the PBT business and having in some cases actually entered the market [20].

A shakeout was of course inevitable, and in the ten years since the cited study, it has indeed happened. Now there are only three major suppliers in the U.S. - G.E., Celanese and GAF, although DuPont has announced an intention to reenter the market. Both Celanese, and G.E. are still active in Western Europe, as are AKZO, ATO Chemie, BASF, Bayer, Ciba-Geigy, Huels, Montedison and Rhone-Poulenc. In Japan, Polyplastics, the Celanese affiliate, remains in the fray, along with Dainippon, Teijin, Mitsubishi Rayon, Mitsubishi Chemical and Mitsubishi Petrochemical.

While the market history was unfolding, so also was the product history. PBT products can be grouped into three generations, each associated with a particular time period. The initial product was a glass reinforced lower molecular weight (Intrinsic Viscosity, or IV = 0.75) material, quickly followed by a flame retardant glass reinforced grade. Interest in the base polymer as a molding resin was rather modest, primarily because its performance envelope was moderate. Although an excellent molding resin with a first class surface finish and superior electrical properties, its mechanical properties without glass reinforcement were ordinary and its impact resistance was limited. The advent of higher molecular weight polymer (IV = 1.0) made a considerable difference in practical impact, although notch sensitivity was not that much improved. The first generation of PBT based molding resins can be considered therefore to consist of the base polymer, coated fiberglass, and a flame retardant package. The key to the success of these products was the ability of PBT to wet out glass fibers and obtain high reinforcing efficiency from them, combined with ease of molding and fast set-up. Without these, it is doubtful that the material would have had much success.

By the mid 70's, the anisotropy of predominantly glass reinforced PBT was being seen to limit its utility in areas which could benefit from its properties and moldability. This led to a wave of low warp products which sought to alleviate the effects of fiber glass orientation by using other filling materials. These formulations achieved their effects by randomizing the glass fibers through the presence of conjugate fillers, or by using reinforcing fillers other than glass fibers. In the latter case, anisotropy was reduced by using fibers of different aspect ratio such as Wollastonite, or by using platelets such as mica or glass flakes. In many cases the additives had no special coating or other surface treatment. Again, the key to success was the ability of PBT polymer to wet out these inorganic additives and effectively transfer stress to them. Thus was born the ability to tailor formulations to specific property requirements [21] albeit in a more primitive way than was later provided by the third generation products.

The capabilities of other inorganic additives were relatively rapidly exhausted so the search moved to different ways of modifying the properties of this protean resin. The third (and present) generation products rely on the use of organic additives, generally other polymeric materials, to broaden the performance range of PBT. First among these other additives are polycarbonate (PCO) and PET, but a wide variety of impact modifiers is also used. The first two fall under the general heading of "polymer alloys", a topic which excited a great deal of interest from the late 70's on. Some impact modifiers, because of their particulate form, retain integrity and cannot be considered "alloys".

The literature on the third generation products is extensive, and like much of the other writing in this field, is mostly in the form of patents.

However, some key points do emerge from this large amount of literature. In PBT/PET blends one would expect the occurrence of transesterification in the melt. With sufficient heat history equilibrium will be reached at the property set of the corresponding copolymer. In practice the copolymer melting points, as shown in Fig. 4 [10], are simply not developed within the typical melt residence time experienced in compounding and molding. Thus, commercial blends (major portion PBT) process at PBT temperatures and give parts with a high gloss surface finish.[22],[23]. The phenomenology of this well-known effect has not been studied in detail, but one might speculate that the glossy surface is an indication of a more amorphous surface layer.

FIG. 4
MELTING POINTS OF
PET/PBT COPOLYESTERS

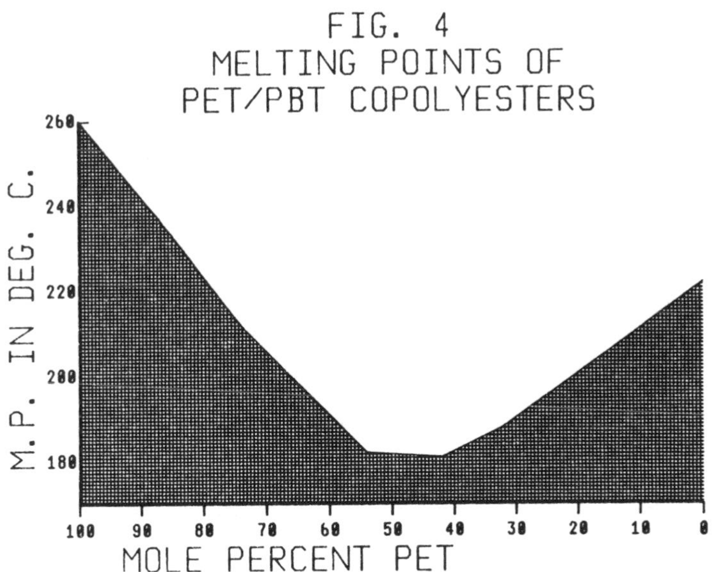

Arguably, the most remarkable property of polycarbonate is its very low notch sensitivity in thin sections. As thickness goes up, the notched Izod impact goes down. PBT/PCO blends exploit this behavior effectively, ameliorating the notch sensitivity which is a drawback of plain PBT. These blends also improve other mechanical properties over those of the base polymer, especially stiffness and heat deflection temperature. Of course, there is no "free lunch" here either, but the price is modest - increased creep with PCO content, associated with an apparent drop in the glass transition temperature [24].

Many elastomers have been blended with PBT, but the greatest interest appears to have developed in the core/shell acrylates [25], [26]. These additives reduce stiffness proportionally with their concentration, but do not much depress heat deflection temperature of the base resin. They do make a major change in notch sensitivity, causing a propagating crack to lose itself in a welter of crazes and other energy absorbing mechanisms.

The third generation technologies described above have by now been tested extensively in the marketplace. G.E. was the first to launch these products with its Xenoy[R] line, the subject of another paper in this symposium [27]. Mobay and Celanese, with Pocan and Duraloy[TM] respectively, were not far behind. At present, the main use area of the third generation products is in automotive applications where their ability to provide toughness, flatness, low warp, paint oven resistance and so on while reducing weight makes them very useful indeed.

The remarkable adaptability of PBT based molding materials is creating for them a future which should last well into the 21st century. Interestingly, PET molding resins should by then present greater competition, because the problems affecting its successful molding appear to have been overcome by a number of companies, first among them being DuPont. Again, this topic will be discussed in another session [28]. Overall, it seems safe to say that just as the development of nylon was the most significant event in engineering thermoplastics in the first half of the 20th century, so will the development of injection moldable polyesters be in the second half.

Celanex is a Registered Trademark of Celanese Corporation

Xenoy is a Registered Trademark of General Electric Company

Duraloy is a trademark of Celanese Corporation

REFERENCES

1. K. D. Thompson, J. Commerce, March 24 1986.
2. V. V. Korshak and S. V. Vinogradova, Polyesters (Pergamon, Oxford 1965).
3. J. R. Whinfield and J. T. Dickson (to E. I. DuPont de Nemours and Co.), U.S. Pat. 2,465,319 (1949).
4. P. J. Flory et al. (to Wingfoot), U.S. Pat. 2,589,688 (1952).
5. H. Batzer and G. Fritz, Makromol. Chem.11, 85 (1953).

6. R. Hill and E. E. Walker, J. Polym. Science, 3, 609 (1948).
7. E. F. Izard, J. Polym. Science, 8, 503 (1952).
8. I. Marshall and A. Todd, Trans. Faraday Society, 49, 73 (1953).
9. D. Zimmerman and R. B. Isaacson (to Celanese Corporation), Brit. Pat. 1285838 (1972).
10. J. G. Smith, C. J. Kibler and B. J. Sublett, J. Polym. Science, 4, 1851 (1966).
11. H. F. Kuhfuss, et al. (to Eastman Kodak), Brit. Pat. 1320520 (1972).
12. Industrial Research Incorporated (1970).
13. Report 73-1, Polybutylene Terephthalate, i (Chem Systems Inc., New York 1973).
14. R. Lovenguth, Proc. IEEE 10th Electrical Insulation Conference, 71 C 38 EI (1971).
15. J. S. Gall (to Celanese Corporation), U. S. Pat. 3,751,396 (1973).
16. R. Lovenguth, Insulation/Circuits, September (1972).
17. S. B. Neff, Automotive Industries, March (1973).
18. Report 73-1, Polybutylene Terephthalate, 3 (Chem Systems Inc., New York 1973).
19. D. McNally and K. A. Murray, Celanese Internal Report (1973).
20. PBT and Related Thermoplastic Polyesters - III, 10 (R. M. Kossoff & Associates, Inc., New York 1976).
21. W. T. Freed and D. McNally, Symposium on Recent Developments in Automotive Plastics, April (A.C.S. Los Angeles 1974).
22. Celanex Bulletin J1A (Celanese Engineering Resins, Chatham, NJ, 1984).
23. Valox Properties Guide (General Electric Company Plastic Operations, Pittsfield, MA.)
24. D. C. Wharmund, D. R. Paul and J. W. Barlow, J. Appl. Polym. Science, 22, 2155 (1978).
25. S. B. Farmham and T. D. Goldman (to Rohm and Haas) U. S. Pat. 4,096,202 (1978).
26. C. Linder, et al. (to Bayer AG), U.S. Pat. 4,417,026 (1983).
27. J. M. Heuscher, Symposium on History of High Performance Polymers, April (A.C.S., New York 1986).
28. E. J. Deyrup, Sumposium on History of High Performance Polymers, April (A.C.S., New York 1986).

INJECTION MOLDABLE PET

Edward J. Deyrup
E. I. Du Pont de Nemours & Company, Inc.
Wilmington, Delaware

ABSTRACT

Glass reinforced injection moldable PET (poly(ethylene terephthalate)) resins have become a major engineering thermoplastic. Discovery of injection moldable PETs will be discussed with particular emphasis on crystallization. General processing characteristics and physical properties are presented as well as recent developments in new PET's with superior processing and/or product properties.

INTRODUCTION

PET was first synthesized by J. R. Winfield in England in the 1940's [1]. PET was used as a fiber for apparel, film for packaging and insulation, and in bottles for beverages long before it was used in any volume as an engineering thermoplastic . The delayed use of PET was in spite of its excellent physical and electrical properties and relatively low cost.

Several factors kept PET from being accepted:

* The most important of these was that PET did not crystallize completely in conventional water heated molds [2].

* Borderline toughness

* Lack of understanding of drying requirements. This accentuated the toughness concern, especially when there was an attempt to overcome low toughness by increasing PET molecular weight.

The first commercial engineering PET resins for injection molding were introduced by Akzo Chemie in Europe and Teijin in Japan. They used conventional nucleating agents such as talc. This did increase crystallization rates but not enough to allow full crystallization in water heated molds. Molders still had to use very high mold temperatures (approximately 130°C) or anneal parts. Sales remained low - an estimated 1 million lbs/yr or less.

In 1971 Celanese introduced PBT, and was shortly followed by General Electric and Eastman. While based on more expensive ingredients than PET, and with a lower end use temperature, it rapidly gained market acceptance. As seen in the patent literature, the emphasis of industrial research shifted to PBT principally because it was easy to mold with conventional water heated molds.

THE PROBLEM OF CRYSTALLIZATION

Increased polymer chain rigidity generally correlates with a slower rate of crystallization as the polymer is cooled down from its melting point. This is seen with a wide range of polyesters with glass transition temperature being used as the measure of polymer chain rigidity and crystallization rates measured at 40°C below the polymer melting point (Figure 1). PET has only two methylene units between aromatic ester groups and therefore has a higher Tg and a slower rate of crystallization at high temperature compared to PBT with its four methylene units.

Copyright 1986 by Elsevier Science Publishing Co., Inc.
High Performance Polymers: Their Origin and Development
R.B. Seymour and G.S. Kirshenbaum, Editors

FIG 1

DEGREE OF SUPERCOOLING (^oC) AND MELT TEMPERATURE
VERSUS GLASS TRANSITION TEMPERATURE (Tg)
SUPERCOOLING (^oC) = MELT POINT -
TIME OF 1/2 CRYSTALLIZATION OF 40 MINS - REF

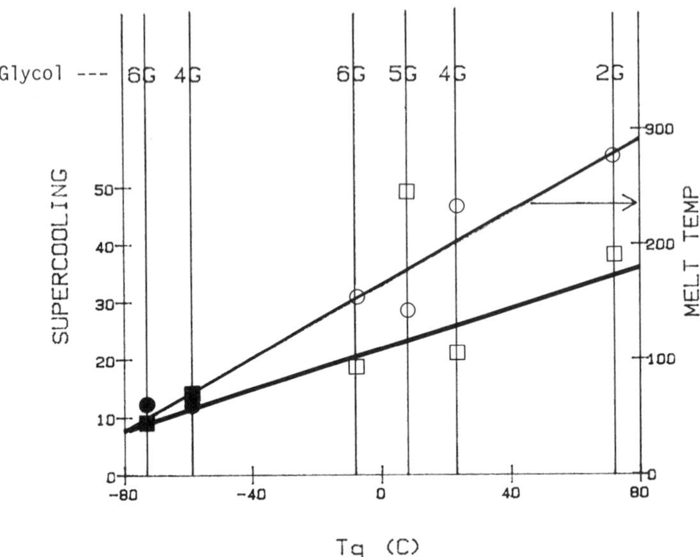

○ MELT POINT - TEREPHTHALATES
□ SUPERCOOLING - TEREPHTHALATES
● MELT POINT - ADIPATES
■ SUPERCOOLING - ADIPATES

This is only part of the problem. A more serious problem is crystallization (or lack thereof) at low temperatures. The injection molding process injects hot melt directly onto a relatively cold mold surface. If the mold is heated by water (instead of oil) - which is normally true and certainly preferred by the molder, the mold surface is generally less than 110°C. When we study crystallization rates directly by differential scanning calorimetry - we cannot begin to reproduce the quench rates involved in injection molding. One method to study crystallization at typical mold temperatures is to reheat in a differential scanning calorimeter a portion of an injection molded sample where the quench (mold) temperature is known . We then measure the magnitude of any exotherm on heating which reflects the amount of primary crystallization that has not occurred at that quench temperature [6],[7],[8].

For all polymers there is a maximum in the rate of crystallization versus quench temperature (Figure 2). As we decrease the quench temperature down from the melt temperature (Tm), crystallization rates continue to become faster until we reach a critical temperature where molecular mobility needed to achieve crystallization becomes the more important factor - and crystallization rate begins to decline. This temperature (Tk), is approximately 180°C for PET [8],[9] and can be related

FIG 2
CRYSTALLIZATION RATE EXPRESSED AS
1/(HALF TIME (mins)) VERUS CRYSTALLIZATION
TEMPERATURE FOR UNMODIFIED PET - REF 11

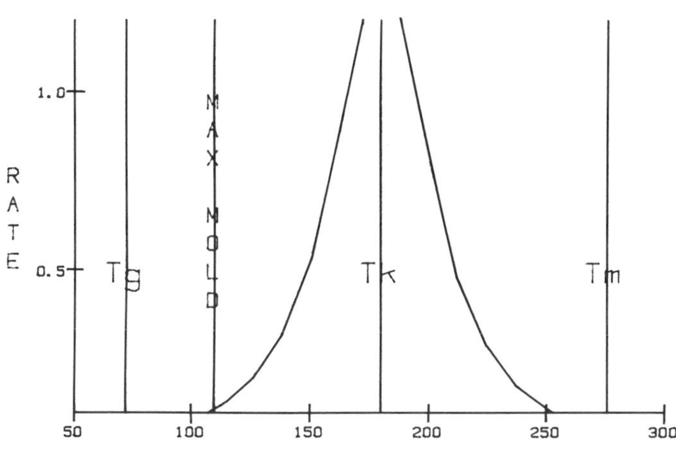

to the glass transition temperature by the approximate relationship Tk=0.5*(Tg+Tm) where all temperatures are in Kelvin. Therefore we would estimate using the values in [4], the temperature where crystallization rate begins to decrease to be 46°C lower for PBT than PET. A consequence is that it is much easier to quench PET (compared to PBT) to a mostly amorphous state.

If we were to injection mold unmodified PET at different temperatures we would obtain a curve similar to Figure 3. At very low mold temperatures, the PET is a glass. It is relatively easy to mold in this region as long as the mold temperature is uniformly below Tg. Parts are also quite tough and physical properties are good as long as the use temperature is below the Tg. Akzo for a period of time sold PET which was to be molded and used in this manner.

As mold temperature is increased to the vicinity of Tg, the unmodified PET has still been quenched to too low a temperature for it to crystallize but if the mold is above the Tg, the physical state of the polymer is a liquid. Understandably, it is difficult to eject a high viscosity liquid - which is why it is very difficult to mold unmodified PET on commercial cycles when the mold temperature is in this region. As the mold temperature is increased to substantially above the Tg, we begin to get some crystallinity especially in the center of the molding where the part cooling rate is retarded. However the resin is still difficult to mold because part surfaces are still mostly amorphous. The part also has a very rough surface if the polymer is reinforced with glass fiber. We believe this is because crystallization is still proceeding when the cavity pressure decays to zero; material then can shrink back leaving on the surface an outline of glass fibers.

FIG 3
CRYSTALLINITY VERSUS CRYSTALLIZATION
TEMPERATURE FOR UNMODIFIED PET
REF (11)

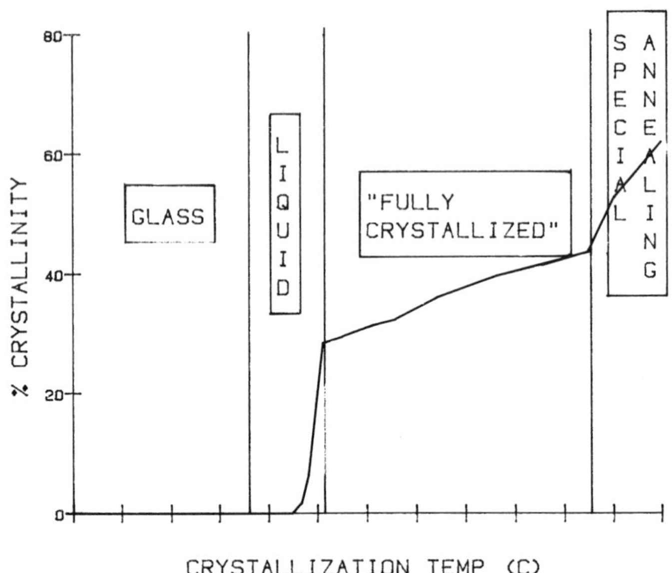

CRYSTALLIZATION TEMP (C)

As we go still higher in mold temperature, we eventually reach a point of complete primary crystallization. The only gain with further increases in temperature will be secondary crystallization - which occurs with all crystalline thermoplastics. Parts are very glossy even with the presence of glass fibers. Toughness is lower because the surface is semicrystalline instead of the more ductile amorphous form of the polymer. One way to differentiate this region is to determine whether there is an exotherm on heating. If absent then the sample is "fully crystallized" which corresponds to about 30 percent crystallinity. Alternatively, if the mold surface is highly polished, high - uniform gloss of glass reinforced moldings is an excellent indication of being "fully crystallized".

Because the part surface insulates the interior of the molding, crystallinity will be greater at the interior and lower at the surface (Figure 4). Appearance, dimmensional stability, cycle performance, solvent resistance, and toughness will strongly depend on surface crystallinity.

Addition of conventional nucleating agents (such as talc) do not make enough of a difference in the minimum mold temperature needed to obtain "full crystallization" (Figure 5).

A SOLUTION TO THE CRYSTALLIZATION PROBLEM

In 1978, Du Pont introduced Rynite® polyester engineering

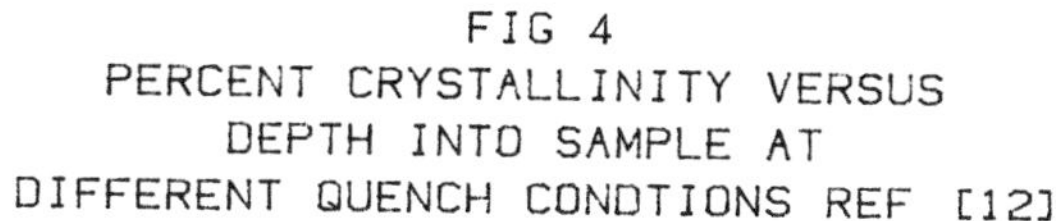

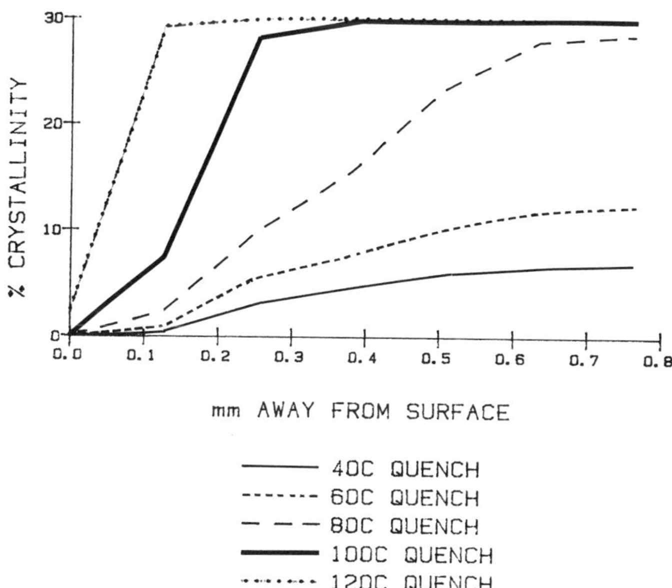

thermoplastic resin, a modified PET based on a proprietary two part crystallization system [5],[6] . This initiated a new worldwide wave of patent literature. During 1981 and 1982 alone at least 69 patents on this subject were issued to 19 different companies. Worldwide sales of injection moldable PET quickly grew to a multimillion lbs/yr.

This crystallization system, which now is advocated by almost all new patents covering PET engineering resins, has two parts. (I): a source of active sodium and/or potassium. An example is sodium neutralized poly(ethylene/methacrylic acid). And (II): an additive to increase mobility of PET to permit crystallization at low mold temperatures. This additive also lowers the Tg.

The sodium source is believed to donate ions ultimately to the PET end groups which are the active species for initiation of crystallization [13],[14]. This has been classified by one author as chemical nucleation as opposed to the conventional nucleation found with additives such as talc [14].

Initial Du Pont compositions contained as compound II the dibenzoate of neopentyl glycol - which lowered the glass transition about 12°C. But this subsequently was replaced by less volatile materials.

The commercial significance of this new technology was that this modified PET, which was desirably combined with the glass reinforcement to

FIG 5
PERCENT CRYSTALLINITY VERSUS MOLD TEMPERATURE
AS DETERMINED FROM DSC OF 1/16 INCH INJECTION
MOLDED BARS

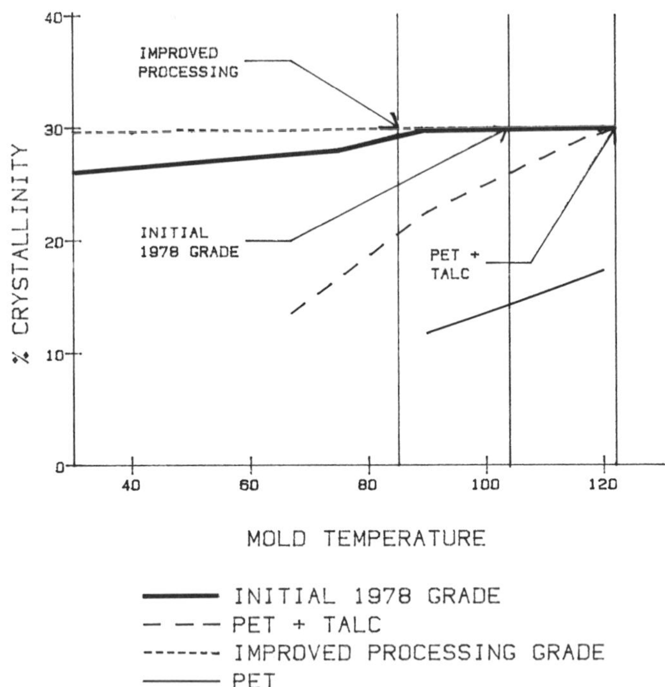

ARROWS POINT TO MINIMUM MOLD TEMPERATURE
FOR "FULL CRYSTALLINITY" FOR VERY THIN PARTS

obtain engineering type properties, produced a resin from which molders were able to mold "fully crystallized parts" for the first time using conventional water heated molds (as illustrated by Figure 5). Even at low mold temperatures where the surface is not "fully crystallized", overall crystallinity remains high. In this graph, the vertical lines represent the minimum mold temperature that can be used to obtain fully crystallized surfaces with molds in which there is a very rapid rate of heat transfer. Frequently, mold temperatures can be lower than indicated because heat transfer is poorer and the part cools less rapidly. For example thicker parts generally can be molded with cooler molds because the actual quench rate is slower.

Recently Du Pont has introduced compositions which permit even lower mold temperatures (Figure 5) these have been found especially useful for attaining fully crystallized parts in thinner sections.

IMPORTANCE OF MELT HYDROLYSIS

All polyesters hydrolyze in the melt rapidly. PET, principally because it is molded at higher melt temperatures (295 versus 250°C), requires more care in drying and maintaining that dryness than does PBT. Any moisture will react completely within the minimum molding times and temperatures used to injection mold. This lowers molecular weight and thereby decreases toughness. Molecular weight loss was more of a problem with the higher molecular weight resins that were used in compositions available prior to 1978 (Figure 6). Increasing PET molecular weight is not a practical route to increased toughness because the drying requirements increase. PET and PBT suppliers expended a considerable effort in educating molders on how to dry thermoplastic polyesters and maintain dryness through the proper use of hopper dryers. Because of its high melting point and low moisture content PET can be adequately dried continuously in hopper driers with an average temperature of 135°C for 4 hours average hold-up time [15].

REINFORCEMENT

Essentially all injection moldable PET is sold as glass reinforced grades - although other fillers are sometimes present. Without glass, PET does not have sufficient high temperature strength and modulus to demand a significant premium over non-engineering polymers like ABS and polypropylene. However, the combination of glass reinforcement and

FIG 6
INHERENT VISCOSITY AND NUMBER AVG MOLECULAR WIEGHT VERSUS PERCENT H2O IN FEED OF INJECTION MOLDING MACHINE STARTING VISCOSITY ON LEFT AXIS

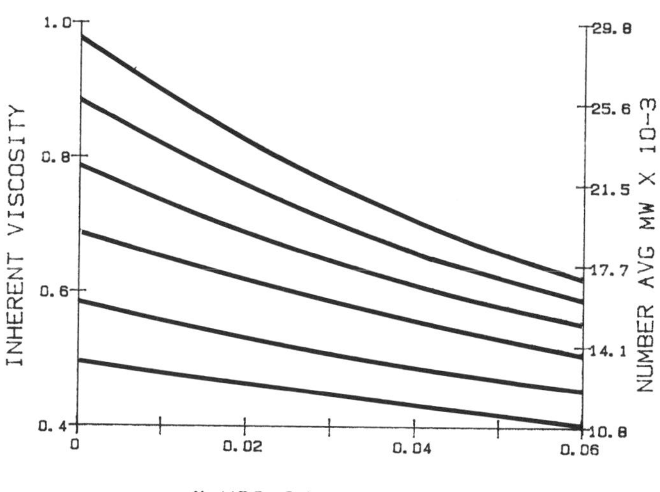

FIG 7
TENSILE STRENGTH (PSI) VESUS
WEIGHT PERCENT GLASS AT 23C AND 150C
REF (15)

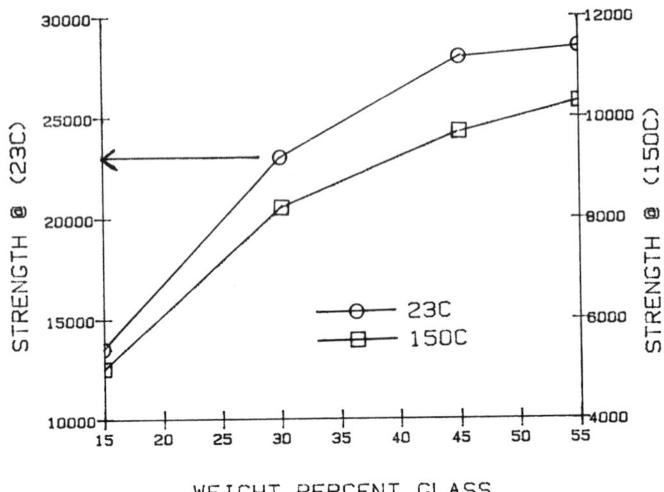

FIG 8
FLEXURAL MODULUS (PSI) VERSUS
WEIGHT PERCENT GLASS AT 23C AND 150C
REF (15)

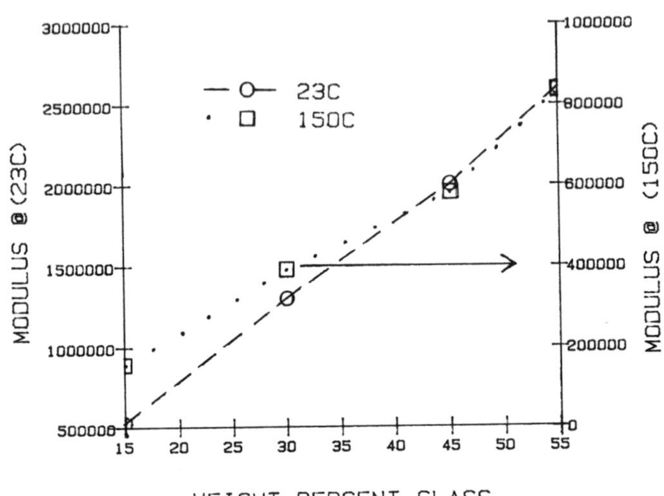

crystallinity provides excellent strength and modulus (Figures 7,8) both at low and high temperatures. Commercial grades span from 15 to 55% glass fiber.

In any injection molded part, glass fibers are anisotropically oriented. Orientation and properties will depend upon depth into the part [16],[17]. But at the surface (the region of highest shear) the primary orientation is with the flow direction. In thinner parts where a greater percentage of the fibers are aligned with the flow direction there tends to be warpage, non-isotropic orientation coupled with crystallization results in non-isotropic shrinkage. This in turn can curl the part.

After initial development of glass reinforced grades, grades were introduced with mica. Mica tends to resist the tendency to curl because mica has wide thin platelets.

Toughness

When we reinforce polymers with glass, some measures of toughness - which are primarily strength related - as are tensile strength, Kc, notched Izod, etc increase [16]. Other measures of toughness become relatively low. In particular elongation which is typically over 30 percent falls to under 5 percent with low loadings of glass. Parts actually can fail in the field from low elongation even at relatively low test speeds. Differences in elongation of 2 versus 3 versus 5 percent etc frequently translate into major differences in part performance. This is in contrast to unreinforced polymers where elongations well over 10 are encountered. When elongation is high (>10), failure is not encountered at low test speeds and differences in values are only meaningful if they correlate to the much lower elongations observed at higher test speed such as are found in impact tests.

While the traditional notched Izod values increase with glass reinforcement, these values can sometimes be misleading if viewed in terms of our experience with unreinforced polymer. First of all the test is normally done so that the crack must propagate through or around the glass. A much lower energy is required if the crack propagates along the direction of orientation of fibers. This frequently is the case in field failures. In addition some of the energy consumption comes from fiber pulling out of the polymer matrix and at that point part damage is visually extensive.

Historically one of the areas of weakness of glass reinforced thermoplastic polyesters relative to their polyamide analogues has been their lower toughness as measured by elongation and unnotched impact. Elongations of 2-3 versus 4-5, unnotched impacts of 15-20 versus over 25 ft-lbs/in. The difference is frequently magnified because nylon often has absorbed some moisture which makes it even more ductile. Sometimes the polyester would not have been thoroughly dried before molding which would exacerbate the difference. There were also applications where the toughness of glass reinforced polyamide was inadequate.

Toughness was even more of a problem with pre 1978 PET compositions. As a consequence all modern injection moldable PET's are toughened to some extent by adding a dispersed low modulus second phase. This increases ductility to that of standard commercial PBT's.

More recently the technology of toughening has advanced to the extent that compositions introduced in 1984 have elongation and Izod impact strength which exceed what is available in glass reinforced PBT without a major sacrifice in modulus (Figures 9,10).

FIGURE 9

TOUGHNESS VS STIFFNESS

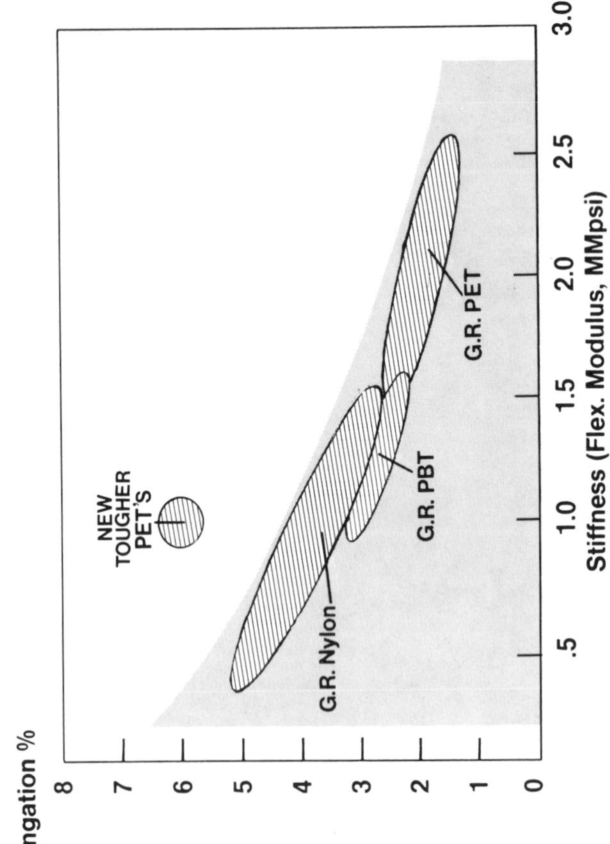

FIGURE 10

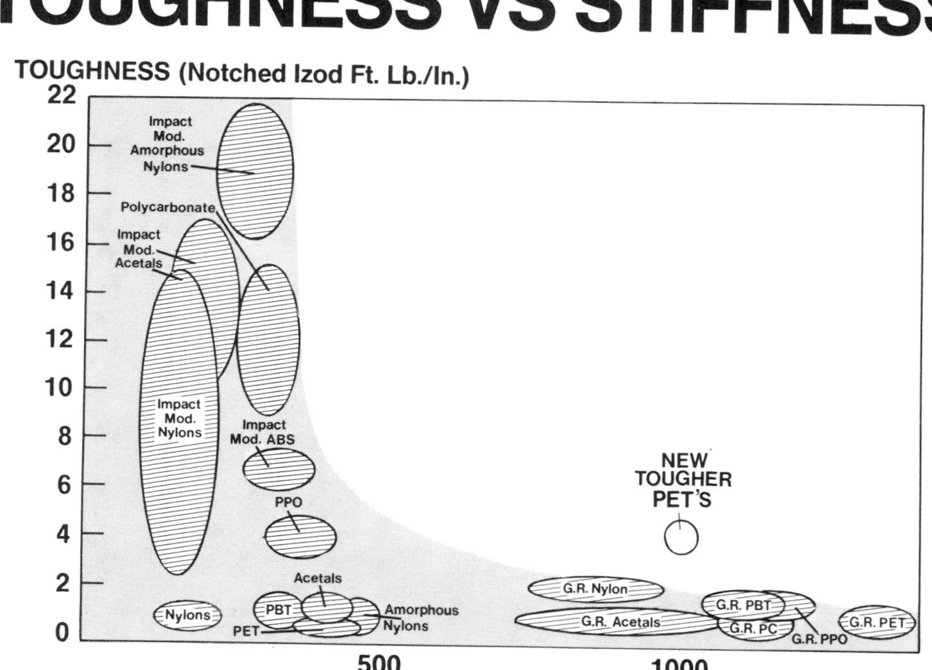
TOUGHNESS VS STIFFNESS
TOUGHNESS (Notched Izod Ft. Lb./In.)
22
20
18
16
14
12
10
8
6
4
2
0
Impact Mod. Amorphous Nylons
Polycarbonate
Impact Mod. Acetals
Impact Mod. Nylons
Impact Mod. ABS
PPO
Acetals
Nylons
PBT
PET
Amorphous Nylons
G.R. Nylon
G.R. Acetals
NEW TOUGHER PET'S
G.R. PBT
G.R. PC
G.R. PPO
G.R. PET
500
1000
STIFFNESS (Flex. Modulus, Mpsi)

APPLICATIONS

A key attribute of thermoplastic polyesters is that their mechanical and electrical properties as well as their molded dimmensions do not change with ambient humidity. With glass reinforcement they have reasonably high temperature strength and modulus, and they are unaffected by most solvents. The overall world-wide market for thermoplastic polyesters can be categorized into 6 areas (Figure 11). Examples for automotive would be: ignition parts, window lift brackets, side roof rails, sunroof frames, and headlamp retainers. Electrical: connectors, coil bobbins, motor housings, and transformer housings. Furniture: chair arms and shells. Appliances: glue guns, typewriter printer gears, and oven handles. Industrial: pumps, pneumatic positioner.

Outlook

Based on properties and economics as well as initial sales growth it is reasonable to assume that injection moldable PET will eventually reach parity in sales with PBT, and possibly surpass it. This will take a long time, since PBT is a well established, dependable and versatile engineering resin. There are strengths for each. For example it is easier to make compositions with high loadings of glass fibers from PET than PBT because of the initially lower viscosity of PET somewhat compensates for the increase in viscosity from the additional glass fiber. Both polymers can be modified with various minerals, tougheners, etc. to accomplish a variety of blends varying in properties and molding performance.

FIG 11
WORLDWIDE THERMOPLASTIC MARKET
BY INDUSTRY

SALES IN MILLIONS
OF POUNDS

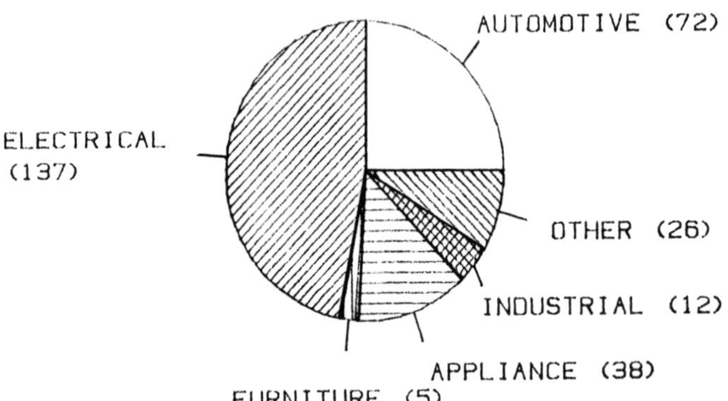

While there are many sources of supply of PET, we expect that the manufacturers of injection moldable PET will be limited for several reasons. An important one is that manufacture is much more complicated than just adding glass fiber. Several ingredients must be monitored to achieve quality. Care must be exerted to use a source of PET which gives constant properties. For example, recycled PET can contain high levels of DEG (diethylene glycol) which will reduce melting point, crystallinity, and high temperature modulus (Figure 12).

ACKNOWLEDGEMENTS

I would like to acknowledge the contributions of W. E. Garrison and T. M. Ford in the preparation of this paper.

FIG 12
EFFECT OF DEG (DIETHYLENE GLYCOL) ON
HDT (HEAT DISTORTION TEMPERATURE - 264PSI)
AND MELTING POINT (CENTIGRADE)

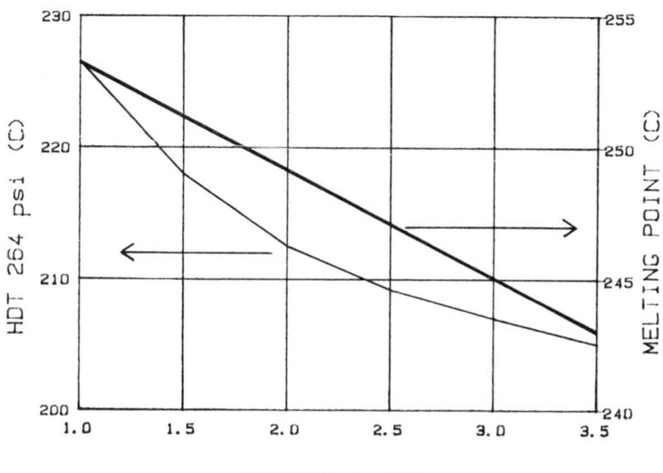

REFERENCES

1 Patent US 2465319

2 Modern Plastics, March 1984, p21

3 E. L. Lawton, Polymer Engineering and Sci, Vol 25 - No 6 (1985), 348-354

4 M. Gilbert and F. J. Hybart, Polymer, Vol 15 (1974), pp 407- 412

5 Patent GB 2015013 E. J. Deyrup

6 Patent GB 2015014 E. J. Deyrup

7 T. M. Ford and J. L. Hecht, SPE ANTEC

8 L. H. Palys and P. J. Phillips, J Poly Sci., Poly Phys. Ed.,Vol 18(1980), pp 829-852

9 M. Gilbert and F. J. Hybart, Polymer 1972 Vol 13 (1972),p327 to 332

10 D. W. Van Krevelen, Properties of Polymers, 2nd Edition,Oxford, 1976, p 433

11 K. Mayhan, 1965 Ph D. Thesis -A Study of the Crystallization Kinetics of Poly(ethyleneterephthalate)

12 J. D. Muzzy, D. G. Bright, and G. H. Hoyas, Polymer Engineering and Science, Vol 18 (1978 mid May), No 6, pp 437 - 442

13 D. Garcia, J. Poly Sci., Polymer Phys. Ed., Vol 22 (1984) 2063-2072

14 R. Legras and J.P. Mercier, Nature, Vol 304 (Aug 1983), pp432-434

15 Du Pont literature for Rynite

16 K. Friedrich, Report CCM-83-18, Center for Composite Materials, University of Delaware, 1980

17 C. Lhymn, J. M. Schultz: J. Material Sci (1983)

18 S. Fakirov et al., Makromol. Chem, Vol 182 (1981), pp 185-197

19 Data by W. E. Garrison - Du Pont

HISTORY OF POLYARYLATES

LLOYD M. ROBESON AND JAMES M. TIBBITT
Union Carbide Corporation, Specialty Polymers and Composites Division, P.O. Box 670, Bound Brook, New Jersey

INTRODUCTION

Polyarylates are an emerging class of new engineering polymers. At one point polyarylates were described as linear polyesters of dihydric phenols [1]. In this paper, however, they are described as wholly aromatic polyesters derived from aromatic dicarboxylic acids and diphenols. Reactions of aromatic dicarboxylic acids and glycols have been known for years [2] as well as diphenols (e.g. acetate ester) with aliphatic diacids [3,4]. The reaction of aromatic dicarboxylic acids and diphenols were first noted by Conix [5] in 1957. The acid chloride and the diacetate routes were noted as feasible synthesis procedures to obtain polyarylates of high molecular weight. Levine and Temin [6] (1958), Eareckson [7] (1959), and Morgan and Kwolek [8] (1959) were also early contributors to the technical literature regarding the synthesis of polyarylates. In the same time period, Russian investigators (Korshak, Vinogradova, Lebyedyeva, and others), were also quite active with many references [9-16] in the Russian technical literature. These papers were summarized in a book by Korshak and Vinogradova that constitutes one of the first reviews of the synthesis of polyarylates [1].

In this review of the history of polyarylates, only those products of an aromatic dicarboxylic acid (or derivative) and a diphenol (or derivative) will be considered. Structurally similar aromatic polyesters based on p-hydroxybenzoic acid have received considerable interest due to the resultant liquid crystalline behavior of these materials. They, however, will not be covered in this paper. The primary polyarylate discussed herein is based on Bisphenol A (2,2-bis(4-hydroxyphenyl)propane) and tere/isophthalic acids. These monomers are large volume commercial products and are utilized in the primary commercial polyarylates available today.

HISTORY OF SYNTHESIS PROCEDURES

The direct condensation polymerization of aromatic diacids and diphenols has not been demonstrated to be a feasible route for obtaining high molecular weight polyarylates. However, three monomer variations have been shown to yield acceptable high molecular weight products. These variations include the acid chloride process in which the aromatic diacid is first reacted to form the acid chloride (e.g. phosgene) followed by condensation with the diphenol. The second route involves the preparation of the diphenyl ester of the aromatic diacid followed by reaction with the diphenol. The third process involves preparation of the diacetate of the diphenol followed by reaction with the aromatic diacid. The historical background of these reaction schemes will be discussed.

The earliest references to the acid chloride route came from Conix [5], Eareckson [7], Morgan and Kwolek [8], and a number of Russian investigators [9-16], specifically Korshak and Vinogradova. The procedure noted by Eareckson [7] involved interfacial polycondensation where the aqueous solution of the dialkali salt of Bisphenol A was contacted with a solution of aromatic acid chlorides in a water-insoluble solvent. An emulsifier was added and the solutions were rapidly mixed followed by coagulation of the

Copyright 1986 by Elsevier Science Publishing Co., Inc.
High Performance Polymers: Their Origin and Development
R.B. Seymour and G.S. Kirshenbaum, Editors

resultant polymer. Conix [5] prepared a series of polyarylates including those based on Bisphenol A and tere/iso phthalates. The procedure noted was interfacial polymerization with small amounts of quaternary ammonium or sulfonium compounds added as effective catalysts. Molecular weights as high as 275,000 were reported using this procedure. Morgan and Kwolek [8] investigated the unstirred interfacial polymerization of the dialkali salt of Bisphenol A with terephthaloyl chloride. It was observed that the polymer film formed in the organic phase. Tetraethylammonium chloride was found to greatly accelerate the rate of film formation. In the same time period, Korshak et al. [13] reported on the interfacial polymerization of isophthalic acid chloride and Bisphenol A and observed that the yield was dependent upon the intensity of stirring the reaction mixture. Further studies by Korshak et al. [15] investigated the effect of emulsifying agents on the polymerization of Bisphenol A and isophthalic acid chlorides.

Another acid chloride process utilized was similar to the Schotten-Baumann ester-forming process. Polymer was produced by adding at least a stoichiometric amount of acid acceptor (e.g. triethylamine or pyridine) to a solution of the dihydric phenol and the aromatic acid chlorides [17]. Other variations of this process involve generation of the acid chlorides in situ (via phosgene addition) [18]. Another variation of the acid chloride process disclosed somewhat later [19] involved addition of solid calcium hydroxide to a solution of dihydric phenols and aromatic acid chlorides with triethylamine as a phase transfer catalyst. A summary of the various acid chloride processes is illustrated in Figure 1.

FIGURE 1

ACID CHLORIDE SOLUTION PROCESSES

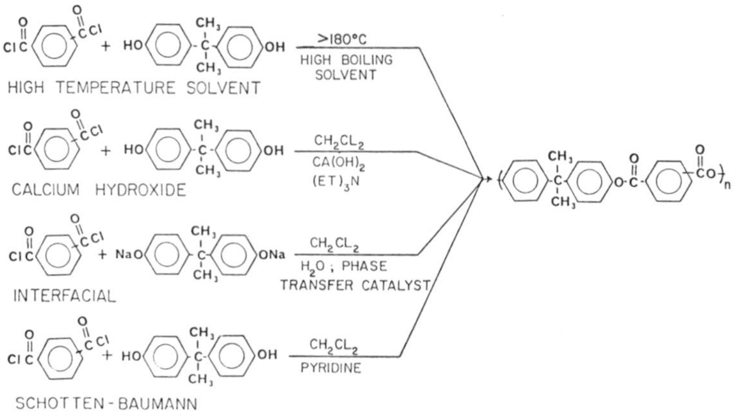

High temperature reactions of diphenols and aromatic acid chlorides in high boiling solvents have also been noted. Korshak and Vinogradova [9] investigated this reaction in a biphenyl solution at 120-230°C. A similar process was reported by Keck where Bisphenol A was reacted with iso and terephthaloyl chlorides at elevated temperatures in a high boiling solvent such as dichloroethylbenzene [20]. The polymer could be isolated by either coagulation or extruder devolatilization.

Korshak and Vinogradova noted the potential of producing polyarylates from the diphenyl ester of a dicarboxylic acid and a dihydric phenol and noted that a similar procedure had been successfully utilized to prepare aromatic polycarbonates [1]. The route now referred to as the diphenyl ester synthesis route was initially described by Blaschke and Ludwig [21]. This process involves the reaction of the diphenyl ester of tere/isophthalic acids with dihydric phenols (e.g. Bisphenol A) at temperatures exceeding 230°C. Catalysts have been noted to accelerate the reaction. Phenol must be removed from the reaction in order to obtain high molecular weight. One method noted to allow for phenol removal at high viscosities was to conduct the final stages of polymerization in a vented extruder [22].

The third method of polyarylate synthesis is commonly referred to as the diacetate process where the diacetate of Bisphenol A (or other dihydric phenols) is reacted with tere/isophthalic acids at high temperatures (>250°C). This involves continuous removal of acetic acid to achieve high molecular weights. Conix [5] and Levine and Temin [6] were among the first to describe the successful synthesis of polyarylates using this process although earlier references were noted for the reaction of the diacetate of hydroquinone and aliphatic carboxylic acids [3,4]. Later references of interest include Beavers et al. [23], Conix [24], and Hamb [25]. Catalysts and heat transfer fluid addition have been used to accelerate the reaction and facilitate acetic acid removal respectively [26,27]. The non-acid chloride processes for polyarylate are illustrated in Figure 2.

FIGURE 2

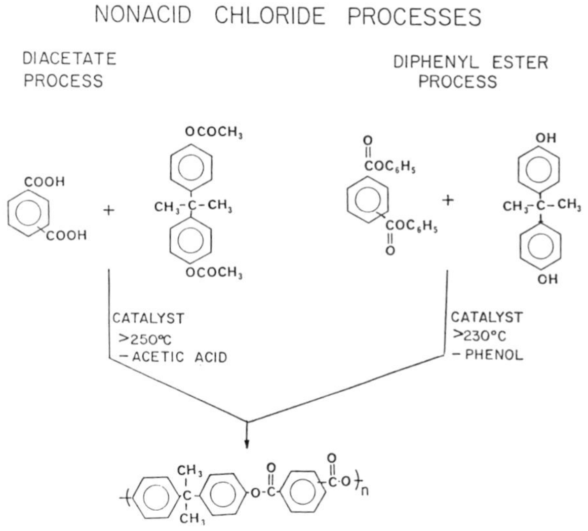

NONACID CHLORIDE PROCESSES

DIACETATE PROCESS

DIPHENYL ESTER PROCESS

CATALYST
>250°C
– ACETIC ACID

CATALYST
>230°C
– PHENOL

HISTORY OF COMMERCIAL DEVELOPMENT

Korshak and Vinogradova noted in their review on polyesters (1965) that polyarylates offer properties such as high softening point, good dielectric constant, high stability to the action of chemical reagents, and capability to form strong films. They hypothesized that polyarylates would find wide application in the near future. Bier [28] in a review article in 1974 discussed the synthesis and properties of polyarylates. Attractive mechanical and electrical properties were noted along with favorable economics and feasible synthesis routes. It was noted that injection moldability was limited whereas the feasible extrusion and cast film markets were small. Bier noted that these drawbacks prevented them (Dynamit Nobel) from entering the polyarylate business. Interestingly, the same year (1974) Unitika announced the commercial production of U-polymer (a polyarylate based on Bisphenol A and tere/iso phthalates. The first United States entry (1978) into the polyarylate business was Ardel D-100 based on Bisphenol A and tere/iso phthalates. In 1979, Hooker Chemical Company announced Durel polyarylate (structure was not disclosed) with a lower heat deflection temperature which implies a different structure than Ardel polyarylate. Both Ardel and Durel polyarylates are injection moldable thus indicating that the primary shortcoming mentioned by Bier [28] had been overcome. The first commercial entry in Europe (1979) was APE (aromatic polyester) from Bayer. This polyarylate was stated to be based on Bisphenol A and tere/iso phthalates.

PROPERTIES OF POLYARYLATE

Eareckson [7] noted that the polyarylate based on Bisphenol A and tere/iso phthalates was soluble in chloroform and methylene chloride. Films were observed to be strong and very tough even down to -70°C. They were noted as having good heat stability and ultraviolet stability even though yellow discoloration occurred in a short time upon exposure to UV. The Bisphenol A isophthalate and Bisphenol A terephthalate structures were found to be highly crystalline. The glass transition values ranged from 180°C at high iso content to >200°C at high tere content.

In one of the early comprehensive review articles on the properties of polyarylates, Bier [28] mentioned that the polyarylate based on Bisphenol A exhibited attractive electrical and mechanical properties. Injection moldability was considered to be a limitation, but extrusion was possible. Bier stated that the hydrolysis resistance was poor (in the same range as polycarbonates based on Bisphenol A).

The property profile of the first commercial polyarylate entry (U-polymer: 1974) was reported by Sakata [29]. Polyarylate was noted to be a tough, transparent, high heat distortion material. One of the unique features noted for polyarylate was the excellent flexural recovery properties. The melt viscosity was noted to be high but in the range for injection moldability.

Domine [30] reported on the property profile of Ardel D-100 polyarylate (the first domestic commercial entry). In addition to the remarks of Sakata above, Domine mentioned the excellent retention of properties of polyarylate exposed to UV. The combustion characteristics determined by a number of different tests were observed to be quite favorable relative to most other non-flame retarded polymers (e.g. V-0 at 1/16", limiting oxygen index = 34, low flame spread index, and low smoke density). The creep resistance of polyarylate was shown to be much better than polycarbonate but inferior to polysulfone.

Freitag and Reinking reported on the property profile of Bayer's commercial polyarylate [31]. They observed that polyarylate was less sensitive to specimen thickness than polycarbonate in the notched izod impact test. They stated that polyarylate is quite similar to polycarbonate in hydrolytic degradation by alkalis, dilute acids, and hot water. The injection moldability of the Bayer polyarylate was discussed in some detail. Proper drying was emphasized as necessary to prevent hydrolysis during injection molding.

The UV stability of polyarylate was first reported in detail by Cohen et al. [32]. The aromatic ester group of polyarylate was shown to undergo a photochemical Fries rearrangement to yield an o-hydroxybenzophenone structure. This structure is a well-known UV stabilizer. When exposed to UV, polyarylate rearranges at the surface yielding a thin coherent skin which is opaque in the ultraviolet but transparent to visible light. This skin thus yields a protective coating preventing the interior from UV degradation. This rearrangement is accompanied by a yellowing of the surface.

The property profile (mechanical, electrical, thermal, and combustion) of Ardel D-100 (Bisphenol A polyarylate) is listed in Tables I through IV, respectively. The dynamic mechanical data are illustrated in Figure 3 for Ardel D-100 showing the high glass transition temperature and consistent modulus over a large temperature range.

TABLE I. Mechanical and physical property data for polyarylate (Ardel D-100).

Flexural Modulus (psi)	310,000
Flexural Strength (psi) at 5% strain	11,000
Tensile Modulus (psi)	290,000
Tensile Strength (psi)	9,500
Tensile Elongation at Yield (%)	8.0
Tensile Elongation at Rupture (%)	50.0
Tensile Impact Strength (ft-lb/in.^{2})	140
Notched Izod Impact Strength (1/8 in.) (ft-lb/in. of notch)	4.2
Dart Impact Strength (ft-lb) 1/8 in. plaque; 3/4 in. radius dart	>68
Density (g/cc)	1.21
Refractive Index	1.61
Water Absorption (%) Immersion	
24 hrs at 72°C	0.27
30 days at 72°C	0.71
Mold Shrinkage (in./in.)	0.009

TABLE II. Electrical property data for polyarylate (Ardel D-100).

Dielectric Constant (60 Hz)	2.73
(10^3 Hz)	2.71
(10^6 Hz)	2.62
Dissipation Factor (60 Hz)	0.0008
(10^3 Hz)	0.005
(10^6 Hz)	0.02
Dielectric Strength (v/mil)	400
Arc Resistance (sec)	125
Volume Resistivity (ohm/cm)	1.2×10^{16}

TABLE III. Thermal properties of polyarylate (Ardel D-100).

Heat Deflection Temperature (1/4" bar) 264 psi	174°C
Glass Transition Temperature	188°C
Continuous Use Temperature	
Mechanical Properties	130°C
Impact Properties	120°C
Electrical Properties	130°C
Injection Molding Range (stock temperature)	350-395°C

TABLE IV. Combustion characteristics of polyarylate (Ardel D-100).

UL-94 Rating 30 mil	V-1
60 mil	V-0
125 mil	V-0
Autoignition Temperature	545°C
Limiting Oxygen Index, 1/8 in.	34%
Radiant Panel Flame-spread Index 1/16 in. thick	1.9
Smoke Density, vertical, flaming mode 1/16 in. thick	2.5 N/cm
Ds @ 1.5 min.	2.3
Ds @ 4.0 min.	34.0
Ds max.	106
Time to Ds max.	11 min.
Time to Ds = 16	3 min.

These numerical flame-spread ratings or flammability ratings are not intended to reflect hazards presented by these or any other materials under actual fire conditions.

FIGURE 3

DYNAMIC MECHANICAL DATA FOR POLYARYLATE ARDEL D-100

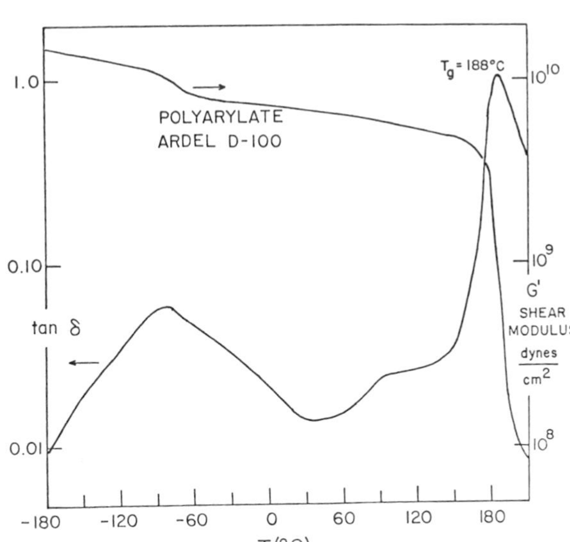

Another property recently reported in detail for polyarylate is in the polymer blend area. An early French patent [33] noted that Bisphenol A polyarylate and poly(ethylene terephthalate) yield interesting mixtures. Further studies by Kimura et al. [34] and Robeson [35] found that with modest ester-exchange polyarylate and poly(ethylene terephthalate) form miscible mixtures (thus transparent with Tg values between the constituent values). Polyarylate and poly(butylene terephthalate) were noted to be miscible without ester-exchange [36]. Miscibility of three different cyclohexane dimethanol based polyesters with polyarylate was observed under minimum time and temperature exposure conditions [35].

APPLICATIONS OF POLYARYLATE

The property balance of polyarylate clearly qualifies it for a number of engineering polymer applications. Korshak and Vinogradova [1] first hypothesized the future commercial importance of polyarylate. Domine [30] reported potential in lighting, appliances, tinted glazing, connectors, solar energy equipment, and spring clips. Freitag and Reinking [31] added housings for electrical and electronic appliances, capacitor cups, switches, incandescent and fluorescent lamp fixtures, lamp enclosures, headlamp reflectors, and transparent lamp diffusers. In a more recent review by Maresca and Robeson [37], transparent signs, safety lamp housings, traffic light lenses, aircraft interior panels, mass-transit panels, safety equipment housings, fire protection equipment (e.g. helmets), snap-fit connectors, fasteners, hinges, and springs were added to the above mentioned potential applications.

REFERENCES

1. V.V. Korshak and S.V. Vinogradova, "Polyesters", Pergamon Press Ltd., New York, 1965.
2. W.H. Carothers and J.W. Hill, J. Am. Chem. Soc., 54, 1559 (1932).
3. J.G.N. Drewitt and J. Lincoln, Brit. Pat. No. 621, 102; assigned to British Celanese Ltd., April 4, 1949.
4. E.R. Wallgrove and F. Reeder, Brit. Pat. No. 636,429; assigned to Courtaulds Inc., April 26, 1950.
5. A.J. Conix, Ind. Chim. Belg., 22, 1457 (1957); also Ind. Eng. Chem., 51(2), 147 (1959).
6. M. Levine and S.C. Temin, J. Polym. Sci., 28, 179 (1958).
7. W.M. Eareckson, J. Polym. Sci., 40, 399 (1959).
8. P.W. Morgan and S.L. Kwolek, J. Polym. Sci., 40, 299 (1959).
9. V.V. Korshak and S.V. Vinogradova, Izv. Akad. nauk SSSR, otdel. khim. nauk., 636 (1958).
10. S.V. Vinogradova and V.V. Korshak, Dokl. Akad. nauk. SSSR, 123, 849 (1958).
11. V.V. Korshak and S.V. Vinogradova, Vysokomol. soyed., 1, 834 (1959); also Vysokomol. soyed., 1, 838 (1959).
12. V.V. Korshak and S.V. Vinogradova, Vysokomol. soyed., 1, 1482 (1959).
13. V.V. Korshak, S.V. Vinogradova, and A.S. Lebyedyeva, Vysokomol. soyed., 2, 61 (1960).
14. V.V. Korshak, S.V. Vinogradova, and A.S. Lebyedyeva, Vysokomol. soyed., 2, 977 (1960).
15. V.V. Korshak, S.V. Vinogradova, and A.S. Lebyedyeva, Vysokomol. soyed., 2, 1162 (1960).
16. V.V. Korshak, S.V. Vinogradova, P.M. Valetskii, and Y.V. Mironov, Vysokomol. soyed., 3, 66 (1961).
17. W.A. Hare, U.S. Pat. No. 3,234,168; assigned to E. I. duPont de Nemours and Co., February 8, 1966.
18. E.P. Goldberg, U.S. Pat. No. 3,030,331; assigned to General Electric Co., April 17, 1962.
19. K. Ueno, U.S. Pat. No. 3,939,117; assigned to Sumitomo Chemical Co. Ltd., February 17, 1976.
20. M.H. Keck, U.S. Pat. No. 3,133,898; assigned to Goodyear Tire and Rubber Co., May 19, 1964.
21. F. Blaschke and W. Ludwig, U.S. Pat. No. 3,395,119; assigned to Chemische Werke Witten. G.M.b.A., July 30, 1968.
22. K. Eise, R. Friedrich, H. Goemar, G. Schade, and W. Wolfes, German Offen. No. 2,232,877; assigned to Werner & Pfleiderer and Dynamit Nobel A-G, January 17, 1967.
23. E.M. Beavers, M.J. Hurwitz, and D.M. Fenton, U.S. Pat. No. 3,329,653; assigned to Rohm & Haas Co., July 4, 1967.
24. A.J. Conix, U.S. Pat. No. 3,317,464; assigned to Gevaert Photo-Producten, May 2, 1967.
25. F.L. Hamb, J. Polym. Sci., 10, 3217 (1972).
26. W.J. Jackson, Jr., H.F. Kuhfuss, and J.R. Caldwell, U.S. Pat. No. 3,684,766; assigned to Eastman Kodak Co., August 15, 1972.
27. G.W. Calundann, U.S. Pat. No. 4,067,852; assigned to Celanese Corp., January 10, 1978.
28. G. Bier, Polym., 15, 527 (1974).
29. H. Sakata, 32nd ANTEC of SPE, p. 459, San Francisco, May 13-16, 1974.
30. J.D. Domine, 37th ANTEC of SPE, p. 655, New Orleans, May 7-10, 1979.
31. D. Freitag and K. Reinking, Kunststoffe, 71(1), 46 (1981).
32. S.M. Cohen, R.H. Young, and A.H. Markhart, J. Polym. Sci., Part A-1, 9, 3263 (1971).
33. French Pat. No. 1,392,883; assigned to N.V. Onder Zoekingsinstitut Research, February 8, 1965.

34. M. Kimura, G. Salee, and R.S. Porter, J. Appl. Polym. Sci., 29, 1629 (1984).
35. L.M. Robeson, J. Appl. Polym. Sci., 30, 4081 (1985).
36. M. Kimura, R.S. Porter, and G. Salee, J. Polym. Sci., Polym. Phys. Ed., 21, 367 (1983).
37. L.M. Maresca and L.M. Robeson, "Polyarylates" in "Engineering Thermoplastics: Properties and Applications," ed. by J.M. Margolis, Marcel Dekker, Inc., New York, 1985.

THE HISTORY OF ACETAL HOMOPOLYMER

KENNETH J. PERSAK, RICHARD A. FLEMING
E. I. du Pont de Nemours & Co., Inc.
Wilmington, Delaware

ABSTRACT

The Du Pont Company offered the first commercial acetal homopolymer, a high molecular weight polymer with a molecular structure of repeating carbon-oxygen links, under the trademark Delrin® acetal resin in January 1960. The commercialization of Delrin® followed 10 years of chemical research and development stimulated by work with pure monomeric formaldehyde and an improved polyformaldehyde made by polymerization in a nonsolvent.

A marketing program launched by Du Pont 2-1/2 years prior to commercialization provided plastics processors and end users with the knowledge, data and technical assistance necessary for proper use of the new acetal polymer.

The Asahi Chemical Industries Co., Ltd., Tokyo, Japan, commercialized an acetal homopolymer in 1972 under the trademark Tenac® polyacetal resin [1].

This review positions acetal homopolymer in civilization's long continued search for new materials. And, beginning with the decade of the 1940's it traces history through early 1986 in terms of needs, key management decisions, the product line and technical marketing concepts. It concludes with some highlights and a look to the future.

This history of necessity highlights Du Pont, which was the only supplier of acetal homopolymer for over a decade, until the advent of a second supplier in 1972. With only one supplier, and even now only two, official statistics on capacity, production, sales or market data are not available.

CIVILIZATION'S SEARCH FOR BETTER MATERIALS

A vital thread in the fabric of man's history has been the persistent search for better materials. The Stone Age, Bronze Age and Iron Age describe progressive successes in the search for better materials by an advancing civilization.

Although iron and small quantities of steel had been produced before the birth of Christ, it was man's continuing need for discovering "more and better" that led to improved processes for making iron and steel. In Central Europe in the 14th century the search led to the blast furnace process for making iron. Then in the 18th century the crucible process was developed, making it possible to produce steel in large quantities.

These historical events illustrate a basic fact well known to the present day researcher: successful new or improved materials result from an awareness of a need, from searching, discovery and commercialization.

Published 1986 by Elsevier Science Publishing Co., Inc.
High Performance Polymers: Their Origin and Development
R.B. Seymour and G.S. Kirshenbaum, Editors

After commercialization the search continues for improvements in terms of performance, quality, manufacturing productivity and cost.

And so it was that in the 19th century these factors, particularly the awareness of needs, led to the synthetic chemical research which marked the beginning of the Plastics Industry [2] [3].

Acetal homopolymer was the result of R. N. MacDonald's pioneering discovery in 1952 of a stable, high molecular weight polymer of formaldehyde [4]. MacDonald's search and discovery has antecedents in: Butlerov's description of formaldehyde polymers in 1859, Du Pont's entry into the plastics industry in 1915 [5], Staudingers' monumental work on macromolecules beginning in 1922, Carother's work which led to the discovery of nylon and its commercialization in 1938 and Barkdoll's work on the purification of monomeric formaldehyde and its chemistry in 1948.

HISTORY BY DECADES

1940 to 1950

World War II's vast consumption of metal created a need for alternate materials. This was followed by the pent-up post war demand for consumer products. One need for alternatives to metal involved materials for use in electrical components and moving mechanical parts. In 1945, Du Pont responded by offering nylon molding resin for these applications. Management recognized that successfully marketing nylon for use in traditional metal applications required a new Technical Marketing effort.

This led to the establishment of Du Pont's Plastics Technical Service Laboratory in 1946 located in Arlington, N. J. and also to the formation of an end use development group.

One early concept, based on field development work, was realization of the necessity of designing the part based on the properties of the engineering resin itself, rather than simply duplicating the design that worked best for metal construction. To implement this concept required generating considerable design data. This experience greatly influenced the scope and content of the acetal homopolymer characterization effort carried out in the next decade.

Near the end of this decade the experience gained from field application developments with nylon and other plastics revealed a need for polymers offering more metal-like properties, especially higher strength and stiffness and less sensitivity to moisture than nylon.

At the end of this decade, management authorized fundamental research into the preparation of pure monomeric formaldehyde and the study of its chemistry.

1950 to 1960

Early in 1950 work on formaldehyde monomer purification led to some interesting results; higher purity led to significant increases in the otherwise relatively low molecular weights achieved in laboratory polymerization. The manpower effort was expanded to improve control over the rate of polymerization and to reduce the undesirable content of low molecular weight polymer. By 1952, a controlled continuous polymerization using a polymeric tertiary amine initiator in a liquid hydrocarbon medium produced a high molecular weight polyformaldehyde. The relatively instable raw polymer was stabilized by end capping in solution using acetic anhydride.

In a key management decision, work on various other projects was deferred in order to concentrate effort on commercializing a polyformaldehyde product. In late 1952 a Task Force was assigned to design a plant process; define the additive system and the initial product line; evaluate molding and extrusion; characterize end use properties and acquire design data which could be used to optimize part design. The latter echoes the concept of designing the part based on the properties of the engineering resin rather than a metal, and is an example of how work on acetal homopolymer benefited by earlier experience with nylon. The Task Force operated a semi-works unit at Du Pont's plant in Belle, W. Va. This unit provided some process experience and supplied polymer for in-house testing and the pre-commercial field program which began later in the decade.

Another important mangagement decision was to establish a new, larger, more customer-oriented Plastics Technical Service Laboratory in Wilmington, De., closing the original facility in Arlington, N. J. The new laboratory began operation in late 1954. Its proximity to Du Pont's Experimental Station, also in Wilmington, De., facilitated early processing tests and development work on the new polymer, which was coded "Polymer F".

In October 1956 a U.S. Patent was issued to Du Pont on a high molecular weight, thermally stable polyformaldehyde [4]. In an unusual move, Du Pont management decided in December 1956 to mount a large pre-commercial sales development effort, 2-1/2 years before the scheduled mid-1959 commercialization date.

This development team of chemists and engineers were responsible not only for field application development but also to ensure that molders and extruders would be ready to use the new product. "Polymer F" was recoded "Resin 91-X". In October 1957, the authorization to build a commercial plant to produce Resin 91-X was given. It was to be at Du Pont's Washington Works in Parkersburg, W. Va.

As mentioned, the experience gained with nylon revealed the need for a new polymer to compete with metal in engineering applications. The product team assigned to Resin 91-X decided that to compete with metal, melt flow characteristics would need tight control if molders were to be able to produce parts to engineering dimensional tolerances heretofore normal for metals, but not plastics. Similarly, molecular weight variation and distribution would need tight control to satisfy end users requirements for tight limits on property values.

These needs were incorporated in the plant process design and in establishing product specifications.

In 1958 Du Pont's Experimental Station obtained its first screw molding machine. This type of machine was recommended and promoted for use with Delrin®. Favorable customer test results with various polymers aided in industrial acceptance of this new type of equipment. By the middle of the next decade virtually all new molding machines sold in the U.S. were screw machines rather than the old ram type [3]. The melting efficiency, good streamlining, and uniformity of melt temperature of screw machines greatly improved molding performance with polyformaldehyde polymer.

A series of papers by Du Pont authors on thermally stable high molecular weight polyoxymethylenes, molecular structure, fine structure, physical properties and on solvents indicate that these resins, which may be described as containing acetal backbone linkages, are characterized by unusually high crystallinity [6].

Early in 1959 Du Pont was granted the trademark Delrin® acetal resin, replacing the code Resin 91-X. Pre-commercial sales of Delrin® 500X NC-10 and Delrin® 100X NC-10 began at $1/lb.

In mid 1959, because of instrumentation problems, production start-up and official commercialization of Delrin® was delayed until January 1, 1960. Nevertheless during the latter half of 1959 there was widespread press and periodical coverage announcing Du Pont's discovery and the impending commercialization [7] [8].

1960 to 1970

Du Pont announced the commercialization of Delrin® acetal resin on January 1, 1960. The truckload price was $0.95/lb. until June 1, 1960 when it was decreased to $0.80/lb. Prices in effect at the beginning of each decade through 1986 are listed in Table I [9].

The product line consisted of natural color Delrin® 500, a molding grade resin with a number average molecular weight of ca 41,000; Delrin 100, a molding resin of ca 58,000 molecular weight for applications requiring maximum impact strength; and Delrin 150, an extrusion grade analogue of Delrin® 100.

Colors consisted of 16 Standard Colors and 42 Service Colors. The molding grade pellets were surface lubricated to reduce friction of the pellets against the torpedo in ram type molding machines.

The standard package was a heat sealed, aluminum foil lined, multi-wall paper container holding 50 lbs. of product. The moisture proof container was used to reduce the effect of moisture pickup on melt flow. This made it easier to mold parts to the desired tight dimensional tolerances.

In addition to the introductory product line Du Pont provided extensive technical literature provided at the time of commercial introduction. The technical marketing concept that Delrin® would compete with metals in engineering applications greatly influenced the content of the two major publications; the Product Handbook and the Molding Manual.

Both publications contained information on Delrin® to facilitate the proper application and molding of engineering parts. In 1961 the Product Manual evolved into a Design Handbook. The Design Handbook contained property data, long term effects of environment, gear and bearing design guides and data on dimensional considerations.

The Design Handbook also provided specifics about the main technical attributes of Delrin®, i.e., tensile strength, stiffness, creep

TABLE I. U.S. truckload price of Delrin® 500 natural at the beginning of each decade through 1986.

January 1 Decade	Price $/Lb.
1960	0.95
1970	0.65
1980	1.04
1986	1.64

resistance, frictional properties, fatigue, chemical and moisture resistance. For the first time it described properties and performance of a plastic in traditional terms understood by metals designers. The Molding Manual provided comprehensive information on estimating cycles, mold design, equipment considerations, dimensional tolerances and on the quality control of molded parts. The Molding Manual, along with Material Safety Data Sheets, provide information on safety considerations for processing Delrin and toxicity information on formaldehyde. Both the scope and depth of these two publications set a standard for future literature on engineering plastic materials.

Another key decision, made in 1960, was that Delrin® would be marketed worldwide. This led to the formation of a new Du Pont Plastics team in Geneva, Switzerland to introduce Delrin into the European market, along with coordinated introduction in the East Asia, South Pacific and South America markets. A product policy was that the Delrin trademark, product codes and product quality would be uniform worldwide. As a result, international end users could be confident that Delrin® delivered anywhere in the world would meet the same product specifications.

In the first year after commercialization worldwide sales of Delrin® exceeded 1 million pounds. Sales the next year, 1961, were 4 million pounds.

In the following year, 1962, the Celanese Plastics Co. introduced "Celcon" acetal copolymer [10].

The acetal homopolymer molecular structure offers benefits of higher tensile strength and stiffness. Acetal copolymers feature resistance to strong bases, a quality useful in specific applications. A Du Pont publication compares acetal homopolymer and copolymer in terms of molecular structure, properties and injection molding performance [11].

To support the rapidly growing market for Delrin® in Europe, a compounding facility came on stream in 1963 at Du Pont's Dordrecht Works in Holland. Initially it produced finished product from capped polymer made in the U.S. The facility was expanded in 1965 and 1968 to more than double the initial capacity [12].

In 1963 there was a market need for homopolymer with better melt flow for use in thin wall, multi-cavity molds. Delrin® 900 was commercialized with a number average molecular weight of ca 32,000. With a lower molecular weight than that of Delrin® 500, Delrin® 900 provided improved melt flow and good tensile properties with some sacrifice in toughness [13]. Delrin® 900 was the forerunner of future high flow, fast cycle and high productivity compositions.

Also in 1963 the U.S. Navy had a need for a special bearing material. The search led to the development by Du Pont of a composition of 20% highly oriented fiber of Teflon® in Delrin® which met Navy requirements. In 1964 similar compositions were commercialized for industrial applications requiring low friction, low wear, no stick-slip, along with good mechanical properties. Two compositions were commercialized based on Delrin® 100 and 500 respectively, coded Delrin® DE-113 and DE-313. Subsequently they were coded Delrin® 100 AF and Delrin® 500 AF.

Another composition containing 20% milled glass fiber in Delrin® 500 was introduced as an experimental product coded Delrin® 570X. It became fully commercial in 1965.

As noted previously in 1958, experimental screw molding machines were under study with encouraging results. By 1964, new molding machines in the U.S. were virtually all of the screw design. The screw molding machine design produces better quality melt with fewer potential hold-up spots. It improved the molders ability to process all acetals, but particularly aided the molding of the new Teflon fiber filled Delrin® compositions.

In 1964 and 1965 a break-through came in the the understanding of how to mold acetal homopolymer to obtain parts of maximum properties and tight dimensional tolerances [14] [15].

During the remainder of the decade there was little change in either pricing or the product line.

In the 1964-1966 period a Federal Specification and an American Society for Testing and Materials (ASTM) specification on acetal materials were issued, and were of considerable assistance to the industry in specifying acetal material [16].

By 1966 the Delrin® production capacity at the Washington Works plant had been tripled versus the 1960 capacity.

Emphasis in the latter half of the decade was on implementing the Worldwide Marketing Concept by providing technical literature in many languages, conducting product training for overseas Du Pont representatives and holding technical seminars for customers and Plastics Engineering groups [17].

A new technical marketing concept, called Total Design, was introduced. This concept described a process where the design versatility and properties of Delrin® enabled design engineers to incorporate the functions of several metal parts into one part. Total Design can reduce the cost of parts by reducing the number of parts, with the associated benefits in procurement, assembly and quality control.

By the end of the decade, Delrin® acetal homopolymer was recognized worldwide as an engineering plastic material. Primary markets were automotive, appliance, consumer products, hardware, industrial, electric, plumbing and stock shapes [18].

<u>1970 to 1980</u>

Early in 1970 the Washington Works Delrin® plant capacity was expanded by ca 33% responding to strong worldwide demand.

A market need for a moderately priced low friction and low wear material led to the commercialization in 1971 of Delrin® 500CL. This new composition did not contain fluoropolymer or silicone.

In mid 1972 Du Pont started up a Delrin® compounding facility in Mexico to better serve this market area.

A major event in the history of acetal homopolymer was the announcement of commercialization of Tenac® acetal homopolymer by Asahi on September 1, 1972. Since then, market development with Tenac® in Japan has successfully competed with metal in typical engineering applications. Asahi Chemical Co. has been sensitive to market needs and their search for new compositions has led to a broad product line. Details on the product line and a recent Asahi development will be covered in the following section on the decade of the 1980's.

As the Total Design concept developed, it led to more and more cost studies of parts of Delrin® versus metal. Du Pont developed various techniques for estimating the cost of molded parts. Computer programs were used effectively; however, a popular method for field use was a convenient manual system [19]. The cost studies revealed in a systematic manner what was already generally known: if molders could reduce the cost of parts of Delrin® they could compete more effectively with metal.

This led to a laboratory search for faster molding compositions which were commercialized in 1972 and 1973. The original codes of Delrin® 8010 and Delrin® 8020 were subsequently changed to Delrin® 900F and Delrin® 500F. These compositions coupled with the latest molding technology resulted in cycle reductions up to 40%.

A polymer plant came on stream in 1973 at Du Pont's Dordrecht, Holland site. This new facility produced capped polymer and combined with the existing extrusion compounding equipment provided the European market with the first fully integrated plant for manufacturing acetal homopolymer.

As the market increased the molding requirements of some parts became very demanding in terms of thin walls, fast injection, long production runs and very tight dimensional tolerances. Under these demanding conditions in some molds a polymer deposit formed on the surface of the mold. The deposit could reduce heat transfer enough to adversely effect cycle time and/or the thickness of the deposit could result in difficulty holding dimensional tolerances.

The need for a low mold deposit resin for demanding molding jobs led to the discovery of a new stabilizer system...the first since commercialization. Based on this technology Delrin® 500D and Delrin® 900D were introduced in 1978. Both compositions have been used successfully in some of the most demanding applications.

At the close of this decade the production capacity for Delrin® at the Dordrecht plant was increased by 35% to supply the strong demand in the European market.

<u>1980 to Early 1986</u>

Asahi published a paper in 1981 on the polymerization of formaldehyde carried out using a solution method and a gas supply method using various initiators at various temperatures [20].

In the previous decade the background is given for the development of fast cycle compositions and low mold deposit resins. Subsequent research effort developed new compositions which provide the combined benefits of fast cycle and low mold deposit. The new compositions were commercialized in 1982 as Delrin® 500HP and Delrin® 900HP.

These new compositions are further examples of how the acetal homopolymer has been improved through the years, resulting in a product line tailored to satisfy specific market needs. Another recent example is the 1982 introduction of a plumbing grade resin coded Delrin® 200PL. It is intended for applications in intermittent hot water up to 180°F.

In the production area Du Pont announced, in 1982, a 35% expansion of the Delrin® facility at the Washington Works plant. This expansion was accomplished to ensure adequate supply for the North American and overseas markets.

A major discovery by Du Pont was an acetal homopolymer/elastomer alloy, Delrin® 100ST, that provides outstanding toughness and fails in a ductile mode in notched impact tests. Izod impact values are 15-20 ft-lb/in (800-1067 J/m) [21]. Commercialization of the new product was announced in early 1983. This new technology led to the commercialization of Delrin® 500T, a composition offering the molding advantages of Delrin® 500 with the toughness of the high molecular weight Delrin® 100.

Delrin® 1700HP, commercialized in 1985, is the latest addition to the family of fast cycle, low mold deposit compositions. With a number average molecular weight of ca 28,000, Delrin® 1700HP is a very high melt flow acetal for use in high productivity molds involving thin walls, high shear rates, long flow paths and fast cycles.

Asahi Chemical recently announced a new technology for the super drawing of Tenac® polyacetal resin. A 20 fold draw in length results in a 20 fold increase in tensile strength and a 10 fold increase in modulus. This technology is the subject of several patents [22].

REFLECTION

In looking back over 26 years of commercial experience it is appropriate to comment on product quality. Significant improvements have been made compared with the commercial Delrin® of 1960 in terms of melt stability, molecular weight distribution and control, the level and uniformity of physical properties and the color and colorability of natural color product. These product improvements were achieved by persistent chemical and engineering effort on the chemistry of the process, modernization of equipment, compounding technology and automated measurement and control systems.

In closing the history of this decade it may be helpful to summarize the product lines of Asahi and Du Pont.

As of early 1986 the two product lines are listed by functional categories in Table II [23] [24]. Various UV stabilized compositions, which both suppliers offer, and colors are not included.

TABLE II. Acetal Homopolymer Product Lines

TENAC®	DELRIN®
General Purpose 3010, 4010, 5010, 7010	General Purpose 100, 500, 900
High Productivity 5050, 7050, 7054	High Productivity Fast Cycle: 100F, 500F, 900F Low Deposit: 100D, 500D, 900D Fast Cycle and Low Deposit: 500HP, 900HP, 1700HP
Low Friction, Low Wear LT200, LT804, LT805 FA405, FS410	Low Friction, Low Wear 100CL, 500CL 100AF, 500AF
Higher Stiffness GA5-10, GA5-20, GA5-25	Higher Stiffness 570
Greater Toughness 4012, 5012	Greater Toughness 100ST, 500T

TABLE II. Acetal Homopolymer Product Lines Continued

TENAC®	DELRIN®
Anti Static and Lubricated EF500	Improved Extrusion 150SA, 550SA
Electric Contact, Deposit Free 4015, 5015	Hot Water, Plumbing 200PL
Electroplating PT300	

HISTORICAL HIGHLIGHTS

Looking at the history to date of acetal homopolymer the following are key events:

- Staudingers and Carothers work on, and understanding of, the nature of high molecular weight polymers.
- The ground breaking of molded nylon as an engineering material and the benefit of that experience in all phases of the Delrin® venture.
- MacDonald's discovery of how to make stable, high molecular weight polyformaldehyde followed by the management decision to launch a 50 man research force to commercialize acetal homopolymer.
- The decision to mount a 17 man sales development and field processing effort 2-1/2 years before commercialization.
- Construction and successful start-up of the first acetal homopolymer plant at the Du Pont Washington Works.
- Worldwide marketing concept which established Delrin® acetal homopolymer as an engineering plastic with a full polymer plant in Europe and compounding facility in Mexico.
- Asahi Chemical Company's introduction of Tenac® acetal homopolymer and the offering of a broad product line.
- Du Pont's discovery of super tough Delrin® 100ST and Asahi's new technology for the super drawing of Tenac®, along with many less dramatic improvements in acetal homopolymer over the past 2-1/2 decades is evidence that opportunities continue to exist for future new developments.

A LOOK TOWARD THE FUTURE

Theoretically, just a few years ago, it would have been impossible to achieve super toughness in a highly crystalline material such as acetal homopolymer. That theory was disproved with the discovery of super tough Delrin® 100ST. With continued high quality research effort the future will hold even more technological breakthroughs.

Experimental work suggests an opportunity to develop lower odor compositions; further enhancement of performance in hot water; new stabilizer systems to benefit molding and end use performance; mechanisms for modifying fine structure and molecular configuration.

The history of this product demonstrates that the needs of the marketplace provide the incentive for searching and discovering the new products which will be commercialized in the coming decade.

ACKNOWLEDGEMENT

The authors wish to thank the following for their support and assistance in preparing this history. In Du Pont, R. L. Ward for review and suggestions, for information and assistance R. N. MacDonald, P. N. Richardson, D. McFall, C. H. T. Tonsbeek, A. D. Seibel, K. A. Jackson, F. Mannis, R. L. Anderson, P. Sample, J. Y. DuTour, T. Matsushima and copy review by L. E. De Shields.

Asahi Chemical, Shinichi Ishida.
Typist, Nancy Bader.

REFERENCES

1. S. Ishida, Asahi Chemical Industries Co., Ltd., Company Communication (1985).
2. M. Kaufman, The First Century of Plastics (ILIFFE Books Ltd., London 1963).
3. J.H. DuBois, Plastics History U.S.A. (Cashners Books, Boston 1972).
4. R.N. MacDonald, U.S. Patent 2,768,994, October 30, 1956.
5. Du Pont Management Newsletter, June 1968.
6. C.E. Schweitzer, R.N. MacDonald, J.O. Punderson, J. Appl. Polymer Sci., 1, 158.
   T.A. Koch, P.E. Lindvig, J. Appl. Polymer Sci., 1, 164.
   C.F. Hammer, T.A. Koch, J.F. Whitney, J. Appl. Polymer Sci., 1, 169.
   W.H. Linton, H.H. Goodman, J. Appl. Polymer Sci., 1, 179.
   R.G. Alsup, J.O. Punderson, G.F. Leverett, J. Appl. Polymer Sci., 1, 185.
7. H. Solow, Fortune, August 1959.
8. Modern Plastics, November 1959, January 1960.
9. F. Mannis, Du Pont Communication.
10. M. Sittig, Polyacetal Resins (Gulf Publishing, Houston 1963).
11. Du Pont Publication, Acetal Homopolymer Vs. Acetal Copolymer, E-59941 (1983).
12. C.H.T. Tonsbeek, Du Pont Communication.
13. K.J. Persak, L.M. Blair, Kirk-Othmer: Encyclopedia of Chemical Technology, 1 (1978).
14. Molding Manual, Delrin® Acetal Resin.
15. Du Pont Plastics Technical Center, Geneva, Switzerland, TRD-30 Report on Delrin®.
16. Federal Specification L-P-392a, (1964); ASTM2133-66 (1966).
17. Designing Acetal Parts, Plastics Design and Processing, June 1966.
18. Modern Plastics, January 1970, 66.
19. How You Can Lower Cost, pp. 12-13, Du Pont Publication, E-18056, (1978).
20. S. Ishida, J. Appl. Polymer Sci., 26 (1981).
21. E.A. Flexman, Modern Plastics, February 1985, pp. 72-76.
22. Japan Patents, Toku Kou Show 57-148616 (1982), Toku Kou Show 57-109651 (1983).
23. S. Ishida, Asahi Communication 1985.
24. Delrin® Design Module III, E-62619, pp. 5.

ACETAL COPOLYMER, A HISTORICAL PERSPECTIVE

THOMAS J. DOLCE AND FRANCIS B. McANDREW
Celanese Engineering Resins, Inc., 86 Morris Avenue, Summit, NJ

INTRODUCTION

Polymers with the repeat units $\leftarrow CH_2 - O \rightarrow_n$, known as polyacetals, have been known since the early 1920's. It was not until the late 1950's, however, that research efforts resulted in high molecular weight linear, stable thermoplastic materials. Whereas these developments were directed towards homopolymers of formaldehyde at DuPont, the research at Celanese Corporation was aimed towards copolymers using trioxane, a cyclic formaldehyde trimer, as the major monomer.

In 1922 Hammick and Boeree[1] observed an insoluble white powder formed during sublimation experiments with trioxane. This material proved to be a low molecular weight, unstable polyacetal homopolymer. In 1957 workers at Celanese investigated ring-opening polymerizations of trioxane with other cyclic ethers which led to the first successful commercial acetal copolymer[2]. In 1960 a pilot plant was completed at Corpus Christi, TX for process development studies and for preparation of material for product development activities. From this pilot unit Celanese commenced limited customer sampling in 1961 and eventually brought on stream a full scale commercial unit in 1962 at Bishop, TX. This product is sold under the trademark Celcon® acetal copolymer. Through its affiliation with Hoechst AG in Germany and Daicel, Ltd. in Japan (forming the two companies Ticona Polymerwerke GmbH and Polyplastics, Ltd., respectively) this copolymer is manufactured and sold worldwide. Since then two other companies have introduced acetal copolymers, viz., BASF in the 1970's and Mitsubishi Gas Chemical Co. in 1981.

POLYMERIZATION

Trioxane polymerization is cationically initiated and can be depicted as follows:

$$R^+ + \text{(trioxane)} \rightleftharpoons R\text{-}\overset{+}{O}\text{(trioxane ring)} \rightleftharpoons ROCH_2OCH_2^+ + CH_2O$$

An induction period during which no visible polymer is formed accompanies this polymerization. Kern and Jaacks in 1960[3] showed that this induction is associated with the liberation of free formaldehyde which must build to a certain concentration before polymer is observed. At this point an

®Celcon is a registered trademark of Celanese Corporation.

Copyright 1986 by Elsevier Science Publishing Co., Inc.
High Performance Polymers: Their Origin and Development
R.B. Seymour and G.S. Kirshenbaum, Editors

equilibrium is reached between monomeric formaldehyde and the growing chain. By deliberately adding formaldehyde or other aldehydes[4] to the polymerization mixture, this induction time was eliminated or reduced. Propagation likely occurs both by addition of free formaldehyde and opened trioxane rings to the growing polymer chain. Traces of impurities containing active hydrogens can induce chain transfer and termination. Molecular weight can thus be controlled by addition of such materials to the polymerization mixture.

Commercial polyacetal copolymers contain 0.1 to 15 mole percent of a cyclic ether, commonly ethylene oxide or 1,3-dioxolane. Typical catalysts for this reaction are BF_3 or its ether complexes. In 1964, Weissermel and coworkers[5] showed that in the copolymerization of trioxane with ethylene oxide, the latter was almost completely consumed before any visible polymer was observed. During this stage of the polymerization, soluble prepolymers of ethylene oxide could be isolated[6]. These prepolymers consisted primarily of oligomers with mono-, di-, and tri-ethylene oxide units. Celanese workers in 1980[7] verified also the presence of cyclic ethers, predominately 1,3-dioxolane and 1,3,5-trioxepane, as part of the reaction mixture. These are likely formed as reaction products of ethylene oxide and monomeric formaldehyde generated from the opening of the trioxane ring.

These findings are inconsistent with the essentially random nature of the final copolymer. Weissermel et al[5] proposed a rearrangement mechanism which accounts for the redistribution of the comonomer units. Transacetalization is the term given to this rearrangement. In 1972, Cherdron[8] described this redistribution as the result of "the reaction of the polymeric carbonium ions with oxygen atoms of the same or of another polymer molecule (intramolecular and intermolecular transacetalization)". He depicted the mechanism as follows:

1. Intermolecular Transacetalization*

```
                          CH2-CH2-OCH2-O~~                      CH2-CH2-OCH2-O~~
                          |                                     |(+)
~~OCH2-OCH2(+)      +     O                      ~~OCH2-OCH2-O
                          |                                   --+--
                          CH2-OCH2-CH2-O~~                      CH2-OCH2-CH2-O~~

                      CH2-CH2-OCH2-O~~
                      |
--> ~~OCH2-OCH2-O                        +  (+)CH2-OCH2-CH2--O~~
```

(cont'd.)

*Reprinted from J. Macromol. Sci.-Chem., A6(6), p.1078 (1972) by courtesy of Marcel Dekker, Inc.

2. Intramolecular Transacetalization

$$\sim OCH_2\text{-}OCH_2\text{-}OCH_2\text{-}OCH_2\text{-}OCH_2\text{-}OCH_2\text{-}OCH_2\text{-}OCH_2^{(+)}$$

$$\longrightarrow \sim OCH_2\text{-}OCH_2\text{-}OCH_2^{(+)} + \text{cyclo-}(CH_2\text{-}O\text{-}CH_2\text{-}O\text{-}CH_2\text{-}O\text{-}CH_2\text{-}O\text{-}CH_2\text{-}O) \quad \text{Pentoxane}$$

The ultimate result of this rearrangement is a copolymer with randomly distributed comonomer units (in this case 1,3-dioxolane) and with a narrow molecular weight distribution.

END GROUP STABILIZATION

The success of polyacetals as engineering thermoplastics lies not only in the capability of achieving high molecular weight linear polymers, but also in the ability to end cap the polymer produced. The polymer obtained from the reactor normally will terminate in a hemiacetal, $-CH_2\text{-}O\text{-}CH_2\text{-}O\text{-}CH_2\text{-}OH$. This molecule is thermally unstable and will fairly easily unzipper to free formaldehyde when exposed to heat especially in the melt. Likewise, this hemiacetal end-group is unstable to alkaline media. Reaction with acetic anhydride will convert the hemiacetal to a more stable acetate ester group. This will achieve a more thermally stable molecule but offer no protection against alkaline attack.

A unique scheme has been devised to achieve thermally stable end groups in polyacetal copolymers. It makes use of the occasional presence of a stable $(-CH_2\text{-}CH_2\text{-}O\text{-})$ unit in the polymer chain as well as the susceptibility of the hemiacetal group to either thermal or alkaline degradation. Subjecting the raw polymer to a controlled thermal treatment will remove the unstable ends until a $\text{+}C\text{-}C\text{+}$ unit is reached. Techniques to accomplish this have been described in the patent literature[9] assigned to Celanese investigators in 1963. By this method, neat polyacetal copolymer from the reactor was subjected to heat and the kneading action of a roll mill. After several minutes, the unstable end-groups were removed and a stable polymer resulted. More recently (1981), this technique has been refined by researchers at Mitsubishi Gas Chemical Co. who performed this controlled thermal unzippering in a vented extruder[10]. By either approach, this must be carried out in the presence of thermal stabilizers, especially antioxidants, to prevent random oxidative attack of the polymer chain.

Alternately, the unstable hemiacetal portion can be removed hydrolytically in the presence of alkaline media. Patent literature disclosures published in the mid to late 1960's describe this procedure being done either homogeneously, i.e., in solution[11], or heterogeneously[12]. In either case, thermally or hydrolytically, a stable end-capped polymer results.

STABILIZATION

In addition to protecting the acetal copolymer from degradation from an unstable end-group, it is necessary to provide protection from attack by oxidative and acidic means. Conventional hindered phenols provide adequate stabilization from oxidative environments both during processing in the melt and subsequent usage. In 1966, the patent literature describes one such antioxidant[13].

Acidic species can result from traces of catalyst residues or from oxidation of released formaldehyde to formic acid. Nitrogen containing compounds, such as amidines, soluble polyamides, and amines, efficiently neutralize acidic species. In the mid 1960's, Celanese was assigned U.S. 3,313,767 describing one such class of compounds[14]. Epoxides and mildly basic salts also provide stabilization for this mode of degradation. Interestingly, many of these same compounds also scavenge free formaldehyde, thus preventing its oxidation to formic acid.

In the polyacetal copolymer, the presence of an occasional $-(CH_2-CH_2)-$ will prevent complete catastrophic depolymerization of a polymer chain which has been attacked either by oxygen or acid. A homopolymer chain without this stabilizing influence will completely degrade. Thus, acetal copolymers are inherently more stable than homopolymers.

Other stabilizers, especially those providing protection against ultraviolet radiation may be added to the resin. Colorants, lubricants, plasticizers, glass fibers, minerals, etc. are common additives extending the utility of acetal copolymers.

COMMERCIAL PRODUCTION

Typically in the commercial manufacture of polyacetal copolymer molten trioxane, comonomer, chain transfer agent and catalyst are fed into a reactor capable of providing shear and mixing action and with the means to remove heat generated by the exothermic polymerization reaction. The solid raw polymer is removed, quenched to neutralize any traces of catalyst and washed free of unreacted monomer. The unstable chain ends are usually removed hydrolytically in a basic media. Subsequently, the polymer is washed, dried and compounded with stabilizers and other ingredients and packaged for shipment.

STANDARD ACETAL COPOLYMER PRODUCTS

In this and in the following sections, the acetal copolymer products mentioned are those developed by Celanese and marketed under the Celcon trademark.

When Celanese made its first product offerings to the plastics marketplace in 1962, it was with linear acetal copolymers, one having a 9.0 melt index (MI) for injection molding and the other having a 2.5 melt index for extrusion and, to a minor extent, injection molding.

For extrusion applications, high viscosity polymers are normally used because they have good melt strength and resist sagging. On the other hand, for injection molding medium to low viscosity polymers are preferred to facilitate easy packing of mold cavities. The initial offering of the 2.5 and 9.0 MI resins thus covered both plastic processing methods, and showed the capability to tailor polymer molecular weights ($\bar{M}_N$ of 50,000 and 35,000, respectively) for the end use application. With time, the need for lower molecular weight resins became apparent as a means to reduce cycle times and provide for improved flow for parts having long, thin sections. Hence, 14 and 27 MI resins followed and very recently even one with an MI of 45. At the other end of the spectrum, a 5.0 MI resin was also added to fill the gap between the 2.5 and 9.0 MI resins.

All of the copolymers described above have essentially the same thermal and mechanical properties, with the key exception being the effect that molecular weight changes have on impact (toughness) related properties. As is seen with most thermoplastics, there is a relationship between polymer ductility and molecular wieght. Thus, a 2.5 MI copolymer exhibits much greater elongation and impact resistance than a 27 MI one, and the latter more than a 45 MI resin. End-use requirements for acetal copolymer parts will frequently dictate the particular melt index of the resin to be used. A summary of properties of these standard grades is given in Table 1.

EARLY SPECIALTY ACETAL PRODUCTS

Acetal Terpolymer

Shortly after the introduction of acetal copolymer, a need was identified for a grade offering improved rheological properties over the 2.5 MI resin for certain profile extrusions and extrusion blow molding. To meet this need, an acetal terpolymer was developed that provided the desired very high viscosity at low shear rates. The route selected was to add a low level of polyfunctional termonomer to the copolymer polymerization which would introduce long chain branching in the resultant terpolymer[15]. This branching imparts the desired high melt strength during processing (a compilation of properties is found in Table 1).

Glass Reinforced Copolymer

Almost from the start of marketing acetal copolymer the need for a fiber glass reinforced resin became apparent. The key advantage of this resin would be in enhanced flexural modulus or stiffness. The initial offering contained 20% fiber glass. This resin provided a doubling of flexural modulus over the 375,000 psi displayed by the basic copolymer but with a minor loss in tensile strength. Studies revealed that the less-than-desired enhancement of mechanical properties was due to poor adhesion between the fiber glass and the polymer matrix. After significant research, a chemical coupling system was identified and with 25% fiber glass there was a tripling of flexural modulus and a doubling of tensile strength[16]. This resin is used primarily where stiffness and strength without bulk are required, and where the many desirable properties of the acetal copolymer are needed to meet end-use service conditions (see Table 1).

TABLE 1. Properties of acetal copolymer products.

	Resin Grades				
Property	M25	M90	M270	U10[a]	GC-25A[b]
Specific Gravity	1.41	1.41	1.41	1.41	1.62
Melt Index, g/10 minutes	2.5	9.0	27.0	1.0	----
Melting Point, °C	165	165	165	165	166
Tensile Strength, psi	8,800	8,800	8,800	8,700	16,000
Elongation, %	75	60	40	67	2-3
Flexural Modulus, psi x 10^5	3.75	3.75	3.75	3.50	10.5
Notched Izod Impact Strength, ft-lbs/in.	1.5	1.3	1.0	1.7	1.1
Heat Deflection Temperature					
@ 66 psi, °C	157	157	157	157	166
@ 264 psi, °C	110	110	110	110	163

(a) Terpolymer for extrusion applications.

(b) Glass reinforced (25%) copolymer.

RECENT FORMULATION DEVELOPMENTS

Over the last 10 years or so, many new acetal copolymer formulations have emerged, all developed to meet identified market needs. All of these newer formulations came at great expense in research effort because the acetal copolymer, with its high level of crystallinity, does not lend itself to easy modification. Each different type of resin system required the development of extensive associated technology for compounding, processing and molded part assembly, decoration, etc. The more interesting recent resin developments are described below and in Table 2.

Mineral-Filled Resins

Due to anisotropic shrinkage frequently experienced in molding unmodified acetal copolymers, there existed a need to minimize this to reduce part warpage that frequently occurs. Minerals of various types are used to achieve this usually at the expense of toughness-related properties. By the judicious selection of mineral filler by type and particle size, and by employing chemical coupling agents, it is possible to minimize the tendency of the acetal copolymer to shrink anisotropically, thus reducing the part warpage while still retaining impact strength. Celanese has developed four mineral-filled resins based on the 9.0 and 27.0 MI copolymers, each at two different levels. These resins are used where control of molded part dimensions is extremely critical for assembly or functional reasons. As one would expect, the mineral filler decreases tensile strength but increases flexural modulus.

Antistatic Resin Grades

Compared with many polymers (such as the polyolefins), acetal copolymer offers good resistance to the build-up of static electrical charge. However, for certain applications (audio and video equipment, devices used in hospital operating rooms or where oxygen is used) it is desired that resin parts have minimal propensity to become charged. Grades of acetal copolymer have been developed which, due to the inclusion of an antistatic compound to their formulation, tend not to build-up static charges. For example, in a laboratory test in which a 6000 volt charge is imposed on an unmodified acetal copolymer part, it requires 200 seconds for the static charge to decay to half its original value. With the antistatic grades, the half-life is reduced to less than 20 seconds. Formulations currently available are based on the 27 and 45 MI resins.

Low Wear Grades

A major application area for acetal copolymer is in gears, bearings and material handling equipment such as chain links and other conveyor components. While unmodified acetal copolymer generally works well in such applications, there are very demanding applications that require improvements in the wear characteristics of the basic acetal copolymer. A number of grades of wear resistant acetal copolymer are now available involving the addition of one or more low wear additives such as polydimethylsiloxane (silicone oil), polytetrafluorethylene (PTFE), molybdenum disulfide, various mineral oils and proprietary compounds. All of these to varying degrees tend to reduce part wear significantly and show up in standardized laboratory tests as having lower coefficients of friction and greatly enhanced pressure-velocity (PV) limits.

High Impact Resins Grades

As noted earlier, the impact resistance of acetal copolymers is molecular weight dependent. However, even the highest molecular weight acetal copolymers commercially available lack adequate impact resistance for certain end-uses. This is due primarily to the very crystalline nature of the resin. To broaden the range of utility of acetal copolymers, formulations have been developed which offer significantly improved practical impact resistance. This is accomplished through the inclusion of impact modifiers (usually elastomers) at various levels. While the impact resistance is greatly enhanced, the tensile strength and flexural modulus of the resin is reduced. The level of reduction is generally dependent on the amount of impact modifier used in the formulation. In 1985, Celanese introduced Duraloy™ 1000 series, a grade of acetal copolymers which exhibited no break notched Izod impact at room temperature and excellent resistance to impact at low temperatures as well.

Electroplatable Resin Grades

Chrome plated plastic parts are widely used in the automotive, appliance, plumbing and other industries. While there were early efforts to chrome plate acetal copolymer, it was only recently that adequate adhesion between the resin and the metal coating was obtained. Recent breakthroughs in the acid etching step and in molding technology to obtain defect-free parts have provided an electroplatable acetal copolymer grade.

Weather Resistant Resin Grades

Acetal copolymer parts in their natural form will, after long-term exposure to sunlight, develop a white film and/or minor crazing on exposed surfaces. The white film is degraded copolymer of low molecular weight. The addition of carbon black or UV stabilizers will provide significant improvements in performance in this respect, with combinations of carbon black and UV stabilizers offering maximum UV protection. Certain pigments, such as titanium dioxide, also offer improved UV protection for the acetal copolymer. Special weather resistant grades in a wide variety of colors have also been developed.

End-Use Areas and Future Trends

Because of acetal copolymer's high crystallinity, its excellent dimensional stability, chemical resistance, and mechanical properties, and its predictable performance in service, it has found broad application in replacing metals. Some of the most significant end-use application market segments are the automotive, plumbing, personal products and materials handling equipment industries.

Research and development programs are underway to make improvements in the performance limits of acetal copolymer grades currently in the marketplace, as well as to develop completely new grades to extend the range of resin utility to new markets and application areas. The future will witness the introduction of a number of new grades which, to a large extent, will be due to the development of new stabilizers, additives, and blend components.

™Duraloy is a trademark of Celanese Corporation.

TABLE 2. Properties of Celcon® acetal specialty products.

PROPERTY	RESIN GRADES							
	M90	MC90	MC90HM	AS270	LW90	TX-90 Plus	EP90	WR90
Grade	General Purpose	Mineral-Coupled	Mineral-Coupled	Anti-Static	Low-Wear	High Impact	Electro-Platable	Weather Resistant[a]
Specific Gravity	1.41	1.48	1.60	1.41	1.43	1.37	1.43	1.41
Melt Index, g/10 min.	9.0	8.5	6.5	27.0	9.0	7.0	9.0	9.0
Tensile Strength, psi	8,800	7,700	6,400	8,800	8,500	6,000	8,200	8,800
Elongation, %	60	55	50	50	50	200+	70	40
Flexural Modulus, psi x 10^5	3.75	4.30	5.0	3.75	3.70	2.20	3.50	3.75
Notched Izod Impact Strength, ft-lbs/in.	1.3	1.2	1.2	1.1	1.2	2.5	1.1	1.2
Heat Deflection Temperature, @ 66 psi,°C	157	154	150	158	----	135	----	158
@ 264 psi, °C	110	93	97	110	97	80	103	110
Static Charge Half Life, 6000 VAC, seconds	200+	----	----	<20	----	----	----	----
Gardner Practical Impact, in-lb.	20	40	60	----	----	240+	----	----
PV Limit @ 200 ft/min.	20,000	----	----	----	30,000	----	----	----
Time to Wear Failure,min. @ 50 psi & 300 ft/min.	10	----	----	----	>120[b]	----	----	----

(a) Offers much better resistance to chalking than the unmodified copolymer.

(b) Sample did not wear-fail in the 120 minute period.

®Registered Trademark of Celanese Corporation.

REFERENCES

1. D. L. Hammick and A. R. Boeree, J. Chem. Soc. 121, 2738 (1922).

2. U.S. Pat. 3,027,352 (Mar. 27, 1962), C. Walling, F. Brown, and K. Bartz (to Celanese Corp.).

3. W. Kern and V. Jaacks, J. Polym. Sci. 48, 399 (1960).

4. U. S. Pat. 3,445,433 (May 20, 1969), F. B. McAndrew, (to Celanese Corp.).

5. K. Weissermel, E. Fischer, K. Gutweiler, and H. D. Hermann, Kunstoffe 54(7), 410 (1964).

6. U.S. Pat. 3,641,192 (Feb. 8, 1972), F. B. McAndrew, G. W. Polly, and W. E. Heinz (to Celanese Corp.).

7. G. L. Collins, R. K. Greene, F. M. Berardinelli, and W.H. Ray, J. Polym. Sci., Polym. Chem. Ed. 19, 1597 (1981).

8. H. Cherdron, J. Macromol. Sci. Chem. A6(6), p. 1078 (1972).

9. U.S. Pat. 3,103,499 (Sept. 10, 1963), T. J. Dolce and F. M. Berardinelli (to Celanese Corp.).

10. U.S. Pat. 4,301,273 (Nov. 17, 1981), A. Sugio and coworkers (to Mitsubishi Gas Chemical Co.).

11. U.S. Pat. 3,174,948 (Mar. 23, 1965), J.E. Wall, E.T. Smith, and G.J. Fisher (to Celanese Corp.).

12. U.S. Pat. 3,318,848 (May 9, 1967), C.M. Clarke (to Celanese Corp.).

13. U.S. Pat. 3,240,753 (Mar. 15,1966), T. J. Dolce (to Celanese Corp.).

14. U.S. Pat. 3,313,767 (Apr. 11, 1967), F.M. Berardinelli and T.J. Dolce (to Celanese Corp.).

15. U.S. Pat. 3,686,142 (Aug. 22, 1972), W.E. Heinz and F.B. McAndrew (to Celanese Corp.).

16. Ger. Offen. 2,121,559 (Dec. 2,1971), F. M. O'Neill and A. L. Baron (to Celanese Corp.).

A PATH TO ABS THERMOPLASTICS

WILLIAM A. PAVELICH
Borg-Warner Chemicals
Technical Centre
Washington, WV

INTRODUCTION

The Cycolac$^{(R)}$ brand of ABS thermoplastics is produced by Borg-Warner Chemicals from acrylonitrile, butadiene, and styrene monomers; hence, the abbreviation ABS. The first commercial products were introduced in 1954 and have evolved into a family of materials. A pattern of blending and alloying that had been established before Cycolac resins became part of the product line was continued when ABS became available as a blend component.

The Borg-Warner Corporation began business in 1928 in Illinois. Components for agricultural implements and for the automotive and tractor age were developed. A certain amount of integration was accomplished, and by 1934 the corporation was able to diversify. The Marsene Corporation of Gary, Indiana, was acquired to provide an entry into chemical manufacturing. They brought a product mix that included paper coatings, adhesives, and chemically modified natural rubber (polyisoprene). One of these modifications was the addition of HCl to yield rubber hydrochloride, and another dealt with the reactions of "cyclizing" polyisoprene [1]. The junior author of this 1927 article on rubber reactions, W. C. Calvert, will appear again in connection with reactions of synthetic rubber to create impact modified materials. Product support involved blending various materials to combine the properties of the components in the blend. The blend of cyclized rubber and polyisobutylene (Butyl rubber) was prepared and found to be an excellent electrical insulator. Wire coatings were developed from this composition, and the product became a vital element in the radar projects of the late thirties and early forties. For our purpose, the important point is that a useful blend of polymers had been created and commercialized.

DISCUSSION

In the late forties, work began to improve the properties of a new thermoplastic blend. It had been revealed [2] in 1948 - 50 that poly(styrene-co-acrylonitrile), or SAN, could be blended with Buna N, a copolymer of butadiene and acrylonitrile, or Buna S, a copolymer of butadiene and styrene, to get useful thermoplastics. These materials were impact resistant, with Izod impact values of 2 to 3 foot-pounds. The commercial use of these materials was hindered by the lack of low temperature impact strength. The rubber technologists of the Marbon Division (as the Marsene Corporation had been named on assimilation into Borg-Warner) knew that polybutadiene remained "rubbery" at lower temperatures than the copolymers cited above. However, blending experiments showed that polybutadiene and SAN were incompatible. The polymerization of SAN could be accomplished in solution, in bulk, or

Copyright 1986 by Elsevier Science Publishing Co., Inc.
High Performance Polymers: Their Origin and Development
R.B. Seymour and G.S. Kirshenbaum, Editors

in an emulsion medium. Since polybutadiene was polymerized in emulsion, the latex was employed as a medium for emulsion copolymerization of styrene and acrylonitrile. It is interesting to note that there is no mention of grafting in Calvert's patent [3] filed in 1953, but his 1959 patent [4] filed in 1957 does contain this concept. In any case, the resins isolated from the emulsion copolymerization had uniformly dispersed rubber domains. The physical properties were as good or better than the mechanical blends of SAN and Buna rubber. What had been achieved was a two phase polymer system with a SAN continuous phase and a dispersed rubber phase. The rubber dispersion was stabilized by the SAN that was grafted to the polybutadiene emulsion particles. This SAN acted at the interface between the two incompatible polymers, much like a surfactant. The polybutadiene had been crosslinked during its polymerization and subsequent exposure to the SAN polymerization. Now these particles retained their size and shape during compounding and processing. Again, the patent gives us an insight into the state of technology, in that the impact strengths were rated as excellent, good, fair, and poor. The test is described as a sharp blow with a sheet of plastic on the corner of a stone slab. This "whack" test became one of the regular screening tools of Thomas Grant, one of Borg-Warner Chemical's most prolific inventors.

After a number of laboratory experiments (estimated at more than 10,000), an ABS composition was accepted for commercialization in 1953. Field trials and additional laboratory work (more than 50,000 formulations) produced a salable material in 1954. This new plastic was designated Cycolac(R) thermoplastic resin, a name that could easily be associated with cyclized rubber even though that material was not in the system. It was a tough, rigid, chemically resistant material that could be processed over a wide range of conditions into intricate shapes. The specification of Cycolac L grade of ABS as the material of construction for portable radio housings by Radio Corporation of America was a banner accomplishment. The advertising showing the radios being dropped from buildings was a flashy way of stating the toughness of this product.

Listening to customers was an important attribute of the field team. As new property balances were requested by a customer, the information was forwarded to the laboratory at Gary. A response in the form of a new grade of Cycolac resin was usually possible. Cycolac plastics were being used for an ever growing list of applications. In 1956, the Gary plant was becoming too small to serve the needs being discovered. Accordingly, a facility was authorized to be built at Washington, West Virginia. This facility had a nameplate capacity of twelve million pounds of ABS which was then the world's largest such plant. In 1957, this new plant became operational with the new Cycolac T grade being put into production. At the time this grade was introduced, it was designated T for "tough". In later years, it would be alleged that the "T" stood for "telephone". Not only did this grade become the standard for general purpose molding grade ABS, it was specified for use in Western Electric's telephone housings. The automotive industry discovered Cycolac ABS, and ordered large volumes of parts molded from this material. Appliance housings, small tool housings, pipe and pipe fittings, and business machine housings were all being molded or extruded from this new plastic. The extrusion of Cycolac plastic into sheet which could then be vacuum formed, thermoformed or pressure formed created markets for large, relatively simple shapes such as luggage, machine cases, refrigerator liners, and boat hulls.

Meanwhile, research was being carried out to define the range of polymer systems that could be obtained. It was obvious that the composition was a prime variable. Constraints arose when the full range from 0 to 100% of any component was considered. As styrene and acrylonitrile copolymerize, styrene monomer is typically consumed faster [5] than acrylonitrile. The monomer composition in the reactor then changes during the reaction and a mixture of polymer compositions is obtained. The properties of these mixtures are not desirable. A composition of 66:34 weight ratio of styrene to acrylonitrile, or approximately 1:1 moles styrene to acrylonitrile, was found to be relatively easy to control in the reactor. Also, one cannot have an impact resistant material without some butadiene, say greater than five weight percent. The system is not a terpolymer, however. The butadiene is added to the reactor as a polymer, along with styrene and acrylonitrile monomers. Polymerization causes SAN to be grafted to the rubber to produce a dispersible domain. It is also a requirement that the polybutadiene exist as a separate phase of a desired size. This size is critical to the impact behavior, so the domain size and shape must be stable through compounding and processing into consumer articles. The rubber particle size that would be useful in the SAN matrix was developed through the technical activities under Howard Irvin, as well as the need for a grafted SAN to rubber ratio that gave good impact behavior. A native Hoosier, Walter Frazer, had joined the Marbon at Gary and provided an early published discussion [6] of the ABS structural requirements needed to obtain desired physical properties. Research activities have continued to define the relationships between the various facets of molecular structure, composition, and system morphology and the impact strength, melt flow, modulus and other physical properties that were required for a particular application. Contemporary commercial practices have been described by Kulich, Kelley, and Pace [7].

As more applications were developed, one of the requests that came in was for a material that would withstand higher use temperatures without changing shape or dimensions, i.e., that had a higher heat deflection temperature (HDT). The Calvert patents had already shown the efficacy of alpha-methyl styrene in this regard. Therefore, this monomer was introduced as a partial replacement for the styrene, both in the graft and in the matrix phases. Depending on the level of alpha-methyl styrene that was used, the HDT could be raised to acceptable levels. These new high heat versions of Cycolac ABS were introduced to the automotive market in 1958 and won widespread use. As in many things, there were penalties: a higher HDT requires a higher processing temperature, first to flux the plastic, and second, to offset the viscosity increase that comes along with the stiffer molecule that results from the addition of a methyl group to the backbone.

Another question that was raised was how to make Cycolac ABS clear. The opacity of the plastic was inherent because the rubber domains that made it tough were also scattering light rather than allowing light to pass through in a straight line. A way to make the system clear would be to make the domains smaller so they had less scattering power. That way lay the destruction of the impact resistance. Changing the refractive indices of the components of the system so the various phases were less different optically was seen as an alternate approach. A fourth monomer, methyl methacrylate, was employed to accomplish this optical change, and Cycolac CIT resin, a "clear", impact resistant thermoplastic was achieved.

The blending skills of the Marsene ancestors had not been shelved. Although polyvinyl chloride (PVC) was known both as rigid and as flexible compounds, the processing characteristics were not entirely satisfactory. The sensitivity of PVC to thermal degration made it necessary to add plasticizers to make it easier to process. Flexible versions of PVC required even more plasticizer. Unfortunately, low molecular weight plasticizers lower the heat deflection temperature of the compound. Cycolac resin and PVC were blended in order to achieve enhanced impact strength, higher heat deflection temperature, or higher tensile strength while preserving flame resistance and chemical resistance [7]. It was found that these blends were easier to process than expected. Further work established the miscibility of SAN and PVC. This provided two directions for products: the flame retarded alloy sold under the name Cycovin(R) plastics, and a family of impact modified PVC's. there was less lowering of the heat deflection temperature when SAN was employed as a plasticizer, but these products were not universally utilized. Blending only the poly[butadiene-g-(styrene-co-acrylonitrile)] with the PVC gave enhanced impact resistance, just as it did with SAN, without decreasing the HDT noticeably. The definition of the appropriate size of rubber domains, S/AN ratios and levels was determined in conjunction with PVC producers. The series of modifier resins known as Blendex(R) has evolved from these findings. Additional work by Thomas Stemple has created a line of flame retarded Cycolac plastics based on lower molecular weight chemicals rather than PVC.

Many of the Cycolac ABS applications are metal replacements. In some of these, the metal look was retained for cosmetic reasons. Automotive grills, wheelcovers, mirror housings, knobs, <u>etc.</u> are examples of this practice of chrome plating an ABS part. A base metal can, of course, be electroplated with a chromium film, but making a plastic conductive is difficult, expensive, and often impairs the physical properties. It was discovered that a nickel film could be deposited chemically on Cycolac plastic parts. This film now provided an electrical path to plate copper, more nickel, and chrome on the part. Major obstacles to this application were developing a receptive surface through the use of appropriate washing systems, etching the plastic surface, and then coping with the two widely different thermal coefficients of expansion of the plastic and metals. In order to make this application succeed, Borg-Warner Chemicals found itself the owner of a process for electroless plating. This process was licensed to platers to assure them and their customers of a quality part. Eventually, this peripheral activity was sold, but the plating grades of Cycolac thermoplastics are still available.

The introduction of polycarbonates presented a new material to consider as an alloying candidate. This material had clarity, impact resistance, abrasion resistance, and HDT to bring to the party. While one might have hoped to get a clear alloy, this didn't happen. The expected HDT increase was obtained [8], and the product was introduced as Cycoloy(R) thermoplastic. The compatibility of these two polymers was shown to be related to the S/AN ratio of the ABS [9]. It is known that polymers as a class do not mix readily, since there is little entropy of mixing to be gained when macromolecules are involved. This has not stopped experimentalists from making blends. These blends are not always useful, but they are tried. For example, it is known that ABS and polycaprolactone can be mixed. This has not made for a marketable material, however.

There have been failures in the development of ABS products over the years. In the early 70's it was proposed that plastics could be used as beverage bottles. They could be recycled and they would have a higher energy content than glass. A major market would have been carbonated beverages. The first hurdle was to contain the carbonation. It was learned that the diffusion was a two step process: first, to saturate the plastic with carbon dioxide, then, to diffuse to the environment. When conventional compositions were blown into bottles, filled, and capped, the beverage went "flat" in a very short time. New compositions were developed in the high acrylonitrile range which had the requisite barrier behavior toward water and carbon dioxide. Impact modification was defined and achieved. This material was to be known as Cycopac(R) resin, and was submitted for evaluation as a bottle. Many of you remember 1974 as the year OPEC was formed. Others among you may remember it for the enhanced activity to reduce vinyl chloride in the environment. It had been recognized from the beginning that a beverage bottle material would have to be kept to a low monomer level because of the potential organoleptic contribution. The problems of achieving, maintaining and certifying to zero monomer levels seemed inordinate, and the Cycopac plastic program was abandoned. "Once bitten, twice shy", and the emergence of polyester containers has not been adequate to renew interest in this product and application.

The day came when the Marbon Division, like its parent, felt it could grow by diversification. An apparently outstanding example of an acquisition was Safety Guide, a manufacturer of the reflective markers used to indicate highway lanes in the dark. These markers can be injection molded and glued to the highway. Both Cycolac plastics and adhesives for this use had been developed by Marbon or Marsene. Other similar acquisitions which could use Cycolac ABS to make useful articles were made, but they are not a part of today's Borg-Warner Chemicals. Nevertheless, there has been continued growth in applications for ABS, and additional production facilities have been established. In 1964, a joint venture, Ube Cycon was opened in Japan. In 1966, a wholly owned subsidiary was opened in Amsterdam, Netherlands, and a facility was begun in Oxnard, California. In 1967, plants in Cobourg, Ontario, Canada and Ottowa, Illinois were added. In 1981, a new plant came on stream at Port Bienville, Mississippi.

SUMMARY

In retrospect, it would seem that Borg-Warner Chemicals has used a consistent strategy in developing Cycolac brand ABS thermoplastics: they have found ways to combine two different polymers, and have studied the properties of those blends or alloys so as to make them perform better by using the knowledge so gained. This has been demonstrated in the beginning by the grafting of compatibilizing polymer onto synthetic rubber to get a stable dispersion of that rubber in a matrix of another polymer. Research activities were then carried out to establish the structure-property relationships which governed the flow-impact-modulus balance that was appropriate to an application. The customer was involved to the extent that suggestions for new performance characteristics were considered a challenge to create a new grade. These company traits created the present family of Cycolac plastics which fit into the world of communications, automobiles, recreational vehicles, tools, appliances, business machines, and other aspects of our daily lives. Today, the new Prevex(R) brand of thermoplastics, an alloy based on polyphenylene ether resins has been added to the Borg-Warner Chemicals product line. It will be interesting to see how it develops in the light of this tradition.

ACKNOWLEDGMENT

The author wishes to thank Borg-Warner Corporation for permission to publish this history.

REFERENCES

1. H. A. Bruson, L. B. Sebrell, and W. C. Calvert, Ind. Eng. Chem. 19, 1033 (1927).
2. L. E. Daly, U. S. 2,435,202 (1948), and U. S. 2,505,349 (1950) (to U. S. Rubber Company).
3. W. C. Calvert, U. S. 3,238,275 (to Borg-Warner Corporation) 1966.
4. W. C. Calvert, U. S. 2,908,661 (to Borg-Warner Corporation) 1959.
5. R. G. Fordyce and E. C. Chapin, J. Am. Chem. Soc. 69, 581 (1947).
6. W. J. Frazer, Chem. Ind. (1966) 1399.
7. D. M. Kulich, P. D. Kelley, and J. E. Pace, Encyclopedia of Polymer Science and Engineering, John Wiley and Sons, New York, vol. 1, p. 388.
8. T. S. Grabowski and H. H. Irvin, U. S. 3,053,800 (to Borg-Warner Corporation) 1962.
9. T. S. Grabowski, U. S. 3,130,177 (to Borg-Warner Corporation) 1964.
10. J. D. Keitz, J. W. Barlow, and D. R. Paul, J. Appl. Polym. Sci. 10, 3131 (1984).

Cycolac(R), Cycovin(R), Cycoloy(R), Blendex(R) and Prevex(R) are registered trademarks of Borg-Warner Corporation.

STYRENE-MALEIC ANHYDRIDE-VINYL MONOMER TERPOLYMERS AND BLENDS

RAYMOND B. SEYMOUR
Department of Polymer Science
University of Southern Mississippi
Hattiesburg, MS

Styrene and its polymers deserve a special place in polymer science history. The polymerization of a liquid obtained by Simon by the pyrolysis of storax in 1839 was the first recorded polymerization of a vinyl monomer (1). Since he confused the polymer with an oxidation product, he called it styrene oxide. Blythe and Hofmann (2) suggested the name metastyrol and Kopp (3) showed that this product and cinnanol, obtained by the distillation of cinnamic acid to be identical. Erlenmeyer (4) described this product as vinylbenzene.

As greater interest in this monomer developed, new methods, such as the dehydration of phenylethyl alcohol (5) and the dehydrogenation of ethylbenzene (6), were used to obtain styrene monomer. However, in spite of considerable work by leading chemists of the 19th and 20th centuries, there was little information available on the polymerization of styrene prior to Staudinger's synthesis of polystyrene in the 1920's (7).

Of course, as a result of the need for styrene monomer for the production of SBR, much information was developed and the annual sales of polystyrene and its plastic derivatives grew to 2.5 million tons in the year 1985. However, in spite of its ease of fabrication by injection molding and its low cost, it was brittle and moldings deformed below the temperature of boiling water.

The ductility was increased by blending with elastomers to produce high impact polystyrene (HIPS) (8). The ductile product, called ABS terpolymer, was also produced by Calvert who made blends of NBR and copolymers of styrene and acrylonitrile (SAN). These products were discussed by W. A. Pavelich of Borg Warner in the preceeding chapter. Both HIPS and ABS are produced in large quantities throughout the world.

It is interesting to note that the first styrene copolymer was a copolymer of a nonpolymerizable monomer viz, maleic anhydride and styrene (SMA) (9). This copolymer (SMA), which was called a heteropolymer by Wagner-Juaregg is resistant to the temperature of boiling water but is readily hydrolyzed by hot aqueous alkaline solutions. The product of hydrolysis viz, styrene-maleic acid copolymer retains its molecular weight (10). The viscosity of this water soluble polymer may be controlled by crosslinking with divinylbenzene (11).

Another styrene copolymer with heat resistance superior to polystyrene, is a copolymer of styrene and fumaronitrile (12). Fumaronitrile, like the corresponding anhydride (maleic anhydride), does not form homopolymers but copolymerizes with styrene to produce a copolymer with as much as 40 percent fumaronitrile. Because of the bulky groups present, this copolymer is only partially alternating i.e., the sequence is not BABA but BABAA where B represents the fumaronitrile monomer and A represents styrene.

Monsanto attempted to commercialize the fumaronitrile copolymer under the trade name of Cerex. However, in spite of its superior thermal properties, the copolymer contained a residual fumaronitrile which is a potent vesicant. Hence, plans for the commercialization of Cerex were abandoned in the early 1950's.

Copyright 1986 by Elsevier Science Publishing Co., Inc.
High Performance Polymers: Their Origin and Development
R.B. Seymour and G.S. Kirshenbaum, Editors

Nevertheless, a terpolymer of styrene-maleic anhydride and acrylonitrile (S/MA/AN) met the specifications of superior thermal properties but it was difficult to injection mold using the plunger-type presses available in the early 1940's. This objection was overcome by blending with NBR, SBR, or HIPS.

This terpolymer (S/MA/AN) and blends of SMA were investigated, patented and produced in a pilot plant in the early 1940's, under the trade name of Cadon (13). However, because of a determination to commercialize Cerex and the resignation of the copatentees of Cadon, the excellent properties of this pioneer engineering polymer were overlooked until the patent had expired and other firms offered commercial polymers based on SMA.

In addition to being used as a textile assistant, dye paste and dispersant, SMA was reacted with polyethylene oxide (Carbowax) to produce bonded fabrics (14). Heavy metal salts of SMA were also used as controlled release biocides (15).

The tendancy for alternation of monomers in a styrene-maleic anhydride and styrene-acrylonitrile copolymers at moderate temperatures has been attributed to the formation of a charge transfer complex (CTC) between a donor (D) and an acceptor (A). This CTC is readily detectable by UV or NMR spectroscopy. More important, the equilibrium constant decreases as the temperature is increased and this effect can be followed by instrumental analysis. Thus it is possible to extrapolate to a higher temperature at which the CTC does not exist (16). Thus, by proper temperature control, it is possible to produce SMA alternating copolymers, block copolymers of vinyl monomers with both alternating and random SMA (17) and completely random copolymers of SMA (18). Half esters of SMA have been used as viscosity control agents in petroleum crudes (19).

Koppers Company produced SMA-type molding powders under the trade name of Dylark but this business was acquired by Arco which continues to produce these SMA type products. Monsanto has extended the SMA terpolymer investigation to include random copolymers and alternating copolymers in which the copolymers with SMA are acrylonitrile, ethyl acrylate, isobutylene, methyl acrylate, and methyl methacrylate (20). These terpolymers (S/MA/X) and rubber modified Cadon have created an entire new family of engineering polymers (21). New patents have been issued for both glassy terpolymers and rubber modified S/MA/AN (22, 23). As might be expected, the incompatable blends are characterized by two glass transition temperatures (T_g) (24).

New patents have also been issued on rubber modified Cadon (25), SMA grafted on EPDM (26), terpolymers with alpha-methylstyrene (S/MA/AN/AMS) (27,29, 30), S/MA/AN (28), blends of SMA and HIPS (31, 32) and SMA/PPO (33).

The physical properties of typical Cadon and Dylark polymers are shown in the following table (34):

PROPERTY	SMA	RUBBER MODIFIED SMA	20% FIBERGLASS REINFORCED SMA
Tensile strength (psi)	8,100	5,000	12,000
% elongation	1.8	3.4	2.5
Flexible strength (psi)		9,000	19,000
Notched Izod impact resistance (ft. lbs./in.)	0.6	4.5	2.4

The physical properties of typical Cadon and Dylark polymers continued:

PROPERTY	SMA	RUBBER MODIFIED SMA	20% FIBERGLASS REINFORCED SMA
Coeff. thermal expansion (10^{-6}Cm/Cm/^{o}C)	80	65	20
Deflection temp. (^{o}F@264 psi)	226	205	235
Specific Gravity	1.05	1.15	1.24
% water absorption	2.1	0.25	0.1
Dielectric strength		450	

REFERENCES

1. E. Simon, Ann. 31 265 (1839).
2. J. Blythe, A. W. Hofmann, Ann. 53 289 (1845).
3. E. Kopp, Compt. Rondu. 21 1378 (1843).
4. E. Erlenmeyer, Ann. 137 353 (1866).
5. S. Sabatay, Bull. Soc. Chim. 45 69 (1929).
6. I. Ostromislensky, M. G. Sheppard, U. S. Patent 1,552,543.
7. H. Staudinger, "Die Hochmol Org Verbind" Springer, Berlin, 1932.
8. R. B. Seymour, U. S. Patent 2,574,438 (1951).
9. T. Wagner-Juaregg, Ber. 63 3213 (1930).
10. I. G. Farbenindustrie, Brit. Pat. 376,479 (1932).
11. R. B. Seymour, U. S. Patent 2,533,635 (1950).
12. R. G. Fordyce, E. C. Chapin, G. E. Ham, J. Am. Chem. Soc. 10 2489 (1948).
13. R. B. Seymour, J. P. Kispersky, U. S. Patent 2,439,227 (1948).
14. R. B. Seymour, J. M. Schroeder, U. S. Patent 2,456,803,4,5,6 (1949).
15. R. B. Seymour, U. S. Patent 2,577,041 (1951).
16. R. B. Seymour, D. P. Garner, Polymer 17 (1) 22 (1976).
17. R. B. Seymour, P. D. Kincaid, D. R. Owen, ACS Adv. Chem. Ser. Washington, 1973.
18. I. E. Muskat, U. S. Patent 3,085,934 (1963), 3,175,335 (1963), 3,338,106, 3,418,292 (1965), 3,415,979 (1969).
19. R. F. Miller, in "The Effect of Hostile Environments on Coatings and Plastics" Chapter 13, D. P. Garner, G. A. Stahl Eds., ACS Symposium Series 229, 1983.

20. W. J. Hall, R. L. Kruse, R. A. Mendelsohn, Q. A. Trementozzi in "The Effects of Hostile Environments on Coatings and Plastics" Chapter 5, D. P. Garner, G. A. Stahl Eds., ACS Symposium Series 229, 1985.

21. O. Olabisi, I. M. Robeson, M. T. Shaw, "Polymer-Polymer Misability" Chapter 5, Academic Press, New York, 1979.

22. Y. C. Lee, Q. A. Trementozzi, U. S. Patent 4,197,376 (1980).

23. Y. C. Lee, Q. A. Trementozzi, U. S. Patent 4,205,869 (1981).

24. W. J. McKnight, F. E. Karasz, J. R. Fried, "Polymer Blends" Chapter 5, D. R. Paul, S. Newman, Eds., Academic Press, New York, 1978.

25. Y. C. Lee, G. A. Trementozzi, U. S. Patent 4,241,695 (1982).

26. Japan Synthetic Rubber KK JA.P. 82-031,913 (1982).

27. Mitsubishi Rayon KK JA.P. 81-081,322 (1981).

28. Mitsubishi Monsanto KK JA.P. 81-041,215 (1981).

29. Mitsubishi Monsanto KK JA.P. 81-041,215 (1981).

30. Monsanto U. S. Patent 4,298,716 (1981).

31. Arco U. S. Patent 4,325,037 (1982).

32. Arco U. S. Patent 4,336,354 (1982).

33. G. E., U. S. Patent 4,108,925 (1978).

34. Modern Plastics 62 (10A) 473 (1985).

HISTORY OF POLYPHENYLENE SULFIDE

H. Wayne Hill, Jr.*
PHILLIPS PETROLEUM COMPANY
BARTLESVILLE, OKLAHOMA

INTRODUCTION

During the last 40 years the plastics industry has experienced tremendous growth, a growth catalyzed and allowed by unprecedented technological advances. These advances have culminated in a new class of synthetic materials often referred to as "engineering thermoplastics". This class of polymers is characterized by unusual toughness and good load-bearing capabilities. These materials have opened many new horizons for designers and have found applications in markets previously dominated by metal, wood, ceramic, glass, etc. One important member of this family of resins is polyphenylene sulfide.

EARLY POLYMERIZATION STUDIES

The early chemical literature contains several references to materials either alleged or assumed to be polyphenylene sulfide. However, material characterization techniques in those early days were not as definitive as they are today. Consequently, structural assignments were not always completely accurate. Several of these references have been reviewed recently.[1] A brief summary of these early investigations follows.

The reaction of benzene and sulfur in the presence of aluminum chloride was studied by Friedel and Crafts[2] in 1888 and by Genvresse[3] in 1897. Materials produced by these workers were characterized very poorly by modern standards. However, the products were oligomeric with molecular weights of about 1000 and too much sulfur to be polyphenylene sulfide. In 1984, Cleary[4] studied the reaction further and reported higher molecular weight material (about 3500) that was mainly poly(thianthrylene sulfide) with extra sulfur linkages. Following the early studies by Friedel and Crafts and by Genvresse, several workers observed by-products when investigating the preparation of diphenyl sulfide and related organo sulfur compounds.[5] These by-products were generally not well characterized, but may have involved oligomeric materials related to polyphenylene sulfide.

In 1948, Macallum[6], a Canadian working in his private laboratory, prepared a phenylene sulfide-type polymer. In his early work Macallum recognized the unusual thermal stability and potential utility of this new polymer. In a sense his work ushered in the investigation of other high temperature polymers in the years to follow. His process for producing phenylene sulfide-type polymers involved the reaction of sulfur, sodium carbonate and dichlorobenzene in a sealed vessel at 275-300°C in the melt.

---

*Present address: Hill Associates, 1236 SE Grevstone, Ave., Bartlesville, OK 74006

Copyright 1986 by Elsevier Science Publishing Co., Inc.
High Performance Polymers: Their Origin and Development
R.B. Seymour and G.S. Kirshenbaum, Editors

$$Cl-C_6H_4-Cl + S + Na_2CO_3 \longrightarrow -(C_6H_4-S_x)_n-$$

He reported that polymers prepared in this manner generally contained more than one sulfur atom per repeat unit (x in the range 1.2-2.3). In addition the polymerization reaction was highly exothermic and difficult to control, even on a small scale.[7] Certainly Macallum's work sparked an interest in polyphenylene sulfide (PPS) and triggered a series of investigations that eventually led to the commercial production of PPS. In 1954 Macallum sold his patents[8] to Dow Chemical Co. where this polymerization scheme was studied further. However, the problems associated with the severe polymerization conditions and control of the exothermic reaction remained largely unsolved.[9] Lenz and coworkers at Dow have studied the mechanism of the Macallum polymerization[10] and the structure of the polymer produced.[11] The structure postulated consists of a crosslinked core to which are attached more or less extended, linear chains.

In 1962 Lenz and coworkers[12] at Dow reported another synthesis of polyphenylene sulfide. This process was based on a nucleophilic substitution reaction involving the self-condensation of materials such as copper p-bromothiophenoxide. This reaction was carried out at 200-250°C in the solid state or in the presence of materials such as pyridine as reaction media.

$$Br-C_6H_4-SCu \longrightarrow -(C_6H_4-S)_n- + CuBr$$

This laboratory process provided a linear polyphenylene sulfide material. Considerable difficulty was encountered in removing the by-product, copper bromide, from polymers made by this process.[13] This difficulty became more formidable on attempted scale-up and PPS was never produced commercially by Dow.

COMMERCIAL POLYMERIZATION PROCESS

The commercial process for producing polyphenylene sulfide was discovered by Edmonds and Hill[14] working in the laboratories of Phillips Petroleum Company. This process involves the production of PPS from p-dichlorobenzene and sodium sulfide in a polar solvent under heat. Since the starting materials for this process are relatively inexpensive, large volume chemicals (p-dichlorobenzene used in moth balls and sodium sulfide as a depilatory agent in the leather tanning industry), commercialization of a high performance material at reasonable cost became feasible.

$$Cl-C_6H_4-Cl + Na_2S \xrightarrow[\Delta]{\text{Polar solvent}} -(C_6H_4-S)_n- + 2NaCl$$

The polymer produced by this process is a linear material containing 150-200 repeat units, giving a molecular weight in the 15,000-20,000 range. The process for producing this material involves the following steps: (1) preparation of sodium sulfide from aqueous sodium hydrosulfide and caustic in a polar solvent, (2) dehydration of the sodium sulfide stream, (3) production of polyphenylene sulfide from sodium sulfide and dichlorobenzene in the polar solvent, (4) polymer recovery, (5) polymer washing to remove by-product, sodium chloride, (6) polymer drying, and (7) packaging.

Polyphenylene sulfide is obtained from the polymerization mixture in the form of a fine white powder, sometimes called "virgin" polymer. Its precise characterization is complicated by its insolubility in most solvents. At elevated temperature, however, it is soluble to a limited extent in some aromatic and chlorinated aromatic solvents and heterocyclic compounds. The inherent viscosity, measured at 206°C in 1-chloronaphthalene is typically 0.16.

While this linear polymer possesses a moderate degree of mechanical strength as it is produced in the polymerization process, it can be converted into a much tougher material by thermal treatment. This conversion was first accomplished by heating the molten polymer in the presence of air.[15] Thus, when the molten polymer is subjected to additional heat in the presence of air, the melt darkens and after a while gels and solidifies. This solid polymer is believed to be crosslinked because it is insoluble in all organic solvents tested, even at elevated temperatures. This process is frequently referred to as a "curing" process and the resulting polymer as "cured" polymer.

While the curing can be accomplished in the melt as described above, a more practical means of curing large quantities of resin in bulk involves curing the finely divided powder at temperatures below the melting point of 285°C in the presence of small amounts of air.[16] Care must be taken to avoid hot spots in the curing vessel so that agglomeration due to melting or softening does not occur. Agglomeration leads to a large particle size and prevents contact of the inside of the polymer particle with air. Since the air plays a vital role in curing, large particle sizes result in a product which is not cured uniformly. If the curing is done in trays in a circulating air oven, the polymer depth should not be over about 1" in order to insure adequate contact with the air. Curing may also be accomplished in an agitated bed with the air stream passing through the bed.

Several changes occur when virgin polyphenylene sulfide is converted to a cured resin by the procedures outlined above. These are: (1) an increase in molecular weight, (2) loss of solubility, (3) substantial increase in toughness, (4) an increase in melt viscosity (decrease in melt flow), (5) a decrease in ultimate crystallinity and (6) a change in color from off-white to brown-black. The magnitude of these changes is dependent upon the degree of cure which in turn is controlled by the time and temperatures employed in the curing process. The behavior of polyphenylene sulfide which allows it to be cured is an extremely important property of the polymer. This is because the cured polymer is much stronger, tougher, and easier to process by injection or compression molding than the virgin polymer.

Polyphenylene sulfide is a unique resin in many ways. Whereas other polymers can correctly be classified as being either thermoplastic or thermosetting in their behavior, polyphenylene sulfide combines the best properties of both these classes of materials. Under normal handling and processing conditions, PPS behaves as a true thermoplastic in that it can be repeatedly melted and solidified with only minor changes in its flow characteristics. Yet, by proper choice of conditions, it can be made to undergo the very complicated and incompletely understood "curing" process. During this reaction there is apparently both polymer chain extension and some degree of crosslinking taking place. Although attempts have been made to elucidate the details of this process,[17] the study is complicated by the insolubility of the polymer in all but a few specialty solvents at elevated temperature. At least three types of reactions are involved in the curing process: (1) disproportionation to produce a higher molecular weight moiety and diphenyl sulfide, (2) oxidative crosslinking, and (3) thermal crosslinking.

2ArS—Ph —Δ→ ArSAr + Ph—S—Ph

Disproportionation

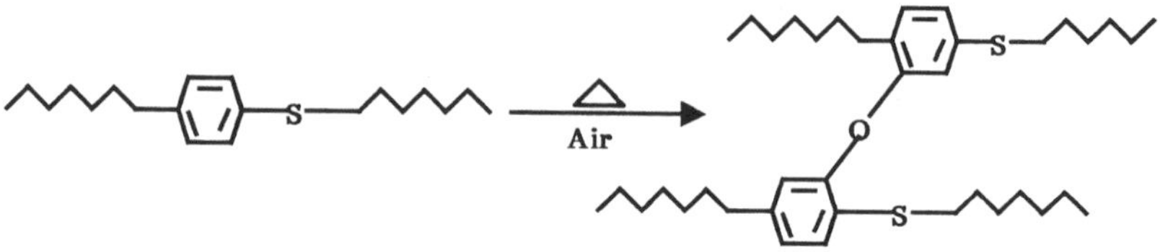

Oxidative Crosslinking

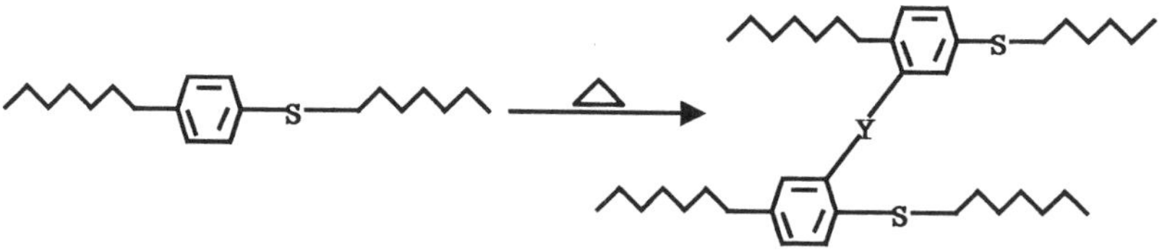

Y = Ar or S

Thermal Crosslinking

Between 1967 and 1973, workers at Phillips refined the new polymerization process, defined physical properties, constructed and operated a pilot plant, established market demand for such a material and designed and constructed a full-scale commercial plant. In late 1972 the world's first polyphenylene sulfide plant came on stream in the Phillips Petroleum Company facility in Borger, Texas. The original plant had a nameplate capacity of 5 MM pounds per year. Subsequently, it has been enlarged and debottlenecked to its present capacity of 12 MM pounds per year. This material is sold under the tradename Ryton® Polyphenylene Sulfide.

PROPERTIES

Polyphenylene sulfide has several fundamental polymer properties which are very important in developing applications for the material and thus assuring commercial acceptance and success. It is a crystallizable polymer with a crystalline melting point of 285°C and a glass transition temperature (T_g) of 85°C. Thus, by proper choice of molding and/or annealing conditions, a crystalline part can be obtained which is dimensionally stable and possesses good mechanical properties.[18] Brady[19] has described a method of estimating crystallinity and related crystallinity to mechanical behavior of the resin. The ability to mold PPS in its crystalline form is very important with respect to usage at elevated temperatures. In filled compositions its Underwriters Laboratories Temperature Indices are in the 200-240°C range, significantly above the T_g of the base resin. However, since T_g is a property of amorphous materials, development of the maximum crystallinity allows the use of filled resins at these higher temperatures. Polyphenylene sulfide is soluble only in aromatic hydrocarbons and chlorinated aromatic hydrocarbons at temperatures above about 200°C.[20] Brady and Hill[21] have demonstrated the excellent chemical resistance of molded parts. Long term thermal stability and chemical resistance have been documented by Hill.[22] Polyphenylene sulfide is inherently non-flammable and has an excellent affinity for a variety of fillers, giving rise to an excellent combination of mechanical properties.[20] The fundamental properties (at least in general terms) of PPS mentioned above were recognized early in the development of this resin. More detailed measurements have continued throughout the years with periodic reports appearing in the literature as noted above.

APPLICATIONS

The initial market for PPS was in the area of coatings. The phenomena of curing were very important in this market. Coatings can be applied by several techniques:[23] slurry coating, fluidized bed coating or electrostatic methods.[24] The slurry coating technique involves spraying an aqueous slurry of the virgin polymer along with various pigments and other additives. The virgin polymer is quite fluid and wets the substrate metal well. This wet coating is then baked in a circulating air oven at about 375°C to dry, coalesce, and cure the polymer to a tough coating. This final coating exhibits excellent chemical resistance and, thus, is used in corrosive environments where, for example, ordinary steel coated with PPS may outperform stainless steel or expensive alloys. Proper formulation with relatively small amounts of polytetrafluoroethylene (PTFE) provides a non-stick cookware coating in a one-coat operation; whereas, many competitive non-stick coatings require both a primer and a top coat. The manufacture and use of these non-stick coatings in food contact service is governed in the U.S. by FDA regulation 177.2490(d) Title 21, Code of Federal Regulations.

The early molding studies on PPS involved the compression molding of highly cured, highly filled formulations. These materials were used in areas such as seals, pistons, and piston rings in severe service where a combination of thermal stability and chemical resistance are required.

As an understanding of curing and careful control of the extent of curing was developed, injection molding began to develop as a market for PPS. Today, this represents the major market for Ryton PPS. The extent of curing can be controlled to give the proper flow behavior for the particular combination of fillers and reinforcements employed. This has led to a large variety of commercial grades of PPS. Table I lists some of the mechanical properties of three common injection molding grades of PPS:[22] (1) 40% glass-filled polymer (Grade R-4), (2) a glass- and mineral-filled polymer (Grade R-8), and (3) a glass- and mineral-filled color compound family (Grade R-10).

The 40% glass-filled material has the lowest density and the highest values for tensile strength, flexural strength, compressive strength, impact strength, and hardness. The glass- and mineral-filled composition offers a good balance of properties at a substantial cost savings, and the glass- and mineral-filled color compounds offer competitive properties and attractive colors at an intermediate cost.

The price/performance ratio offered by polyphenylene sulfide and its compounds has allowed it to find wide acceptance in a variety of areas, including electrical/electronic, mechanical, automotive, appliance, and coatings applications. In terms of performance, polyphenylene sulfide offers good mechanical and electrical properties, high temperature resistance, chemical and moisture resistance, inherent flame resistance, excellent stiffness and creep resistance, easy processing, precision moldability and dimensional stability, and excellent property retention at elevated temperatures. As with most engineering plastics, it is usually no single property which causes PPS to be selected for a given application. Instead,

TABLE I

MECHANICAL PROPERTIES OF POLYPHENYLENE SULFIDE COMPOUNDS

Glass/Mineral

Property	40% Glass Filled	Glass/Mineral Filled	Filled Color (Natural)
Density, g/cc	1.6	1.8	1.97
Tensile Strength, psi	19,600	13,300	10,000
Elongation, %	1.3	0.7	0.5
Flexural Modulus, psi	1,700,000	1,900,000	1,800,000
Flexural Strength, psi	29,000	20,400	17,500
Compressive Strength, psi	21,000	16,000	16,500
Izod Impact Strength, Ft. Lbs./In.			
Unnotched	8.1	2.2	1.7
Notched	1.4	0.5	0.7
Shore D Hardness	92	88	90
Heat Deflection Temp.			
(264 psi), °F	>500	>500	>500

it is generally a combination of two or more properties which are not available in another resin at lower cost.

The largest market areas for PPS injection molding compounds are the electrical/electronic area and the mechanical area. Typical parts involved in the electrical/electronic area include: coil forms, sockets, bolt yokes, motor brush holders, connectors, switches, integrated circuit and capacitor encapsulations, electronic watch bases, and relay components. In the mechanical area the following applications are important: pump housings, impeller diffusers, pump vanes, end plates, valve components, heat shield, oil well valves, oil well sucker rod guides, and pH meter components.

A specific application which utilizes several of the advantages offered by PPS compounds is that of an exhaust gas emission control valve on Toyota automobiles.[25] Requirements for this application included high temperature resistance, chemical resistance, precision fabrication, dimensional stability, and easy, inexpensive manufacture. The item had previously been made of 14 separate machined, stamped and extruded metal parts, many of them stainless steel. The successful replacement is injection molded in three PPS parts which are subsequently ultrasonically welded together. This change allowed considerable savings in materials and assembly costs and with a weight savings of over 0.2 kg per part. The latter is especially important to today's fuel-conscious automobile industry.

In other automotive applications, PPS compounds have been used successfully in such things as crankshaft position sensor housings, carburetion components, lighting reflectors, and thermostat housings. In most of these applications, it is a combination of high temperature and chemical resistance coupled with adequate mechanical properties, precision moldability, and dimensional stability which lead to PPS compounds being the materials of choice.

Another applications problem which was solved with a PPS resin was that of a new, more efficient street light design.[25] Here, the requirements for the reflector were very high temperature resistance, precision manufacture of the sharp-cornered, multifaceted design, high-gloss surface finish, adequate impact resistance, long production tool life, and low overall cost. Other solutions were tried, including hydroformed aluminum, die cast metal and injection molded polysulfone, polyethersulfone, polyphenylsulfone and polyamide-imide. However, a PPS compound was the only material meeting all the design requirements.

Another PPS success story in the electrical/electronics industry concerns use of PPS in televisions.[26] Here requirements included high temperature resistance, precision moldability, good arc tracking resistance, and good dimensional stability. Again, numerous materials were tested, but a fiberglass-reinforced PPS compound proved to be the material of choice for a series of pin cushion corrector coils used in color television sets.

HIGH MOLECULAR WEIGHT POLYMERS

The commercial polymers described thus far have been made with the now traditional Phillips process combined with the curing process. Campbell,[27] also working in the laboratories of Phillips Petroleum Company, developed a new process for producing a high molecular weight polymer directly during polymerization. Thus, the curing step is not involved in this new process. This high molecular weight product is suitable for use as an injection molding resin as produced in the polymerization process. This material is also useful as an extrusion resin for the production of fibers and films. Hill[28] has summarized the properties of these new resins. The polymerization method involves the use of an alkali metal carboxylate as a polymerization modifier and eliminates the need of the curing portion of the conventional process. It is possible to prepare an even higher molecular weight soluble polymer by the incorporation of a very small amount of a trichloroaromatic compound into the polymerization recipe. This modified polymerization process provides resin of higher molecular weight polymer than is produced in the Edmonds and Hill process as evidenced by a higher inherent viscosity (0.29 vs. 0.16) and by a lower melt flow value (ca. 200 g/10 min. vs. ca. 6000 g/10 min).

Comparisons of the properties of injection molded specimens of the high molecular weight polymer and the cured polymer are given in Table II (unfilled resin) and Table III (40% glass-filled resin). In general, the high molecular weight resin provides higher tensile strength, elongation, flexural strength, and impact strength. In addition, when thick sections are molded from the high molecular weight materials, no cracking of parts occurs.

TABLE II

PROPERTIES OF INJECTION MOLDED UNFILLED RESINS[a]

Property	Cured Resin	High Mol. Wt. Resin (Uncured)
Tensile Strength, psi	7,000	11,400
Elongation, %	1.1	21
Flexural Modulus, psi	558,000	494,000
Flexural Strength, psi	15,000	21,300
Izod Impact, Ft. Lbs./In.		
Notched	0.2	0.3
Unnotched	1.5	10.9
Heat Deflection Temp. (264 psi), °F	232	221

a - Properties measured on annealed specimens.

TABLE III
PROPERTIES OF INJECTION MOLDED, 40% GLASS-FILLED RESINS[a]

Property	Cured Resin	High Mol. Wt. Resin (Uncured)
Tensile Strength, psi	13,400	16,800
Elongation, %	0.5	0.8
Flexural Modulus, psi	1,670,000	1,723,000
Flexural Strength, psi	22,200	26,100
Izod Impact, Ft. Lbs./In.		
Notched	0.9	1.1
Unnotched	2.6	5.4
Heat Deflection Temp. (264 psi), °F	462	465
Cracking[b]	Yes	No

a - Properties measured on annealed specimens.

b - The appearance of cracks in 3/4" thick injection molded specimens.

TABLE IV
POLYPHENYLENE SULFIDE FIBERS

Processing Conditions	
Spinning Temperature, °C	300
Draw Temperature, °C	100
Draw Ratio	3.8
Properties	
Denier	230
Tenacity, gpd	2.5-3.0
Elongation, %	25-35
Initial Modulus, gpd	30-40
Knot Strength, gpd	2.0-2.5
Elastic Recovery, %	
2% Extension	100
5% Extension	96-100
10% Extension	86
Boiling Water Shrinkage, %	16
Continuous Use Temperature, °C	175

Fiber spinning conditions and properties of polyphenylene sulfide fiber are given in Table IV. The tenacity, elongation, and knot strength values indicate the toughness of the fiber. More than 50% of the tenacity is retained at 200°C and oven aging suggests a continuous use temperature of 175°C. Tested under controlled conditions, the fiber possesses the nonburning characteristics and chemical resistance typical of polyphenylene sulfide resins. Potential applications may include flue gas filtration, wet filtration, and flame-retardant clothing and upholstery.

Blown films can be produced from these high molecular weight resins on conventional equipment. These films are transparent and amber as produced. If films are biaxially oriented, this transparency is retained on annealing. At a blow-up ratio of 3.0, the tensile strength and elongations are essentially the same in both the machine and transverse directions (Table V). Electrical properties are typical for polyphenylene sulfide resins.

Sheet (20 mils) can be extruded readily from the new high molecular weight polymer. Quenched sheet is transparent and amber in color, while annealed sheet is opaque (Table VI). Tensile strength increases somewhat on annealing and elongation decreases. There is essentially no difference between machine direction and transverse direction in mechanical properties.

ADVANCED COMPOSITES

A relatively recent development in the history of PPS is its use as the base resin along with long reinforcing fibers as an advanced thermoplastic composite. The use of the new high molecular weight polymers with their good flow behavior and the well-known affinity of PPS for a variety of reinforcing fibers are both very important in this latest development. The mechanical properties of PPS thermoplastic composites based on long glass fibers are characterized by extremely high notched Izod impact strengths (5-14 ft. lbs./in.), combined with flexural moduli of 0.9-2.1 x 10^6 psi. These and other mechanical properties are summarized in Table VII.

TABLE V

POLYPHENYLENE SULFIDE FILM PROPERTIES[a]

Property	Value
Thickness, mils	1.2
Tensile Strength, psi	
Machine Direction	7500
Transverse Direction	7000
Elongation, %	
Machine Direction	8
Transverse Direction	6
Elmendorf, g	
Machine Direction	13
Transverse Direction	16
Dielectric Constant (1 KHz)	4.2
Dissipation Factor (1 KHz)	0.0073

a - Blown film; blow-up ratio, 3.0.

TABLE VI

POLYPHENYLENE SULFIDE EXTRUDED SHEET[a]

Property	Quenched[b]	Annealed[c]
Thickness, mils	20	20
Haze, %	46	100
Gloss, %	89	72
Tensile, psi		
Machine	8600	10,800
Transverse	8400	10,100
Elongation, %		
Machine	17	5
Transverse	12	5

a - Extrusion temperature, 316°C.

b - Polishing roll temperature, 79°C.

c - Annealing Temperature, 204°C.

TABLE VII

PPS/GLASS FIBER COMPOSITE PROPERTIES

Property	Glass Fiber Form			
	Chopped Mat		Swirl Mat	Woven Cloth
Glass Content, Wt., %	20	30	40	40
Density, g/cc	1.53	1.59	1.66	1.66
Tensile Strength, 10^3 psi	13	18	23	26
Elongation, %	1.3	1.5	1.9	-
Flexural Modulus, 10^6 psi	0.9	1.0	1.45	2.16
Flexural Strength, 10^3 psi	21	27	34.5	21
Compressive Strength, 10^3 psi	30	35	38	-
Izod Impact, Ft. Lbs./In.				
Notched, 1/8" specimen	5	7	14	8
Unnotched, 1/8" specimen	9	18	25	17
Heat Deflection Temp., 264 psi				
°C	266	273	273	273
°F	511	523	523	523

An even higher level of mechanical properties can be obtained through the use of long carbon fiber as reinforcing agent in the PPS composite. Notched Izod impacts are high as 30 ft. lbs./in., tensile strengths of 75-170 x 10^3 psi, and flexural moduli as high as 18 x 10^6 psi are possible. Table VIII lists typical mechanical properties of the long carbon fiber reinforced PPS composites.

TABLE VIII

PPS/CARBON FIBER COMPOSITE PROPERTIES

Property	Carbon Fiber Form		
	Unidirectional	Cross Plied (0,90,0)s	Woven Cloth
Fiber Content			
Wt., %	66	62	60
Vol., %	60	54	52
Density, g/cc	1.6	1.6	1.6
Void Content, Vol. %	<1.0	<1.0	<1.0
Tensile Strength, 10^3 psi	240	79	75
Flexural Modulus, 10^6 psi	18	9	6.8
Flexural Strength, 10^3 psi	190	100	93
Short Beam Shear, 10^3 psi	10	4	7
Heat Deflection Temp., 264 psi			
°C	280	280	280
°F	536	536	536
Izod Impact Strength, Ft. Lbs./In.			
Notched, 1/8" Specimen	30	-	12.5
Unnotched, 1/8" Specimen	59	-	21.5
Fracture Toughness, G_{1C} (double cantilever beam test method)	3-5	-	-

These PPS composites from fiber mats can easily be fabricated by fast compression molding similar to metal stamping operations. Three- to five-inch flows in the mold are feasible.

Finished parts fabricated from these long fiber composites possess unusually high impact strength and are particularly suited for end use applications involving exposure to hostile chemical and/or high temperature environments. With proper part design, these materials offer improved performance at a competitive finished part price compared to stainless steel and other metal alloys in applications where corrosion resistance is a major concern. Typical applications under development include chemical valve and pump components, housings, plate heat exchangers, oil field components, pipe couplings and fittings, compressor components, etc. Lightweight, inherent flame resistance, and low smoke generation suggest future applications in public transportation arenas including aircraft seating.

Additional information on these PPS advanced composites can be found in publications by Brady, Murtha, Walker, South, and Ma[29] and by Brady and Murtha.[30]

CONDUCTING POLYMERS

While many applications for PPS involve its inherent electrical insulating behavior, it is a member of the family of organic polymers which can be rendered electrically conductive by "doping" with appropriate compounds such as arsenic pentafluoride. PPS is the only commercially available polymer which can be made conductive in this manner. Furthermore, it is the only melt processable resin in the group of such polymers. This behavior was initially described jointly by Chance, et al, at Allied and by Street, et al, at IBM at the 1980 Spring Meeting of the American Chemical Society[31,32,33]. More recently workers at GTE Laboratories have reported selectivity in doping either the amorphous or crystalline regions of a PPS sample which has been preferentially crystallized in a desired pattern.[34,35] These workers have extended this technology to include a metal plating technique by which conductive patterns of a chosen design may be applied to the surface of a PPS sample, returning the PPS itself to its natural, non-conductive state.[36] The implications of this additive plating process to the fabrication of printed circuit boards are obvious.

FUTURE

While this paper is concerned with the history of polyphenylene sulfide, a brief word about the future of PPS is pertinent since that future will be the next chapter in PPS history. The Phillips polymerization process is a versatile one which permits the use of a wide variety of arylene chlorides as monomers. Thus, a wide variety of polyarylene sulfide materials can be prepared and proper choice of monomer permits the tailoring of the resin to obtain specific properties. For example, Campbell and Scoggins[37] have reported the preparation of a series of copolymers based on mixtures of m-dichlorobenzene and p-dichlorobenzene and sodium sulfide. As the meta content increases, the melting point and glass transition temperature (Tg) of the polymers decrease and at 50% and higher meta content the polymers are no longer crystallizable. Other monomer choices can yield polymers with higher melting points and glass transition temperatures than PPS. In due course of time it is likely that other polyarylene sulfide resins will be produced commercially to meet specific end use needs. In addition new grades of PPS involving various combinations of fillers, reinforcing agents, and other additives will undoubtedly be developed for specific applications. Furthermore, as more and more companies and industries become familiar with PPS, new applications for this versatile resin will be developed and the market base will continue to expand. Thus, the future for Ryton PPS is indeed bright and will remain so for many years to come.

SUCCESS FACTORS

The successful commercialization of a new engineering plastic, or for that matter any new product, requires that several favorable factors all must be involved at the same time. These factors are outlined below:

1. The process for manufacturing the proposed product must be both practical and economical. This means that the raw materials must be available at reasonable costs or at least potentially available at reasonable cost.
2. The proposed product must possess a desirable combination of properties/price as evaluated by potential customers.
3. Patent protection for the process or product must be obtainable.
4. Management backing at the research and development level to permit the original exploratory bench-scale research studies, followed by property measurement, application testing and development, pilot plant process evaluation and sample preparation.

5. Management backing at the corporation level is required to provide the capital to build a plant, expense budget to fund the overall development of the new product and patience and dedication to stick with the new project through the initial difficult years to attain the profitable years.

6. Manpower who are dedicated to the success of the proposed product. These people must be available from the initial research studies through the necessary development and pilot plant work, market studies, market development and sales, production and all the support personnel needed throughout the project.

Fortunately in the development of Ryton Polyphenylene Sulfide at Phillips Petroleum Company, all of the above factors were present throughout the project, thus contributing to its success. Once a new product becomes profitable, the same sort of backing is required to continue the growth of the market. However, this backing is generally easier to achieve since everyone likes a winner.

BIBLIOGRAPHY

1. Cleary, J. W., "Poly(phenylene sulfide)" in "Advances in Polymer Synthesis", B. M. Culbertson, Editor, Plenum Press, New York, in press.
2. Friedel, C. and Crafts, J. M., Ann. Chim. Phys., 14 (6), 433 (1888).
3. Genvresse, P., Bull. Soc. Chem. Fr., 17, 599 (1897).
4. Cleary, J. W., ACS Polymer Div. Preprints, 25 (2), 36 (1984).
5. For reviews of early synthetic work see:
   a. Lenz, R. W., Handlovits, C. E., and Carrington, W. K., J. Polym. Sci., 41, 333 (1959).
   b. Gaylord, N. G., Polyethers, in High Polymer Series, Vol. XIII, Part III, Interscience, New York, 1962, p. 31.
   c. Smith, H. A., Encycl. Polym Sci. Technol., 10, 653 (1969).
   d. Hortling, B., Suomen Kemistiseuran Tiedonautaja, Helsinki, 81, 1 (1972).
6. Macallum, A. D., J. Org. Chem., 13, 154 (1948).
7. Macallum, A. D., private communication.
8. Macallum, A. D., U.S. Patents 2,513,188 (June 27, 1950) and 2,538,941 (January 23, 1951).
9. Smith, H. A., and Handlovits, C. E., ASD-TDR-62-372, Report on Conference on High Temperature Polymer and Fluid Research, Dayton, Ohio, 1962, p. 100.
10. Lenz, R. W., and Carrington, W. K., J. Polym. Sci., 41, 333 (1959)
11. Lenz, R. W., and Handlovits, C. E., J. Polym. Sci., 43, 167 (1960).
12. Lenz, R. W., Handlovits, C. E., and Smith, H. A., J. Polym. Sci., 58, 351 (1962).
13. Smith, H. A., and Handlovits, C. E., ASD-TDR-62-322, Part II, Phenylene Sulfide Polymers, 1962, pp. 18-19.
14. Edmonds, J. T., Jr., and Hill, H. W., Jr., U.S. Patent 3,354,129 to Phillips Petroleum Company (November 21, 1967).
15. Edmonds, J. T., Jr., and Hill, H. Wayne, Jr., U.S. Patent 3,524,835 to Phillips Petroleum Company (August 18, 1970).
16. Rohlfing, R. G., U.S. Patent 3,717,620 to Phillips Petroleum Company (February 20, 1973).

17. Hawkins, R. T., Macromolecules, 9, 189 (1976).

18. Hill, Harold Wayne, Jr., and Edmonds, James T., Jr., U.S. Patent 3,562,199 to Phillips Petroleum Company (February 9, 1971).

19. Brady, D. G., J. Appl. Poly. Sci., 20, 2541 (1976).

20. Short, J. N., and Hill, H. Wayne, Jr., Chemtech 2, 481 (1972).

21. Brady, D. G., and Hill, H. W., Modern Plastics, 51 (5), 60 (1974).

22. Hill, H. Wayne, Jr., ACS Symposium Series, 95, 183 (1979).

23. Hill, H. Wayne, Jr., and Brady, Don G., J. Coatings Technology, 49 (627), 33 (May 1977).

24. Blackwell, J. P., Brady, D. G., and Hill, H. W., Jr., J. Coatings Technology, 50 (643), 62 (August 1978).

25. Shue, R. S., "Polyphenylene Sulfide" in "Developments in Plastics Technology - 2", Elsevier Applied Science Publishers, London.

26. Anon., Plastics Design and Processing, 13 (9), 26 (Sept., 1973).

27. Campbell, R. W., U.S. Patent 3,919,177 to Phillips Petroleum Company (November 11, 1975).

28. Hill, H. Wayne, Jr., Ind. Eng. Chem. Prod. Res. Dev., 18 (4), 252 (1979).

29. Brady, D. G., Murtha, T. P., Walker, J. H., South, A., and Ma, C. C., SPE, 42nd Annual Technical Conference, New Orleans, Technical Papers, Vol. XXX, 690 (1984).

30. Brady, D. G., and Murtha, T. P., SPE, 43rd Annual Technical Conference, Washington, D.C., Technical Papers, Vol. XXXI, 1178 (1985).

31. Anon., Chem. Eng. News, March 31, 1980, p. 36.

32. Chance, R. R., Shacklette, L. W., Miller, G. G., Ivory, D. M., Sowa, J. M., Elsenbaumer, R. L., and Baughman, R. H., J. Chem. Soc. Chem. Comm., 1980, 348.

33. Rabolt, J. F., Clarke, T. C., Kanazawa, K. K., Reynolds, J. R., and Street, G. B., J. Chem. Soc. Chem. Comm. 1980, 347.

34. Rubner, M., et al, J. Elect. Materials, 11 (2) 261 (1982).

35. Druy, M., and Tripathy, S., Profile, GTE Laboratories, Inc., March/June, 1983.

36. Rubner, M. F., and Cukor, P., U.S. Patent 4,486,463 to GTE Laboratories, Inc. (December 4, 1984).

37. Campbell, R. W., and Scoggins, L. E., U.S. Patent 3,869,434 to Phillips Petroleum Company (March 4, 1975).

THE DEVELOPMENT OF POLYSULFONE AND OTHER POLYARYLETHERS

R. A.Clendinning, A. G. Farnham, and R. N. Johnson
Union Carbide Corporation, Bound Brook, NJ

INTRODUCTION

Nucleophilic displacement chemistry involving aromatic dihalides and bisphenols, represented the first means by which high molecular weight aromatic polyethers could be produced. This technology has since been the basis for all commercial processes developed for this family of Engineering Polymers. The preparative method involves the nucleophilic polycondensation of a bisphenol salt with an activated aromatic dihalide in an aprotic solvent. Further investigations have shown that the method has a very wide scope and that it can be applied to the preparation of a host of aromatic poly (sulfone ethers), poly (ketone ethers) and other related polyethers.

These materials have shown excellent and unique properties as high performance plastics. These include excellent toughness and good high temperature and hydrolytic resistance and excellent electrical properties.

EARLY RESEARCH ACTIVITIES

For many years, Union Carbide was a major producer of bisphenol A, the acid catalyzed condensation product of phenol and acetone. During the 1950's there were several research programs at Carbide to utilize bisphenol A as a monomer to make a variety of condensation polymers. Phenoxy®, a high molecular weight thermoplastic polyhydroxyether from bisphenol A and epichlorohydrin, was one of the commercial products to come out of this research.

$$HO-C_6H_4-C(CH_3)_2-C_6H_4-OH + CH_2\overset{O}{-}CH-CH_2Cl \xrightarrow{NaOH} \left[-O-C_6H_4-C(CH_3)_2-C_6H_4-O-CH_2-\underset{}{\overset{OH}{CH}}-CH_2-\right]_n$$

Phenoxy®

Reaction of Bisphenol A with Organic Dihalides

In the late 1950's a project was started to make other polyethers from bisphenols, especially bisphenol A, and alkylene dichlorides and aromatic dihalides. One goal was improved hydrolytic stability over the polyesters and polycarbonates known at that time. A second goal was high thermal stability with that of diphenyl ether being the ultimate target.

Copyright 1986 by Elsevier Science Publishing Co., Inc.
High Performance Polymers: Their Origin and Development
R.B. Seymour and G.S. Kirshenbaum, Editors

Throughout the 1950's and early 1960's there were literature reports [1,2] and private communications from academic laboratories indicating that certain organic reactions could be conveniently carried out in dimethyl sulfoxide, DMSO. Often, reaction rates were many times faster in this solvent than in the more commonly used organic solvents. In the case of nucleophilic aromatic substitution reactions, the effect was thought to be due to the complexing of cations by DMSO, thus leaving the corresponding anion unsolvated and hence, very reactive.

A process had been developed for making the disodium salt of bisphenol A in aqueous alcohol which gave a stable salt as the hexahydrate. Reaction of this salt in DMSO with 1,4-dichlorobutane was rapid and gave good yields of polymer but of low molecular weight. Dehydration of the salt prior to reaction also gave low molecular weight product.

$$\mathrm{NaO{-}C_6H_4{-}C(CH_3)_2{-}C_6H_4{-}ONa} \;+\; \mathrm{Cl(CH_2)_4Cl} \xrightarrow{\mathrm{DMSO}} \left[\mathrm{O{-}C_6H_4{-}C(CH_3)_2{-}C_6H_4{-}O{-}(CH_2)_4}\right]_n$$

low mol. wt.

Side reactions which eliminated hydrogen chloride from the dichlorobutane were suspected as the cause for the low molecular weight polymer; this was clearly established with dichloroethane where the major product was not polymer but vinyl chloride.

The desire to avoid the elimination of hydrogen chloride led to the investigation of the reaction with methylene chloride. With the bisphenol A salt, it was found that the polyformal readily formed in good yield and high molecular weight providing a slight excess of methylene chloride and caustic was used.

Numerous polyformals from a variety of bisphenols were prepared [3]. These are tough polymers with a somewhat lower softening temperature than polycarbonates, stable to base but less stable to acid.

Preparation of Polysulfone

Another potential monomer that would not eliminate hydrogen chloride and would contribute to the goal of high thermal stability was 4,4'-dichlorodiphenyl sulfone. This monomer had been investigated earlier at Union Carbide as a precursor to 4,4'-dihydroxydiphenyl sulfone and potentially attractive synthetic schemes had been investigated. The electron withdrawing character of the sulfone group strongly activates the chlorides toward displacement by nucleophiles such as phenoxide anion.

Initial experiments with the hydrated bisphenol A salt in DMSO gave high extents of reaction and yield but the molecular weight was too low to achieve desirable properties. Several variations were tried; the first successful reaction yielding useful molecular weight Polysulfone was a procedure by Dr. R. N. Johnson in December 1961. He used a regulated addition of potassium t-butoxide to a DMSO solution of bisphenol A and dichlorodiphenyl sulfone. While this technique demonstrated the necessity of an anhydrous system in achieving high molecular weight, it was recognized as an economically unattractive process. However, additional lab work soon led to the use of alkali metal hydroxides (first potassium, then sodium) to form the bisphenol salt in situ before the sulfone was added. Water, added with the caustic and formed by reaction between caustic and bisphenol, was removed by codistillation with an azeotrope forming solvent such as benzene, toluene, or chlorobenzene. When anhydrous, the sulfone was added as a solid or as a solution in chlorobenzene and polymerization to high molecular weight allowed to procede at 160°C.

$$HO-C_6H_4-C(CH_3)_2-C_6H_4-OH + 2\ NaOH \longrightarrow Na^+O^--C_6H_4-C(CH_3)_2-C_6H_4-O^-Na^+ + 2H_2O$$

$$NaO-C_6H_4-C(CH_3)_2-C_6H_4-ONa + Cl-C_6H_4-SO_2-C_6H_4-Cl \xrightarrow[160°C]{DMSO}$$

$$\left(-O-C_6H_4-C(CH_3)_2-C_6H_4-O-C_6H_4-SO_2-C_6H_4-\right)_n + 2NaCl$$

Polysulfone

Reaction Variables

Stoichiometry Control: As is true in most condensation polymerizations, stoichiometry control is crucial to attaining high molecular weight. In the case of Polysulfone, sodium hydroxide stoichiometry as well as bisphenol A and dichlorodiphenyl sulfone stoichiometry is crucial. This is illustrated in Figure 1 using purified reagents, where the reduced viscosity (RV) is plotted against the moles of dichlorodiphenyl sulfone per mole of bisphenol A salt. While very high molecular weight can be achieved in this system, the use of commercially available monomers with varying purities restricts the achievable molecular weight somewhat. To obtain useful mechanical properties it is necessary to achieve an RV of 0.4 or higher.

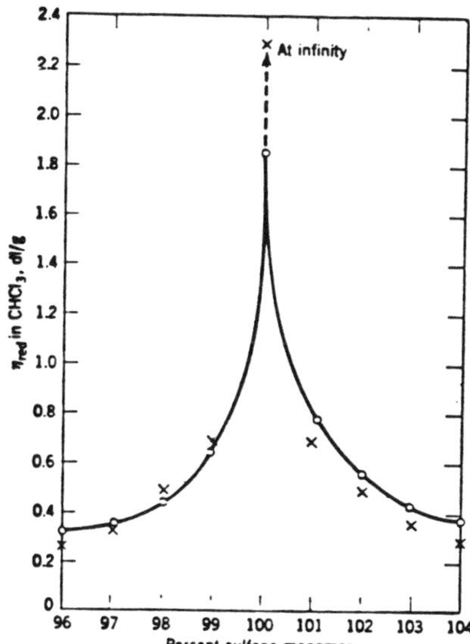

Fig. 1. Preparation of polysulfone with varying amounts of dichlorodiphenyl sulfone. Key: O, experimental; X, theoretical data.

Side Reactions: The major side reaction in the synthesis of polyaryl ethers is due to incomplete dehydration of the polymerization system. This can lead to solvation of the bisphenol salt in the dipolar aprotic-solvent or it can lead to hydrolysis of the bisphenol salt which produces sodium hydroxide. The latter can react with polymer end groups or with the activated dihalo compound such as dichlorodiphenyl sulfone.

$$Cl-C_6H_4-SO_2-C_6H_4-Cl + 2\ NaOH \xrightarrow{DMSO} Cl-C_6H_4-SO_2-C_6H_4-O^-\ Na + NaCl + H_2O$$

Under these polymerization conditions, the reaction does not proceed past the first chlorine. This reaction upsets the stoichiometry; the resulting chlorophenoxide is a chain terminator since this sodium salt is not nucleophilic enough to react in DMSO at 160°C.

In addition to having an essentially anhydrous system, it is also necessary to have an air-free system. Reactions of phenoxides with oxygen are well known and very rapid. Oxygen leads to severely discolored polymers at best and ultimately to difficulty in achieving high molecular weight. This is particularly true in systems involving hydroquinone where the aqueous caustic process leads to badly discolored, low molecular weight products.

Molecular Weight Control: Since in commercial production, it is often desirable to have products of varying molecular weight, means of stopping the reaction at the desired point were essential. In the Polysulfone system this can be accomplished by the addition of a calculated amount of monofunctional reagent such as phenol or, more practically, by the addition of a monofunctional reagent such as methyl chloride at the appropriate point. Methyl chloride converts the growing phenoxide end groups to the methyl ether end group. This not only stops the polymerization; it provides a more color and thermally stable end group than the phenoxide or the phenol.

$$\text{Polymer}-C_6H_4-O^-Na^+ + CH_3Cl \longrightarrow \text{Polymer}-C_6H_4-OCH_3 + NaCl$$

Scale Up: Mechanical and thermal properties of the early lab samples were determined soon after synthesis. The polymer was found to have a glass transition of 190°C, with good impact strength, excellent thermal stability for long periods in air at elevated temperatures, and outstanding hydrolytic stability. Indeed, the two original goals of the program had been achieved. Scale up of the laboratory process in the pilot plant provided sufficient material for fabrication studies. The new polymer could be fabricated on conventional injection molding and extrusion equipment. The high performance characterisitics of the early lab samples were confirmed on the fabricated parts made from pilot plant material.

A marketing study concluded that significant opportunities existed based on the differentiated performance of Polysulfone, namely high temperature resistance and hydrolytic stability. Thus UDEL® Polysulfone was introduced by Union Carbide in 1965 based on a commercial manufacturing plant in Marietta, Ohio.

Scope of Reaction

As the excellent properties of Polysulfone became apparent from the early lab samples, the research program was accelerated and expanded to investigate the polymerization of other bisphenols and dihalides to fully explore the discovery of this new class of high performance polymers [4,5].

Dihalide Reactivity: As expected, the order of halogen reactivity was found to be the same as in the earlier organic synthesis work, i.e., F >>Cl>Br>I. The particular group activating the halogen is also important with the order being $NO_2 \sim SO_2 > C=O > N=N$. Thus with a carbonyl activating group, it may be necessary to use fluorine as the halogen while with the sulfone group chlorine reactivity will be sufficient. Table I illustrates these relationships between activating group and halogen reactivity with the reaction time being an indicating or reactivity.

TABLE I: Reaction of Activated Dihalides with Bisphenol A Salt in DMSO

Dihalide	Alkali Metal	Reaction Time, hrs.	Reaction Temp. °C	RV
Cl-⟨◯⟩-SO_2-⟨◯⟩-Cl	Na	4	135	0.71
F-⟨◯⟩-SO_2-⟨◯⟩-F	K	0.5	25-145	0.94
Cl-⟨◯⟩-C(=O)-⟨◯⟩-Cl	K	18	135	0.16
F-⟨◯⟩-C(=O)-⟨◯⟩-F	K	0.5	135	1.00

In the early stages of polyether development, difluoromonomers were judged to be too expensive to be used in practical economical systems. As a result, higher boiling solvents were investigated in order to achieve reasonable reaction rates with dichloromonomers. Recently, however, difluoromonomers have been used in some new polyarylethers to achieve high molecular weight polymers with unique properties.

Bisphenol Reactivity: Reactivity differences were also noted in the bisphenols with the reactivity controlled as in the dihalo compounds by the nature of the group separating the aromatic rings. Those with electron withdrawing groups between the rings, such as sulfone or carbonyl were much less reactive than those with alkyl groups such as bisphenol A. The electron withdrawing groups render the bisphenol more acidic; hence the corresponding phenoxide is less nucleophilic and therefore less reactive. A correspondingly longer reaction time or higher temperature is thus required to achieve useful molecular weight. The reactivity differences are illustrated in Table II.

TABLE II: Reaction of Bisphenols with Dichlorodiphenyl Sulfone in DMSO.

Bisphenol	Alkali Metal	Reaction Time, hrs.	Reaction Temp. °C	RV
HO–C₆H₄–C(CH₃)₂–C₆H₄–OH	Na	1	160	1.4
HO–C₆H₄–C₆H₄–OH	K	4	165	0.7
HO–C₆H₄–C(=O)–C₆H₄–OH	K	10	165	0.5
HO–C₆H₄–SO_2–C₆H₄–OH	K	10	165	0.3

Choice of Base: Although the early work was done by preparing the dialkali metal salts of the bisphenol in situ, dehydrating them and then adding the dihalo compounds, it was evident that the insolubility and instability of some of the disalts precluded successful reactions, especially on a commercial scale. This has already been alluded to above in reference to the disalt of hydroquinone. Hence, alternatives were examined early in the polyether research program.

The use of alkali metal carbonates as bases in polyether syntheses was the result of a joint effort between the Bound Brook and South Charleston laboratories of Union Carbide [6]. In the carbonate process there is not a separate formation of bisphenol disalt and dehydration step followed by addition of the activated dihalo compound. Indeed, the dihalo compound is charged at the start of the reaction along with the bisphenol, carbonate, and solvents. Smooth formation, without precipitation of disalts, of high molecular weight polymer is achievable even when using bisphenols whose dialkali metal salts were very insoluble and/or unstable in the reaction media. This observation, along with the very slow evolution of water and the non-criticality of the carbonate stoichiometry led us to conclude that the dialkali metal salt was not formed per se in a polymerization employing carbonate as the base. Potassium carbonate is favored over sodium carbonate because it is more soluble and the resulting potassium phenoxide more reactive than the sodium compound. It is thought that the bisphenol mono salt is formed along with potassium bicarbonate which disproportionates under the conditions of the polymerization to potassium carbonate which continues the salt formation and polymerization.

$$HO-C_6H_4-R-C_6H_4-OH + K_2CO_3 \longrightarrow HO-C_6H_4-R-C_6H_4-OK + KHCO_3$$

$$2\ KHCO_3 \longrightarrow K_2CO_3 + H_2O + CO_2$$

The bisphenol monosalt reacts with the dihalo compound to start the chain growth and the resulting phenol ether is available to react with another molecule of potassium carbonate.

$$HO-C_6H_4-R-C_6H_4-OK + X-C_6H_4-SO_2-C_6H_4-X$$
$$\longrightarrow HO-C_6H_4-R-C_6H_4-O-C_6H_4-SO_2-C_6H_4-X$$

Water and carbon dioxide are observed during the course of the polymerization and an azeotrope forming solvent is usually included to co-distill the water as it is formed.

One important difference between the use of strong bases such as sodium hydroxide and the use of weaker bases such as potassium carbonate is that with the latter, there is present throughout the polymerization a finite concentration of phenolic hydroxyl groups which have not yet been converted to phenoxide ions. This is important not only for salt solubility but for the selection of the dipolar aprotic solvent.

Solvent Selection: The choice of dipolar aprotic solvent for aromatic nucleophilic substitution reactions is a very complicated one. Factors involved include: (1) the reactivity of monomers, (2) the solvating power of the solvent, which may enhance [1], [3] the temperature necessary to achieve useful molecular weight in a reasonable time; [4] the solubility of the high molecular weight polymer under the reaction conditions, and [5] the choice of base, either carbonate or hydroxide. Among the dipolar aprotic solvents most frequently used are dimethyl sulfoxide, dimethyl acetamide, N-methylpyrrolidinone (NMP), sulfolane (tetramethylene sulfone), dimethyl sulfone, and diphenyl sulfone.

Dimethyl sulfoxide is generally favored, particularly with processes using sodium hydroxide as base. In carbonate polymerizations, the presence of free phenolic hydroxyl groups at polymerization temperatures of 150-160°C leads to undesirable side reactions involving acid catalyzed decomposition of DMSO which ultimately can lead to stoichiometry upsets. For polymerizations involving carbonate, dimethyl acetamide is often the solvent of choice [7], DMSO is limited by its 180°C boiling point to temperatures of 170°C or lower while DMAc is useful in the 150-160°C range.

When it is necessary to polymerize at a higher temperature, one must resort to a higher boiling solvent such as sulfolane, diphenyl sulfone, or NMP. For example, the more acidic bisphenol, 4,4'-dihydroxydiphenyl sulfone (bis S), is not reactive enough to polymerize with dichlorodiphenyl sulfone to high molecular weight in DMSO or in DMAc. To achieve high molecular weight it is necessary to raise the polymerization temperature to 220-230°C in

$$HO-C_6H_4-SO_2-C_6H_4-OH + Cl-C_6H_4-SO_2-C_6H_4-Cl \xrightarrow[220\text{-}230°C]{\text{Sulfolane}} \left(O-C_6H_4-SO_2-C_6H_4-O-C_6H_4-SO_2-C_6H_4\right)_n$$

sulfolane or 270-280°C in diphenyl sulfone. In this case one can polymerize in the lower boiling solvent by using the much more reactive difluoro compound instead of the dichloro compound. On a commercial basis however, this is ruled out by the cost of the difluoro monomer.

$$HO-C_6H_4-OH + Cl-C_6H_4-SO_2-C_6H_4-Cl \xrightarrow[K_2CO_3]{\text{Sulfolane}} \left(O-C_6H_4-O-C_6H_4-SO_2-C_6H_4\right)_n$$

In this case the polymer is crystalline as made and precipitates at low molecular weight in the lower boiling solvents.

In the even more crystalline polyethers derived from the benzophenone systems rather than sulfone systems, even higher boiling solvents like diphenyl sulfone are required. Thus the polyether from hydroquinone and 4,4'-difluorobenzophenone requires diphenyl sulfone and a temperature of 300-320°C, which approaches the melting point of the polymer (335°C).

$$HO-C_6H_4-OH + F-C_6H_4-\overset{O}{\overset{\|}{C}}-C_6H_4-F \xrightarrow[300\text{-}320°C]{\text{diphenyl sulfone}} \left(O-C_6H_4-O-C_6H_4-\overset{O}{\overset{\|}{C}}-C_6H_4\right)_n$$

This polyetherketone was commercialized by ICI in 1980; it and other polyetherketones were described in the early UCC publication [4].

Polymer Properties

Detailed properties of the commercial polyarylethers introduced by Union Carbide are beyond the scope of this paper but can be found in the accompanying paper by M. E. Sauers, L. A. McKenna, and C. N. Merriam. Provided here is a basic mechanical property profile of the three polyethers commercially manufactured by Union Carbide.

TABLE III. Mechanical Properties of UCC Polyarylethers

Property	Polysulfone	Radel A	Radel R
Glass Transition Temp.	190°C	220°C	220°C
Tensile Modulus	2,482 MPa	2,700 MPa	2,140 MPa
Tensile Strength	70.3 MPa	75 MPa	85.5 MPa
% Elongation at Break	50-100%	40%	60%
Notched Izod Impact Strength	69 J/M	75 J/M	640 J/M
Heat Distortion Temp. 1/8" 264 psi	174°C	202°C	204°C
UL Temp. Index	160°C	(180°C)	(180°C)
Clarity	Transparent	Transparent	Transparent
FDA Status	Single & Repeated Use	Single & Repeated Use	-
Hydrolytic Stability	Excellent	Excellent	Excellent

Present Status

The field of aromatic polyethers made by nucleophilic substitution is now almost 25 years old and it continues to grow. In addition to Polysulfone, Union Carbide has introduced Radel A-400® which has higher heat distortion temperature and better solvent resistance and has announced plans to introduce Kadel R Polyketone resin, a new high temperature crystalline polyaryletherketone.

Union Carbide's nucleophilic displacement reaction has proven to be the process of choice for polyarylethersulfones and ketones as opposed to Friedel-Crafts processes originally utilized by ICI, 3M, duPont, and Raychem.

Acknowledgements

The commercial development of any polymer involves the dedicated effort of many people from a variety of functions including R&D, engineering, manufacturing, marketing, sales, and management. In the early R&D effort, the following individuals made significant contributions: Dr. E. W. Beste, Dr. W. F. Hale, Dr. C. N. Merriam, Mr. E. G. Hendricks, Dr. U. A. Steiner, Dr. L. M. Robeson, and Dr. T. T. Szabo.

REFERENCES

1. J. F. Bunnett and R. E. Zahler, Chem. Rev. 49, 273 (1951).
2. A. J. Parker, Quart. Rev. 16, 163 (1962).
3. R. Barclay, U.S. Patent 3,069,386 (1962); A. G. Farnham, unpublished work (UCC).
4. R. N. Johnson, A. G. Farnham, R. A. Clendinning, W. F. Hale, C. N. Merriam, J. Poly. Sci. A15 2375, 2399, 2415 (1967).
5. R. N. Johnson, A. G. Farnham, U.S. Patent 4,108, 837 (1978).
6. R. A. Clendinning, A. G. Farnham, N. L. Zutty, and D. C. Priest, Canadian Patent 847,963 (1970).
7. R. Viswanathan, B. C. Johnson, and J. E. McGrath, Polymer 25, 1827 (1984).

POLYSULFONE - EARLY MARKET DEVELOPMENT ACTIVITIES

M. E. Sauers, L. A. McKenna, and C. N. Merriam
Union Carbide Corporation, PO Box 670, Bound Brook, NJ

INTRODUCTION AND BACKGROUND

UDEL® Polysulfone, formerly known as BAKELITE® Polysulfone, was introduced via a flurry of promotional activities. The initial introduction was performed via two press conferences, capping several years of research and product development activities. Both conferences were held at Union Carbide's Bound Brook facility. The first conference held on March 25, 1965 consisted of an informal presentation of polysulfone with respect to its physical properties, chemical structure and processing methods. In attendance were a select group of editors representing the following professional publications: Managing Editor of Modern Plastics, Editor of Plastic Technology, and Associate Editor of Materials in Design Engineering.

The main purpose of the second conference held on April 7, 1965 was to announce the development of a new heat resistant structural plastic: UDEL® Polysulfone; and to also announce the construction of a plant to produce this product. In attendance were representatives of the following publications: Chemicals Engineering, Plastics Technology, Modern Plastics, Materials in Design Engineering, Steel, New York Times, Chemical Engineering Progress, Chemical Processing, EEE-The Magazine of Circuit Design Engineering, Iron Age, Design News, Electronic Design, Plastics Design and Processing, and Industrial and Engineering Chemistry.

Accompanying these guests and representing Union Carbide were:

Dr. A. G. Farnham, Co-inventor
Dr. R. N. Johnson, Co-inventor
Dr. R. W. Quarles, R&D Director
Mr. W. T. Higgins, R&D Director
Mr. R. K. Walton, Product Development Manager
Mr. R. K. Dearing, Product Development Engineer

The conference included presentations covering polysulfone's chemical structure, physical properties and manufacturing processes; followed by a tour of the fabrication and laboratory facilities with accompanying demonstrations.

The production facility began producing commerical product in January 1966, and since then UDEL® Polysulfone has gained significant market acceptance.

In 1967 3M introduced another member of the polysulfone family -- Astrel 360.

In 1972 another member of this important family of products was introduced by Imperial Chemical Industries. In October 1976, Union Carbide announced RADEL® R, a polyphenylsulfone, thereby adding a fourth member to the polyarylsulfone family. The last member of this group was introduced by Union Carbide in 1983, RADEL® A Polyarylethersulfone.

Any discussion of markets and applications needs to be preceeded with an understanding of the polymer's properties and potential utility. Prior to the introduction of this polymer, extensive laboratory evaluations were

Copyright 1986 by Elsevier Science Publishing Co., Inc.
High Performance Polymers: Their Origin and Development
R.B. Seymour and G.S. Kirshenbaum, Editors

conducted in order to determine this utility. The following two sections on structure and properties are a result of these studies.

STRUCTURE

UDEL® Polysulfone is prepared by a nucleophilic substitution reaction between the disodium salt of bisphenol A and 4,4'-dichlorodiphenyl sulfone yielding the following structure:

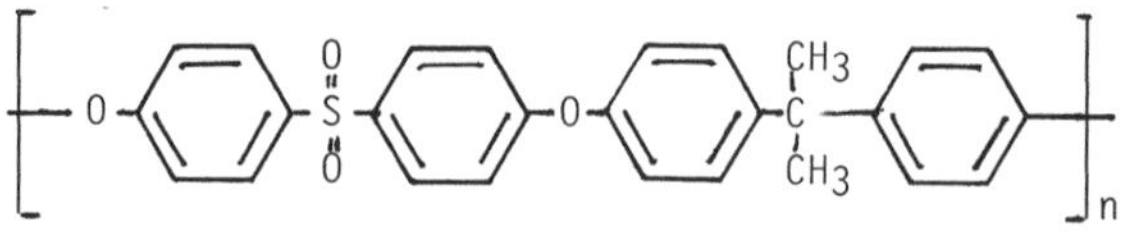

$$\left[-O-C_6H_4-\overset{O}{\underset{O}{\overset{\|}{\underset{\|}{S}}}}-C_6H_4-O-C_6H_4-\overset{CH_3}{\underset{CH_3}{\overset{|}{\underset{|}{C}}}}-C_6H_4- \right]_n$$

The unique feature of the chemical structure is the diarylsulfone grouping (reference figure below)

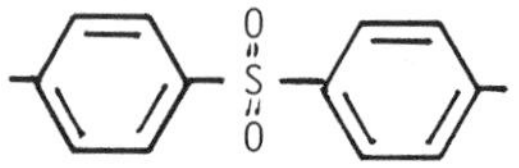

$$-C_6H_4-\overset{O}{\underset{O}{\overset{\|}{\underset{\|}{S}}}}-C_6H_4- \quad \text{diarylsulfone group}$$

This is a highly resonating structure in which the sulfone group tends to draw electrons from the phenyl rings. This resonance is enhanced by having oxygen atoms para to the sulfone group. Oxidation, by definition, is the loss of electrons.

Having the electrons tied up in resonance imparts excellent oxidation resistance to the polymer. Also, the sulfur atom is in its highest state of oxidation and therefore is not susceptible to further oxidation.

The high degree of resonance has two additional effects: (1) it increases the strength of the bonds involved and, (2) it fixes this grouping spacially into a planar configuration. This provides excellent thermal stability to the polymer and provides rigidity to the polymer chain which is retained at high temperatures.

The excellent thermal stability of polysulfone is verified by thermal gravimetric analysis which shows that UDEL® Polysulfone is stable to air up to 450°C (840°F). The excellent thermal stability coupled with excellent oxidation resistance of polysulfones provides excellent melt stability for molding and extrusion.

The ether linkage imparts some flexibility to the polymer chain which gives inherent toughness to the material. Common with all tough, rigid thermoplastic polymers, polysulfone has a second low temperature glass transition at -100°C. This minor glass transition is assigned to the ether linkages.

In these polysulfones the linkages connecting the benzene rings are hydrolytically stable. Therefore, the polymers are resistant to hydrolysis and to aqueous acid and alkaline environments. This is a distinct advantage over the ester linkages present in polycarbonates and thermoplastic polyesters.

TYPICAL PROPERTIES

Polysulfone offers outstanding performance characteristics. It is an excellent structural material in that the properties are maintained over a wide temperature range and are also retained even after long periods of exposure to high temperature.

Continuous use temperature is limited by the glass transition temperature and not by thermal or oxidative degradation of the polymer, thus confirming the stability of the chemical structure.

Union Carbide UDEL® Polysulfone has the structure shown previously. This structure was selected as offering the best combination of properties, ease of fabrication and economics. The traditional form of property presentation at room temperature is shown in Table I.

TABLE I

TYPICAL PROPERTIES OF UDEL® POLYSULFONE

Property	ASTM Method	P-1700
Density, g/cm^3	D1505	1.24
Tensile Strength at yield, psi	D638	10,200
Tensile Modulus, psi	D638	360,000
Elongation at Break, %	D638	50-100
Flexural Strength, psi	D790	15,400
Flexural Modulus, psi	D790	390,000
Izod Impact, 1/8 in., ft-lb/in notch	D256	1.3
Tensile Impact, $ft\text{-}lb/in^2$	D1822	200
Heat Deflection Temperatures @ 265 psi, °F	D648	345
Flammability(1)		
Average time of burning, sec. (P-1720)(2)	D635	5(<5)
Average extent of burning, in. (P-1720)	D635	0.4 (<0.2)
UL Bulletin 94, 1/16 in. thickness (P-1720)	--	V-2(V-0)
1/8 in. thickness (P-1720)	--	V-2(V-0)
1/4 in. thickness (P-1720)	--	V-0(V-0)
Limiting Oxygen Index (P-1720)	D2863	30 (32)
Dielectric Strength, 1/8 in., S/T, v/mil	D149	425
Volume Resistivity, ohm-cm	D257	5×10^{16}
Dielectric Constant, 670 Hz-1 MHz	D150	3.07 - 3.03
Dissipation Factor, 60 Hz-1 MHz	D150	0.0008-0.0034

(1) These numeral flame spread ratings are not intended to reflect hazards presented by this or any other material under actual fire ratings.

(2) A modified P-1700 with improved flammability ratings.

Polysulfone is a transparent, strong, rigid and tough material with a high heat deflection temperature. Its flammability ratings and smoke density ratings are among the best for thermoplastic resins. The electrical properties are very good and remain relatively constant over a wide frequency, having a dissipation factor of 0.005 at the microwave oven frequency.

This data is insufficient to qualify a material as a high performance plastic. Some of the types of data required are answers to the following questions.

What are the changes in properties over a wide temperature range?

What happens to properties on long term exposure to high temperature?

What is the load bearing ability at various temperatures?

How does it perform in environments other than air, such as water?

First, let's consider polysulfone properties vs. temperature. Some highlight properties of UDEL® Polysulfone vs. temperature are shown in Table II.

TABLE II

PROPERTIES vs. TEMPERATURE

Rigidity	Flexural Modulus	
	@ 72°F	390,000 psi
	@ 300°F	300,000 psi
Strength	Tensile Strength at Yield	
	@ 72°F	10,200 psi
	@ 300°F	6,000 psi
Toughness	Notched Izod Strength	
	@ 72°F	1.3 ft-lbs/in
	@ -40°F	1.2 ft-lbs/in
Dielectric Properties	Dielectric Strength (10 mil thickness)	
	@ 72°F	2,200 volts/mil
	@ 300°F	1,600 volts/mil

Polysulfone is a rigid material; even at 300°F it is twice as rigid as polypropylene at room temperature. It is strong, being as strong at 300°F as a general purpose phenolic molding material at room temperature. It is a tough material with a brittle temperature of -150°F. This coupled with a heat deflection temperature of 345°F provides a wide working temperature range of -150°F to 345°F. The electrical properties are relatively constant over this working temperature range.

The retention of properties of UDEL® Polysulfone on exposure to high temperatures for prolonged periods of time and to externally applied stress is shown in Table III.

Table IV shows the maximum continuous working stress levels in air and in hot water or steam.

After two years exposure to 300°F air, the rigidity, strength and heat deflection temperature actually increase slightly. The electrical properties remain essentially unchanged and the notched Izod impact strength drops to 60% of its original value.

TABLE III

PROPERTIES vs. TIME

	Retention of Original Properties After 2 Years at 300°F
Rigidity	110%
Strength	110
Heat Deflection Temperature	105
Toughness	60
Electrical Properties	95

DIMENSIONAL STABILITY

Dimensional Stability	Less than 0.020 in/in elongation in 2 years @ 3000 psi tensile stress @ 210°F
	0.001 in/in elongation after immersion in water @ 100°C for 7 days (equilibrium

TABLE IV

Maximum Continuous Working Stress Levels in Air

72°F (22°C)	3000 psi
140°F (60°C)	2800 psi
210°F (99°C)	2500 psi
260°F (127°C)	2200 psi
300°F (149°C)	2000 psi

Maximum Continuous Working Stress in Hot Water & Steam

72°F (22°C)	2000 psi
140°F (60°C)	1500 psi
230°F (110°C)	500 psi
275°F (135°C)	300 psi

It has excellent dimensional stability at high temperatures under externally applied loads. The excellent oxidation resistance, thermal stability and creep resistance are taken into account in assigning the maximum continuous working stress levels. For example, polysulfone can be used at 210°F under a 2500 psi external stress. Interestingly, the Metals Handbook recommends that zinc die casting alloys should be used up to only 1700 psi external stress at 210°F and under no external stress temperatures above 225-250°F. Polysulfone can be used up to 2000 psi external stress at 300°F.

All these factors are taken into consideration when assigning recommended continuous use temperatures for materials of construction. UDEL® Polysulfone has a maximum use temperature of 300°F-340°F, the same as that for general purpose phenolic molding materials.

Various agencies have evaluated UDEL® Polysulfone and the following approvals or ratings were obtained:

UL (E36098A)	160°C. Thermal rating for all properties except impact, flammability classification 94V-2 and 94V-0 (P-1720).

FAA	Low flame spread index, low smoke density.
NSF	Potable water.
FDA (177.1655)	Sanctioned for food contact applications: repeated use (no limitations) and single service use (reheatable frozen food containers).
Dairy Association (2005)	Accepted for use as contact surface with dairy products.

UDEL® Polysulfone is not perfect. There are certain areas where improved performance is desired. These include:

1. Environmental stress crack resistance.
2. Notch sensitivity
3. Weather resistance.

Environmental stress crack resistance is not chemical resistance where a chemical reaction takes place changing the structure such as hydrolysis or ozone attack in rubbers. Actually, polysulfone has excellent chemical resistance in that it is not attacked by aqueous acids and bases or by oxygen and ozone. Environmental stress crack resistance is a breakdown of physical structure due to a combination of applied stress, temperature and an aggressive environment. In general the aggressive environments for polysulfone are the organic ketones, esters, aromatic hydrocarbons and certain chlorinated hydrocarbons. Resistance to environmental stress cracking is difficult to predict and testing in the actual environment of use is recommended. Use of glass fiber filled polysulfone is the best solution to this deficiency.

All polymers have some degree of notch sensitivity. The polycarbonates are outstanding in having very low notch sensitivity. The notch sensitivity of polysulfones can be eliminated by proper part design, that is by using generous radii or fillets to prevent stress concentration.

The weather resistance of polysulfone is poor. The high degree of resonance that imparts many beneficial properties also makes it an excellent UV absorber. This results in poor weather resistance and makes the conventional means for improvement, UV absorbers, inoperable. The best means for improving polysulfone in the respect is via painting or electroplating.

APPLICATIONS

UDEL® Polysulfone is used in many of its applications because of one or a combination of several of its outstanding properties, including:

Heat resistance
Hydrolysis resistance, repeated steam sterilization
Low flame spread rate
Transparency
Dimensional integrity
Melt processability
Electrical properties

UDEL® Polysulfone can be fabricated by the melt processing techniques available for thermoplastics, injection molding, extrusion and thermoforming. For injection molding, melt temperatures of 650-750°F, mold temperatures of 200-300°F and injection pressures of 15,000-20,000 psi have proved suitable. For extrusion, melt temperatures of 600-700°F are required. For sheet extrusion, the take-off rolls must be capable of being controlled between

300°F and 350°F. Before melt processing, the polysulfone pellets must be dried at a minimum temperature of 275°F for 3.5 hours.

Examples of early application areas and end uses for UDEL® Polysulfone are shown below:

APPLIANCES

Coffee makers
Humidifiers
Hair dryer components
Hot lather dispensers
Steam iron components
Egg cookers
Clothes steamers
Hot chocolate dispensers
Water heater dip tubes
Kitchen range hardware
Microwave oven components

ELECTRICAL & ELECTRONIC

Integrated circuit carriers
Connectors
Coil bobbins
Capacitor film
TV components
Brush holders
Terminal blocks
Business machine components
Printed circuit boards
Alkaline battery cases

AIRCRAFT & AEROSPACE

Passenger service units
Luggage rack bulk heads
Astronauts' outer face mask shields

AUTOMOTIVE

Steering column units
Relay insulators
Pistons in load leveler
Electroplated lamp bezels

MEDICAL - SURGICAL

Respirator parts
Nebulizers
Dialysis components
Instruments
Sterilizable packages
Hospital feeding trays

PROCESSING EQUIPMENT

Milking machine components
Ball valves
Steam table trays
Membranes for reverse osmosis microfiltration
Peterson separator candles

MISCELLANEOUS

Electroplated Polaroid SX-70 camera cases
Digital watch cases
Fuel cell components
Overhead projector transparencies
Electroplated computer print-out wheels
Projector components
Toy corn poppers

As in all businesses, there are some applications that are short-lived and some that represent continuing business for decades. The life cycle of these applications are subject to many diverse pressures. Some of the applications that started almost two decades ago and continue today are:

Coffee makers
Hot water dip tubes
Integrated circuit carriers
Electrical switches
Medical respirator components
Milking machine components
Electrical switches

Some of the applications were low volume and short lived but exciting to be associated with such as the Astronauts Space Shield used during the moon walks. Other applications, like the Polaroid SX-70 camera case, required extensive product development efforts which resulted in significant business.

The need for an expanding product line led to the development of alloys and composites based on UDEL® Polysulfone. Union Carbide introduced a polymer alloy named UCARDEL® designed for specific markets in 1971. At about the same time, Uniroyal Inc. introduced an ABS modified UDEL® Polysulfone named Arylon T for chrome plated applications requiring higher heat resistance than ABS.

Since polysulfones were investigated in the Bound Brook laboratories for a number of years prior to their market introduction, great volumes of data were generated to understand their processing characteristics and their inherent properties; as well as comparing these characteristics with other polymers and metals. As a result of these investigations, a significant number of technical articles and presentations emerged soon after the commercial announcement of UDEL® Polysulfone. An attempt is made to document these activities chronologically.

POLYSULFONE TECHNICAL PUBLICITY

September 1965 — Society of Plastics Engineers (SPE) Regional Conference "A New Thermoplastic Resin - Polysulfone".

September 1965 — SPE Journal article - "Polysulfone-A New Heat Resistant Structural Plastic".

A series of technical papers at regional technical conferences sponsored by SPE and other agencies from September 30, 1965 to March 11, 1966 in Toronto, Montreal, Detroit, Chicago and Newark, NJ on the following topics:

1. "Tooling for Polysulfone"
2. "Polysulfone-A New Engineering Thermoplastic"
3. "Design Criteria for a New Heat Resistant Thermoplastic-Polysulfone"
4. "Techniques for Predicting Injection Molding Behavior"
5. "A New Resin for the Auto Industry".
6. "Extrusion Screw Design"
7. "The Rheology of Polysulfone"
8. "Polysulfone-A Unique Structural Adhesive Resin".

March 1966 — 22nd ANTEC (SPE), "Injection Molding of Polysulfone".

March 1966 — 23rd Annual Western Section Society of Plastics Industry Conference, "Polysulfone-A New Engineering Material".

April 1966 — Rubber Age article, "A New Thermoplastic Resin for the Auto Industry".

June 6-9, 1966 — Molding demonstrations at Society of Plastics Industry Exposition - New York.

January 1967 — Thermal cycling demonstrations of UDEL® Polysulfone performed at 1967 SAE Exposition.

March 1967 — Eastern New England SPE Meeting, "Polysulfone for Extrusion and Injection Molding".

March 1967 — Boston SPE RETEC, "Molding of Polysulfone".

April 1967 — SPI Plastics & Western Industry Meeting, "Polysulfone-A New Thermoplastic for the Aerospace and Electronics Industries".

May 1967 — Society of Automotive Engineers, "Polysulfone-A New Thermoplastic Resin for the Automotive Industry".

May 1967 — SPE Silver Anniversary ANTEC, "Testing for Thermal Endurance".

December 1967 — Houston SPE Meeting, "Polysulfone vs. Zinc Die Casting".

Presentations were made during this period to individual corporations in the form of seminars at their facilities. But, in order to get the message to larger audience including a significant number of injection molders, a new plan was developed and implemented. During 1966 and 1967 regional technical seminars on UDEL® Polysulfone were held around the country. These three hour seminars included properties, processing, economics, design and application discussions. Large audiences turned out all over the country to learn about this new high temperature thermoplastic. Small molders and giant corporations were represented. This may have been our most effective means of communicating the message of polysulfone and Engineering Polymers directly to the fabricators and users of this new product.

Advertising played a large role in the early market development activities of UDEL® Polysulfone. By September 1966 mass mailings of descriptions and photographs of new applications had been distributed. These publicity releases included:

Protected circuit alarm switch
Integrated circuit carrier
Circuit card edge connectors and carrier for I.C. flat packs
Polysulfone film
Polysulfone extruded shapes for prototyping
Automotive throttle control components
Steam iron sight glass
Control plug for Sterco phonograph cartridge
Airline passenger service unit
Television remote control switch assembly

CONCLUSION

Polysulfones represent a class of high temperature, hydrolytically stable polymers that have found wide usage in many industries. Twenty-one years ago when Union Carbide first introduced the first aromatic polymers in this series, it was a new product requiring extensive publicity and technical support. The same holds true today. Continuing product and application development activities result in new usages for these materials. There are many other polymer systems to choose from today; but with the base established over the past twenty years, UDEL® Polysulfone remains in the forefront of these activities.

BIBLIOGRAPHY

T. E. Bugal, Testing for Thermal Endurance: A Case History on Thermoplastic, August 1966, Union Carbide Corporation, Chemicals & Plastics Research & Development, Bound Brook, NJ.

R. N. Johnson, A. G. Farnham, R. A. Clendinning, W. F. Hale and C. N. Merriam, "Poly(aryl)ethers by Nucleophilic Aromatic Substitution. I. Synthesis and Properties", J. Polymer Science, Part A-1, Vol. 5, 2375-2398 (1967).

W. F. Hale, A. G. Farnham, R. N. Johnson and R. A. Clendinning, "Poly(aryl ethers) by Nucleophilic Aromatic Substitution. II. Thermal Stability", J. Polymer Science, Part A-1, Vol. 5, 2399-2414 (1967).

R. N. Johnson and A. G. Farnham, "Poly(aryl ethers) by Nucleophilic Aromatic Substitution. III. Hydrolytic Side Reactions", J. Polymer Science, Part A-1, Vol. 5, 2415-2427 (1967).

L. M. Robeson, A. G. Farnham and J. E. McGrath, "Synthesis & Dynamic Mechanical Characteristics of Poly(aryl ethers)", Union Carbide Corporation, Chemicals & Plastics Research & Development, Bound Brook, NJ.

DISCOVERY AND DEVELOPMENT OF THE "VICTREX" POLYARYLETHERSULPHONES

JOHN B. ROSE
formerly ICI Plastics Division, Welwyn Garden City, now
Chemistry Department, University of Surrey, Guildford, Surrey
GU2 5XH, U.K.

INTRODUCTION

Polyarylethersulphones were reported [1,2,3] in the patent literature as new materials prepared by novel polycondensation reactions during the early 1960s. By this time the basic theory of polycondensation was well established [4] and qualitative theories linking the bulk properties of polymers to their molecular structure had been formulated in some detail [5]. This led polymer chemists engaged in adapting reactions known for monomeric compounds to polymer synthesis to guide their work towards those polymer structures which could be expected to show particularly useful properties. The effects of chain rigidity and polarity on Tg and Tm, increase in either of these parameters leading usually to an increase in the transition temperatures, was well known [6] and incorporating phenylene rings in polymer chains was an obvious method of increasing chain rigidity. However, the predictive power of the knowledge available was limited by its qualitative nature and there was (and still is) a major problem in predicting the softening points of aromatic polymers because the relationships between crystallisability and molecular structure were less well known for these polymers than for their aliphatic analogues.

Another important area of doubt was in relating the toughness of these polymers to their structures, for although the toughness of Polycarbonate was widely appreciated, recognition that introducing aromatic rings into the main chain was the most promising way of obtaining tough polymers with high softening points did not come until later [7]. It was realised that the upper continuous use temperatures of the high melting plastics then available was set by their oxidative stability, the prime example being Nylon 66 with Tm = 265°C but a UL Temperature index of only 100°C. It was expected that polymers consisting of linked aromatic rings devoid of alkyl substituents and aliphatic inter-ring linkages would show good thermal and oxidative stability, but it was not clear that the overriding requirement for thermoplasticity, that the polymer's decomposition temperature must be above the temperature at which it becomes melt processable, say (Tm + 30°C) for crystalline and (Tg + 150°C) for amorphous polymers, could be met with polymers of this type. Low flammability and low smoke evolution on burning are now properties expected from aromatic thermoplastics, but this was not so in 1960.

Thus, by 1960 there was sufficient information available to suggest that in the search for new thermoplastics suitable for continuous use above 125°C attention should be directed towards polymers consisting of linked aromatic rings, but there was little guidance on the choice of inter-ring linkages. It can now be seen that the combination of flexible aryl ether bonds with polar aryl sulphone (or ketone, see next chapter) linkages provides an effective solution to this problem,but (in ICI at any rate) the discovery of these combinations arose mainly from the chemistry of the polycondensation processes investigated, rather than from properties versus structure predictions.

ICI's interests in novel thermoplastics was centered on the Plastic Division in Welwyn Garden City, much of the experimental chemistry being performed in an Exploratory Section set up in 1959 and headed by G.D.

Copyright 1986 by Elsevier Science Publishing Co., Inc.
High Performance Polymers: Their Origin and Development
R.B. Seymour and G.S. Kirshenbaum, Editors

Buckley. The author was a member of this group and became responsible for its technical direction on the promotion of G.D. Buckley in 1961. The temperature limitations of existing thermoplastics were well recognised at Welwyn, where relationships between the structure of polymers and their bulk properties had been under investigation for some time by a well known team under the direction of C.W. Bunn. In this environment, new syntheses of novel aromatic polymers appeared to be a good topic for research and our attention was directed towards the aryl sulphone linkage because of its high polarity and because phenyl sulphone was known to distill unchanged at 379°C [8].

Throughout the exploratory phase of this work the collaborating teams of chemists and physicists leaned heavily for advice on a group of experienced polymer scientists forming the permanent membership of an Exploratory Research Committee whose photographs are shown in FIG. 1 together with those of the late E.G. Williams and H.C. Raine, Chairman and Research Director respectively, of ICI Plastics Division at the time the work was started.

FIG.1. Permanent Membership of the ERC

It is interesting to note that three different companies, ICI in the UK and 3M and Union Carbide in the USA, all independently developed routes to polyarylethersulphones, filing patents within less than a year from November 1962. Each chose different polymers for exploitation: the structures chosen are shown in Table I. Inevitably with such close over-lapping of inventions in the field, there was a clash of interest between Union Carbide, ICI and 3M in the patent area. This was resolved by the US Patent Office Board of Interference Examiners (see Record for Interference Nos. 95,807 and 95,808 pages 983 and 1139) in 1970 and their decision, confirmed by rejection of 3M's appeal to the US Court of Customs and Patent Appeals in 1973, confirmed ICI's priority for the invention of PES. Union Carbide's priority for the invention of Udel has not been questioned in the USA.

TABLE I. Commercial Polyethersulphones.

Polymer repeat unit	Trade Name	Producer	Production from- to-
$-C_6H_4-SO_2-C_6H_4-O-C_6H_4-CMe_2-C_6H_4-O-$	Udel	Union Carbide	1965 onwards
Copolymer of $-C_6H_4-C_6H_4-SO_2-$ and $-C_6H_4-SO_2-C_6H_4-O-$	Astrel 360	3M	1967-1976
$-C_6H_4-SO_2-C_6H_4-O-$	Victrex PES	ICI	1971 onwards
Copolymer of $-C_6H_4-SO_2-C_6H_4-O-$ and $-C_6H_4-C_6H_4-SO_2-$	Victrex 720P	ICI	1973-
$-C_6H_4-SO_2-C_6H_4-O-C_6H_4-C_6H_4-O-$	Radel	Union Carbide	1976-

DISCOVERY OF PES, I.

Two main routes are available for the synthesis of polyarylethersulphones, the polysulphonylation reactions (1) and the polyether syntheses (2):

(1a) $ArH_2 + ClSO_2-Ar'-SO_2Cl \longrightarrow -Ar-SO_2-Ar'-SO_2- + 2HCl$

(1b) $HAr-SO_2Cl \longrightarrow -Ar-SO_2- + HCl$

(2a) $Hal-Ar''-Hal + MO-Ar'''-OM \longrightarrow -Ar''-O-Ar'''-O- + 2MHal$

(2b) $Hal-Ar''-OM \longrightarrow -Ar''-O- + MHal$

Polysulphonylation is an electrophilic aromatic substitution and as aromatic rings are activated to such reactions by ether substituents, reactions (1) proceed particularly well when Ar has the structure II to give polyethersulphones. In the polyether syntheses (2) aromatic halogen is activated by para sulphone linkage(s) to nucleophilic displacement by alkali metal phenoxides, so that the formation of polyethersulphones by using halides where Ar" has the structure III is a preferred variant of these reactions.

I, PES —C_6H_4—SO_2—C_6H_4—O—

II —C_6H_4—O—C_6H_4— III —C_6H_4—SO_2—C_6H_4—

Both types of process were examined by ICI at Welwyn in 1961-2 and at this time polymers of high molecular weight were obtained from the polysulphonylations, but the polyether syntheses gave polymers of only moderate molecular weight [9]. The polysulphonylations were also examined by 3M Corpn. [2], who made their Astrel 360 product in this way, probably via reaction (3) [10]. Union Carbide discovered the polyether synthesis (2a), demonstrated its broad applicability, including the synthesis of PES, and used it to manufacture Udel Polysulfone, the first polyarylethersulphone to become available commercially, via reaction (4) [3,11].

(3) 0.45 $ClSO_2$—C_6H_4—O—C_6H_4—SO_2Cl + 0.45 C_6H_5—C_6H_5 + 0.10 C_6H_5—C_6H_4—SO_2Cl ⟶ 3HCl + [—C_6H_4—C_6H_4—SO_2—]$_{0.55}$ [—C_6H_4—O—C_6H_4—SO_2—]$_{0.45}$

(4) Cl—C_6H_4—SO_2—C_6H_4—Cl + NaO—C_6H_4—CMe_2—C_6H_4—ONa ⟶ —C_6H_4—SO_2—C_6H_4—O—C_6H_4—CMe_2—C_6H_4—O— + 2NaCl

Polysulphonylations

These reactions were examined at Welwyn between 1961 and 1962 as part of a general programme aimed at finding new routes to aromatic polymers. The traditional technique for sulphonylation with sulphonyl chlorides uses at least one mole of Friedel-Crafts halide per mole of sulphonyl chloride and it was realised that this would lead to difficulties in isolating polymers free from inorganic contaminants. However, the preparation of monomeric sulphones using catalytic quantities of $FeCl_3$ had been reported in a war-time German patent [12] and we applied this procedure to reactions (1). Polymers of high molecular weight (reduced viscosity, RV, up to 1.2 for 1% solutions of polymer in dimethyl formamide at 25°C) were obtained from 4-phenoxybenzenesulphonyl chloride or from equimolar mixtures of phenyl ether and bis-4-chlorosulphonylphenyl ether (reactions (5) and (6)) by polycondensation of the molten monomer(s) in the presence of 0.2-2 Wt.% of $FeCl_3$, and polymers free from iron were isolated without undue difficulty [9].

(5) — SO_2Cl → — SO_2 —

(6) + $ClSO_2$ — SO_2Cl → — SO_2 —

It was shown later that the use of small rather than equimolar quantities of $FeCl_3$ or $SbCl_5$ as catalyst also prevented side reactions which increased the evolution of HCl beyond that expected theoretically from the amount of sulphonyl chloride employed and led to inclusion of aberrant structures in the polymer [13]. Reaction (5) was preferred over (6) in the first series of laboratory experiments as it was easy to upset the stoichemistry of reaction (6) by losing a little phenyl ether in the HCl evolved and reaction (6) also gave a significant proportion of insoluble product which was filtered off from solutions of the polymer. The first measurements of properties were carried out on polymer from reaction (5), which was fortunate as we found later that the product from reaction (6) was lacking in toughness. Thus, attention became concentrated on this novel polymer which we called PES. We found it to be thermoplastic giving tough, rigid mouldings which retained useful mechanical and electrical properties up to 200°C, suggesting use as an engineering plastic at temperatures well above those possible with existing thermoplastics. The polymer was amorphous, which in view of its apparently regular structure we found surprising, with Tg = 230°C (at RV = 0.6).

Polycondensation of 4-phenylbenzenesulphonyl chloride gave a partially crystalline polymer, Tm > 370°C, which was intractable, but polymerisation of this monomer mixed with 4-phenoxybenzenesulphonyl chloride, reaction (7), gave a series of amorphous copolymers showing softening points greater than that of PES [14]. A copolymer of this type was eventually produced as a higher softening grade of PES (Victrex 720P see Table I).

(7) SO_2Cl + SO_2Cl → — SO_2 — — SO_2 — + HCl

Polyether Synthesis

Reactions (2) were also examined at Welwyn as part of the programme aimed at new aromatic polymers. At this time reactions of this type had received little prior attention, and the Union Carbide work in this area was not yet published. Nucleophilic substitution of aryl halides was well known to be a difficult reaction requiring catalysis by copper salts, but powerful electron withdrawing groups positioned either _ortho_ or _para_ to the halogen activated it to the reaction [15]. It was also known that the copper catalysed reactions occurred more readily in dipolar aprotic solvents such as dimethyl sulphoxide [16]. We therefore examined the reactions shown in Table II using dimethyl sulphoxide as solvent and 5-10 Wt.% cuprous oxide as catalyst, concentrating attention on bromides as these reacted more rapidly than the chlorides. In this way polymers of moderate molecular weight were obtained from the activated halides but with the other halides the reaction gave only dimers and trimers under our conditions.

TABLE II. Polycondensation of Bis-phenoxides with Aryldihalides, Reaction (2a),[9].

Bis-phenoxide	Dihalide	RV[a]	%Hal.	DP[b]
$NaO-C_6H_4-CMe_2-C_6H_4-ONa$	$Br-C_6H_4-C_6H_4-Br$	0.03	16.7	2
$NaO-C_6H_4-CMe_2-C_6H_4-ONa$	$Br-C_6H_4-O-C_6H_4-Br$	0.05	15.2	2
$NaO-C_6H_4-CMe_2-C_6H_4-ONa$	$Cl-C_6H_4-CO-C_6H_4-Cl$	0.20	1.0	21
$NaO-C_6H_4-CMe_2-C_6H_4-ONa$	$Br-C_6H_4-SO_2-C_6H_4-Br$	0.36	0.5	40
$NaO-C_6H_4-SO_2-C_6H_4-ONa$	$Br-C_6H_4-SO_2-C_6H_4-Br$	0.20	0.9	19

[a] Reduced viscosity = (t. solution-t. solvent)/t. solvent for a 1% solution of polymer in dimethylformamide at 25°C. [b] Calculated from halogen contents

The work by A.G. Farnham and his colleagues at Union Carbide has shown that with the activated halides no catalyst is required to obtain polymers of high molecular weight [11]. We now know that polyethersulphones with RV>0.40 are high enough in molecular weight to show useful mechanical properties and it is clear from Table II that we came near to achieving this, the polymer made from _bis_-phenol A and _bis_-4-bromophenyl sulphone having RV = 0.36. It is possible that the cuprous oxide used as catalyst had a deleterious effect by removing bromine from the system as copper bromide and upsetting the reaction stoichiometry. The polymer made from _bis_-bromophenyl sulphone and _bis_-phenol S was amorphous, confirming the same observation made on polymer of structure I prepared by the polysulphonylation reaction (5).

SELECTION OF THE POLYETHERSYNTHESIS FOR PRODUCTION OF PES

Problems with the polysulphonylation reactions (5) and (6).

The original melt polycondensation was not suitable for development as it gave a foamed mass which occupied a large reaction volume and the use of temperatures above Tg to complete the reaction which caused the formation of some gelled material, probably via homolytic decomposition of sulphonyl chloride groups [17]. The use of solvents was investigated and nitrobenzene found to be suitable so that by 1964 reaction (6) was established in the laboratory using this solvent at 120-140°C to give 100 g. batches of polymer consistently high in molecular weight. Reaction (6) was the preferred polysulphonylation as phenyl ether was available industrially within ICI and bis-4-chlorosulphonylphenyl ether was easily obtained in a pure state from phenyl ether. However, it became clear that moulded specimens of the polymer from reaction (6) failed un-notched in Sharpy type impact tests, whereas polymer from reaction (5), provided that it was high enough in molecular weight, required notching to induce failure. After some investigation this was found to be a real effect and due to structural differences between the two types of polymer. These differences were detected by measurement of ^1H nmr spectra [14] which showed that the polymer from polysulphonylation (5) was virtually identical to the polymer

FIG. 2 ^1H nmr Spectra of Polyethersulphones.

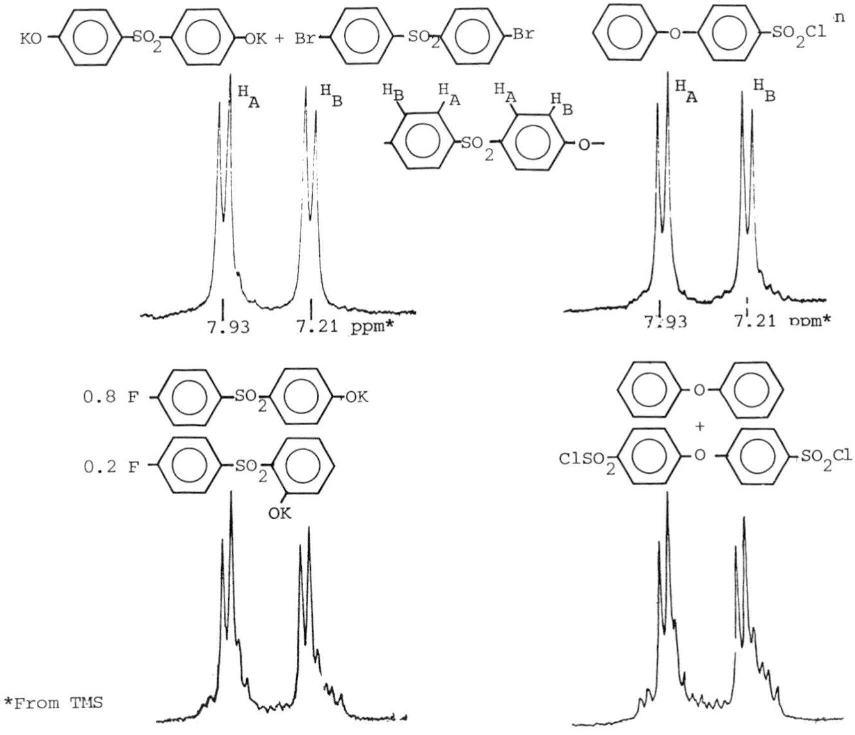

polymers having the all *para* structure I and giving spectra showing an AB quartet (FIG. 2) as I contains only two non-equivalent types of proton. The coupling constant, 8 cycles sec.$^{-1}$, is typical of non-equivalent aromatic protons positioned *ortho* to each other. The polymer from reaction (6) showed a more complicated spectrum, the AB quartet being distorted and several additional peaks appeared on the high field sides of the main resonance signals. Comparison with the spectra of model compounds and eventually with the spectra of co-polymers made by the polyether synthesis (8) [18] to contain known quantities of the *ortho*, *para*-repeat unit, IV, showed that the product of reaction (6) was a copolymer of I and IV containing ca.20% of IV [19]:

(6) $ClSO_2-C_6H_4-O-C_6H_4-SO_2Cl$ + $C_6H_5-O-C_6H_5$ ⟶ $-C_6H_4-O-C_6H_4-SO_2-$ (80%) + $-C_6H_4-O-C_6H_4(SO_2-)-$ (20%) + 2HCl

IV

These structural differences may be due to the electronic effects of sulphone groups, acting via ether linkages, on the orientation of sulphonylation in adjacent rings, for in reaction (5) the sulphone linkages are all formed by substitution at rings linked via oxygen to rings containing sulphone or sulphonyl chloride substituents, while in reaction

(8) x $F-C_6H_4-SO_2-C_6H_4(OK)$ + (1-x) $F-C_6H_4-SO_2-C_6H_4-OK$ ⟶ x $-C_6H_4-SO_2-C_6H_4(O-)-$ (IV) + (1-x) $-C_6H_4-SO_2-C_6H_4-O-$ (I) + KF

(5) $C_6H_5-O-C_6H_4-SO_2Cl$ ⟶ $-C_6H_4-O-C_6H_4-SO_2-$ + HCl

I

(6) only half of the sulphone links are formed in this way the rest resulting from mono-sulphonylation of phenyl ether. It is likely that sulphonylation of phenyl ether gives a substantial quantity of the *ortho*-substituted product, as is known to occur during nitration, while sulphonylation of rings attached via ether links to phenylsulphonyl groups occurs entirely in the *para*-position as appears to occur in the nitration of 4-nitrophenyl phenyl ether [20].

All phenylene rings in the polymer chain are far less readily sulphonylated than phenoxyl rings on chain ends as rings in the chain have a sulphone substituent which is strongly electron withdrawing and deactivates these rings to attack by electrophilic reagents. Thus, attack by sulphonyl chloride occurs much more easily at chain ends and a substantially linear chain results. However, as polycondensation proceeds the concentration of in-chain phenylene rings increases while the concentration of phenoxyl ends decreases and sulphonylation of phenylene rings becomes more likely, for it has been shown by experiments with model compounds that bis-4-chlorosulphonylphenyl ether reacts with in-chain phenylene groups, but at a rate much slower than that at which it reacts with phenoxyl groups [21]. Investigation of polysulphonylations to which an excess of sulphonyl chloride groups were added (as bis-4-chlorosulphonylphenyl ether) showed that repeat units I and IV can be sulphonylated further (reactions (9) and (10)) to give branched chain polymers in which the branch point has the structure V [13]. However with I this reaction is so slow that polymer formed from pure bis-4-phenoxybenzenesulphonyl chloride contains less than 0.4 branch points, V, per 100 repeat units, I. The ortho-para repeat, IV, reacts faster, presumably because there is less steric hindrance to attack at the para position in IV than at the ortho positions in I, so that polycondensation of pure bis-4-chlorosulphonylphenyl sulphone with equimolar quantities of pure phenyl ether gives polymer containing from 0.7 to 1.7 branch points per 100 repeat units, I + IV, the concentration of branches increasing with molecular weight [21].

(9) $-O-C_6H_4-SO_2Cl$ + $-C_6H_4-O-C_6H_4-SO_2-$ (I) —v.slow→ $-C_6H_4-O-C_6H_3(SO_2-C_6H_4-O-)-SO_2-$ (V) + HCl

(10) $-O-C_6H_4-SO_2Cl$ + $-C_6H_4-O-C_6H_4(SO_2-)$ (IV) —slow→ V

Thus polymer from reaction (5) consists essentially of repeat units I but that from (6) contains ca. 20% of repeat units IV and 1-2% of branch points, V. Measurement of impact strength on copolymers made by the polyether synthesis (8) where x ranged from 0 to 0.5 (FIG. 3) shows that the brittleness observed for polymers from (6) may be attributed to the ortho para repeat units in the polymer as the presence of 20% of these is

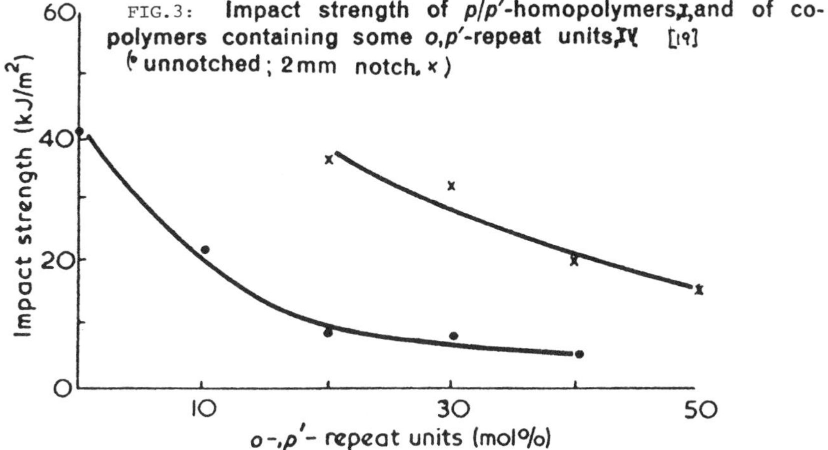

FIG.3: **Impact strength of *p*/*p*'-homopolymers, I, and of copolymers containing some *o*,*p*'-repeat units, IV [19] (• unnotched; 2mm notch, ×)**

sufficient to account for the change from a polymer which is tough, so that un-notched specimens do not fail in the Charpy test, to a brittle material giving un-notched specimens which break quite easily [22]. The presence of branched chains in the polymers from (6) makes this situation worse [22]. Considerable effort was therefore deployed to develop a commercially applicable synthesis of 4-phenoxybenzenesulphonyl chloride via reaction sequences such as (11), but the crude product was always contaminated with bis-4-chlorosulphonylphenyl ether and molecular distillation was required to remove this impurity and obtain a monomer which did not give branched polymer.

(11)

$$\text{Ph-O-C}_6\text{H}_5 + \text{HClSO}_3 \longrightarrow \text{Ph-O-C}_6\text{H}_4\text{-SO}_2\text{OH} + \text{HCl}$$

$$\text{Ph-O-C}_6\text{H}_4\text{-SO}_2\text{OH} + \text{COCl}_2 \longrightarrow \text{Ph-O-C}_6\text{H}_4\text{-SO}_2\text{Cl} + \text{CO}_2 + \text{HCl}$$

Selection of the polyether synthesis

By 1965 the problems with the polysulphonylation routes to PES had become very apparent prompting reappraisal of the polyether syntheses to PES in the light of information about the Union Carbide work on reaction (2a) which had become available in the patent literature. High polymers of I were obtained in the laboratory from reactions (12) and (13) and both were shown by ^{1}H nmr to have the PES structure, I, and both gave tough mouldings. The advantages of these reactions over the polysulphonylations were now clear, as the key monomer, bis-4-chlorophenyl sulphone, was easily made, the most elegant route being reaction (14) [23], and supplies were available commercially.

(12)

$$\text{Cl-C}_6\text{H}_4\text{-SO}_2\text{-C}_6\text{H}_4\text{-Cl} + \text{KO-C}_6\text{H}_4\text{-SO}_2\text{-C}_6\text{H}_4\text{-OK} \longrightarrow 2\ \text{-C}_6\text{H}_4\text{-SO}_2\text{-C}_6\text{H}_4\text{-O-} + 2\text{KCl}$$

(13)

$$\text{KO-C}_6\text{H}_4\text{-SO}_2\text{-C}_6\text{H}_4\text{-Cl} \longrightarrow \text{-C}_6\text{H}_4\text{-SO}_2\text{-C}_6\text{H}_4\text{-O-} + \text{KCl}$$

(14)

$$2\ \text{C}_6\text{H}_5\text{-Cl} + \text{Me}_2\text{SO}_4 + \text{SO}_3 \longrightarrow \text{Cl-C}_6\text{H}_4\text{-SO}_2\text{-C}_6\text{H}_4\text{-Cl} + 2\text{MeHSO}_4$$

Monomer grade bis-phenol S was not readily available and presented some difficulty as the product from reaction of phenol with sulphuric acid (as operated at that time) gave a product heavily contaminated with isomeric sulphones which was difficult to purify. However, work at Welwyn showed that 4-chlorophenyl 4-hydroxyphenyl sulphone was easily obtained in

95% yield by partial hydrolysis of bis-4-chlorophenyl sulphone, reaction (15), so that PES was now available directly from the dichloride [24].

It is established that reaction of activated aryl halides with nucleophilic reagents, e.g. reactions (12), (13), and (15), occur via an intermediate addition complex, VI, and that electron withdrawing groups aid reaction by stabilizing this complex via electron delocalisation as indicated for reaction (16) [25]. Our work shows that with diphenyl

(15) Cl–C$_6$H$_4$–SO_2–C$_6$H$_4$–Cl + 2KOH ⟶ Cl–C$_6$H$_4$–SO_2–C$_6$H$_4$–OK + KCl

(16)

–SO_2–C$_6$H$_4$–Cl + $\bar{O}$–C$_6$H$_5$ ⇌ –SO_2–C$_6$H$_4^-$(Cl)(O–C$_6$H$_5$) ⇌ –SO_2–C$_6$H$_4$–O–C$_6$H$_5$ + $\bar{Cl}$

VI

sulphone derivatives the electronic effects of the sulphone group on the reactivity of a functional group, e.g. halogen or phenoxide, attached to one phenylene ring is modified by substituents in the other [26]. Thus, comparing the rate constants (which are the numbers over the reaction arrows x 10 1./mole./sec.) for reactions (17) to (19) it is seen that either of the two equivalent chlorine atoms is removed from bis-4-chlorophenyl sulphone ca. 1.5 times faster than is the single chlorine from 4-chlorophenyl phenyl sulphone, but when one chlorine has been displaced by $\bar{O}$H the second chlorine is removed, reaction (19), over 30 times more slowly. It

(17) C$_6$H$_5$–SO_2–C$_6$H$_4$–Cl + 2KOH $\xrightarrow{112}$ C$_6$H$_5$–SO_2–C$_6$H$_4$–OK + KCl + H_2O

(18) Cl–C$_6$H$_4$–SO_2–C$_6$H$_4$–Cl + 2KOH $\xrightarrow{179^a}$ Cl–C$_6$H$_4$–SO_2–C$_6$H$_4$–OK + KCl + H_2O

VII   [a] for each Cl   VIII

(19) KO–C$_6$H$_4$–SO_2–C$_6$H$_4$–Cl + 2KOH $\xrightarrow{3.2}$ KO–C$_6$H$_4$–SO_2–C$_6$H$_4$–OK + KCl + H_2O

VIII

appears that the electronic effects of a substituent in one ring is transmitted to the other via the sulphone group, halogen reactivity in one ring being enhanced by the presence of a para-chloro substituent in the other, while the strongly electron donating para-phenoxide pole reduces halogen reactivity in the adjacent ring substantially. The operation of these "bridge" effects may be envisaged as indicated below, the sulphone group's ability to aid formation of the intermediate complex by accepting electrons from the reaction centre, as in VII, being enhanced by electron withdrawal towards chlorine but substantially diminished by electron accession from the phenoxide pole in VIII. Thus, chlorine is displaced from bis-4-chlorophenyl

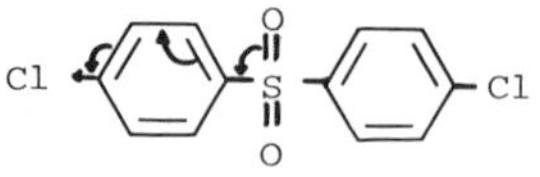

VII 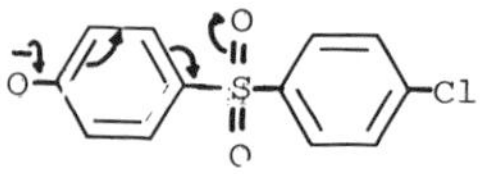 VIII

sulphone 100 times more rapidly than from the chlorophenoxide, VIII, so that reaction (15) gives very high yields of VIII rather than the 50% conversion that would be available (at complete reaction of the KOH) if the halogen reactivities were the same in both reactions [26].

It was found that potassium halophenoxides polymerised rapidly on melting at ca. 300°C, to give polyarylethersulphones which were usually of high molecular weight [18]. The introduction of repeat unit IX into the PES chain by copolymerisation of VIII with X provides an effective way of raising Tg, as this introduces 4-phenylene-4'-sulphonylphenylene groups (as did reaction 7) into the chain [18]. Copolymerisations of this type provided an industrial route to the higher softening, 720 P, grade of PES [22].

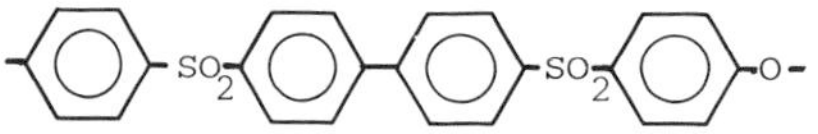

IX 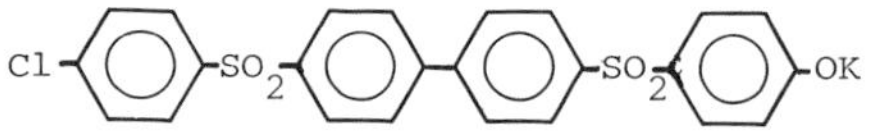 X

DEVELOPMENT OF THE POLYETHER SYNTHESIS FOR MANUFACTURE OF PES

As a result of the work described above a decision was taken in 1966 to scale up the polyether synthesis and semi-technical and pilot plant scale units were built on the Plastics Division Headquarters site at Welwyn Garden City. This was not done without considerable reservations, but the decision to concentrate this development at Welwyn enabled the research team to dominate technological development of the process to monomers and polymers in a way that would not have been possible on a production site. A 100 tonne/year Pilot Plant operating novel technology was commissioned by 1971 and the first "open market" sale of polymer from this plant was made in January of that year.

By 1972, polymer from this plant was freely available for sale in Europe and the product was introduced to the US market in 1974. During 1976 production was transferred to a Pioneer Production Plant built at ICI's Hillhouse Works on the Fylde coast in Lancashire. The initial plant capacity was in excess of 200te/year and this was tripled to almost 1000 te/year by the end of 1981; plant capacity was doubled by further expansion during 1985. Plans to build a new plant at Hillhouse for producing PEEK see next Chapter) and a range of new "Victrex" polymers currently under development were announced by ICI's Advanced Materials Business Group at the end of 1985.

PRODUCT DEVELOPMENT

The initial perception of PES as a high performance engineering thermoplastic suitable for continuous use under stress in air at elevated temperatures has been confirmed and the material now has an Underwriters Laboratories Temperature Index of 180°C; this advance beyond the temperature capabilities of other engineering thermoplastics is indicated by FIG. 4

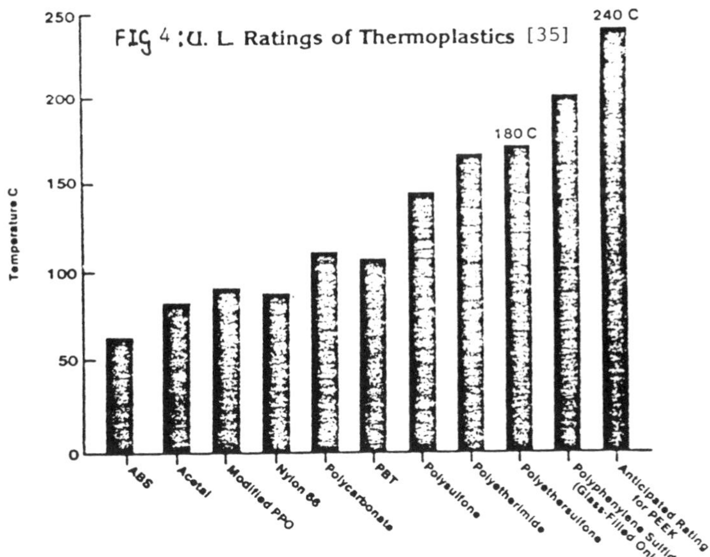

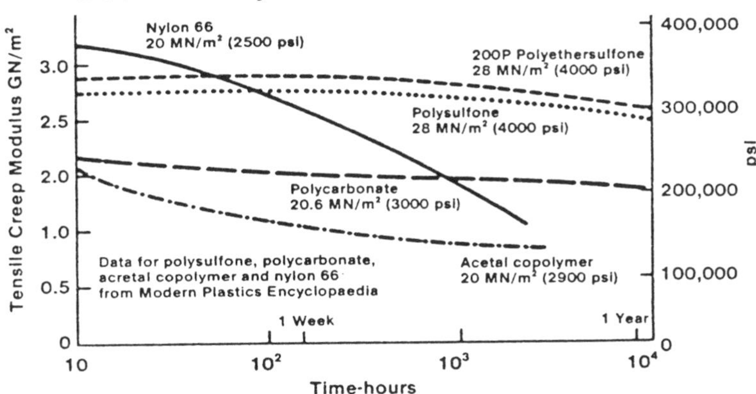

At room temperature PES is rigid and tough with outstanding resistance to creep, FIG. 5, retaining these properties well up to 200°C; good electrical insulation properties are also retained to 200°C [27]. The high resistance to oxidation, FIG. 6, is due to the absence of aliphatic groups from the repeat unit, a structural feature which also confers low flammability (UL94 V-O at 0.43 mm and LOI 34-41) and low smoke emission when burning [28].

Proliferation of Grades

The number of grades has increased as the polymer has been tailored to fit particular application areas by controlling molecular weight and end-group composition, by co-polymerisation and by compounding with fillers,

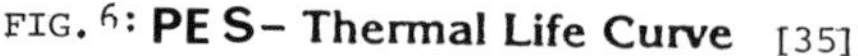

FIG. 6: PES- Thermal Life Curve [35]

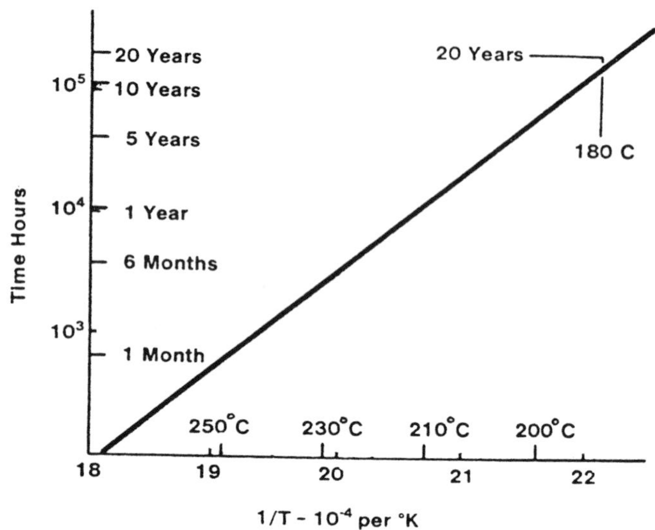

especially glass fibre. Thus, at the start of the marketing program two main grades were offered, 200P designed for injection moulding and a grade of higher molecular weight, 300P, for extrusion or injection moulding of components requiring superior mechanical performance especially with regard to toughness and creep; both grades were supplied as granules. A third grade, 100P, sold as a powder and containing reactive end-groups [29], was developed for use in adhesive formulations for metals and glass where it upgrades the temperature performance of some epoxy based systems, as a lacquer for surface coatings and in the impregnation of glass and carbon fibre [30]. By 1974 the co-polymer 720P (see Table I for structure) had been added to provide a grade with Tg = 250°C, some 25°C greater than that for the homopolymer [30].

Two glass filled grades, first introduced by Fiberite Corpn. [29] were added to the ICI range by 1977 and these showed marked improvement in mechanical properties especially modulus, tensile strength (FIG. 7), and environmental stress cracking, presumably because the increased modulus resulting from filling with glass fibre reduces the strain resulting from a given stress [29]. ICI's 1984 PES trade brochure [31] lists four un-filled grades sold as powder or as granules, a coating/adhesives powder grade equivalent to 100P, and five glass reinforced grades. Further development grades were added in 1985, including one designed for high performance bearing applications and two, both glass filled, for injection moulded circuit boards [27]; no doubt this proliferation of grades will continue. Special formulations ("SUPER SHIELD" [32]) were developed to provide heat-resistant decorative finishes in a range of colours by 1979 and in 1986 ICI Petrochemicals and Plastics Division launched an unoriented PES film under the trade name "STABAR"S [33].

APPLICATION DEVELOPMENT

Detailed discussion of this area is not attempted here as it is beyond the present author's field of competence but it is clear that, after a slow

FIG. 7 : Effect of Glass Fibre on Tensile Modulus and Tensile strength of PES [35].

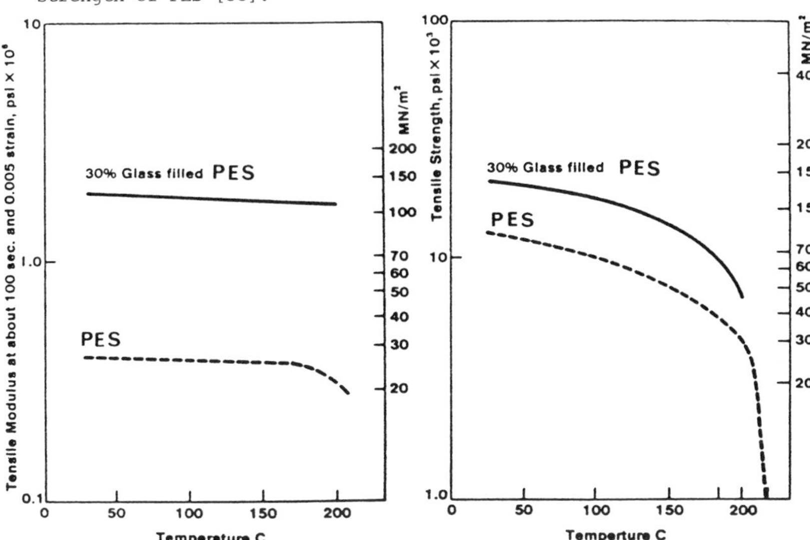

start, a wide range of applications have been developed for PES as injection moulded parts, films, coatings, and composites often in areas served traditionally by metals, ceramics, glass and thermosetting resins. It is also used for making semi-permeable membranes. A full description of PES applications has been given recently by R.B. Rigby [34].

WITH HIND-SIGHT

The late H.C. Raine, FIG. 1, once said that the best recipe for successful exploratory research was to recruit lucky chemists and then ensure that they deserved to be lucky. Looking back over the present history of this project one can see that indeed the project team, and the management, FIG. 8, did enjoy much good fortune. The work was started in 1961, which was at a time of great prosperity for the plastics industry, and this lasted until the critical phases of the project; discovery, product/process selection, and the first part of the process development had been completed. We were fortunate in that our investigation of the polysulphonylation route to PES became concentrated first on polycondensation of 4-phenoxybenzenesulphonyl chloride, reaction (1b), as this gave the tough all para-oriented polymer, I, rather than on the alternative 2-monomer reaction, (1a), which gave a brittle polymer of more complex composition. At first, we failed to realise the value of the polyether syntheses, reactions (2a) and (2b), but because of our early work in this area were able to change quickly to this route when we failed to develop an economic polysulphonylation process to I. Our work on polysulphonylation, which secured ICI's patent priority for I, also led us towards the higher softening co-polymers containing diphenyl groups via the polysulphonylation (7) and then to the co-monomers, e.g. X, required for synthesis of the high softening 720P copolymers by polyether synthesis. Finally, all the teams involved in the discovery of polyethersulphones were fortunate in that these polymers were amorphous, a fact which was not predictable and is as yet not explained, for if the PES or UDEL polymers had been crystalline, as might have been expected from the regularity of their molecular structures, then

PES Research and Development Team

their final melting points would probably have been high enough to push the temperatures required for melt processing above their decomposition temperatures, so that they would not have been useful as thermoplastics. Thus, most of the chances turned up by the researchers were taken so that their good luck was well deserved.

REFERENCES

1. M.E.B. Jones, British Patent 1,016,245 (1962), to ICI; U.S. Patent 4,008,203 (1977).
2. H.A. Vogel, British Patent 1,060,546 (1963), to 3M; J. Polym. Sci., Part A-1, 8, 2035 (1970).
3. A.G. Farnham and R.N. Johnson, British Patent 1,078,234 (1963) to UCC; U.S. Patent 4,108,837 (197).
4. Principles of Polymer Chemistry, P.J. Flory, Cornell University Press, New York (1953).
5. C.W. Bunn in Fibres From Synthetic Polymers, Ed. R. Hill, Elsevier, London (1953).
6. C.W. Bunn, ref. [4] Chapter 12.
7. J. Heÿboer, Br. Polym. J., 1, 3 (1969).
8. Beilstein, 1st Edition, VI, p.301 (1923).
9. B.E. Jennings, M.E.B. Jones and J.B. Rose, J. Polym. Sci., Part C, 16, 715 (1967).
10. H.A. Vogel, British Patent 1,122,192 (1968).
11. A.G. Farnham et. al. J. Polym. Sci., A-1, 5, 2375 (1967).
12. J. Huismann, German Patent 701,954 (1941).
13. J.B. Rose et. al. J. Polym. Sci., Part C, 22, 747 (1969).

14. J.B. Rose et. al., Polymer, 6, 589 (1965).
15. J.F. Burnett, Chem. Rev., 49, 293 (1951).
16. R. Bacon, Proc. Chem. Soc., 113 (1962).
17. P.J. Bain et. al., Proc. Chem. Soc., 186 (1962).
18. J.B. Rose et. al., Polymer, 18, 359 (1977).
19. J.B. Rose, Polymer, 15, 456 (1974).
20. P.B.D. de la Mare and J.H. Ridd, Aromatic Substitution, Butterworths, London, p.161 (1959).
21. J.B. Rose et. al. Polymer, 9, 265 (1968).
22. J.B. Rose et. al. Polymer, 18, 369 (1977).
23. R. Bucourt, J. Mathiew and R. Joly, Rec. Trav. Chem., 78, 527 (1950).
24. D.A. Barr and J.B. Rose, British Patent 1,153,035 (1965).
25. J. Miller, Aromatic Nucleophilic Substitution, Elsevier, London, 1968.
26. A.B. Newton and J.B. Rose, Polymer, 13, 465 (1972); T.E. Attwood, A.B. Newton and J.B. Rose, Br. Polym. J., 4, 391 (1972).
27. M.K. Thompson, Circuit World, 9, [3], 41 (1983).
28. J.B. Rose, Plastics and Rubber Int., 9, [6], 11 (1984).
29. R.B. Rigby et. al., Proc. 31st Ann. Tech. Cont. Reinforced Plastics (Composites Inst., Soc. Plastics Industry, 31, [19-B], 7 (1976).
30. J.B. Rose et. al., ACS Symposium Series No. 4, New Industrial Polymers, ACS, Washington DC, p.63 (1974); Chemtech. 426 (July 1975).
31. Victrex PES, ICI Brochure VS1/0684 (1984).
32. ICI Technical Data Sheets F27/8/1 (1979) and F27/8/2 (1982).
33. ICI Technical Data Sheet SRTD 10.
34. R.B. Rigby in "Engineering Thermoplastics; Properties and Applications" Ed J. Margolis, Marcel Dekker, New York (1985).
35. ICI Data.

DISCOVERY AND DEVELOPMENT OF THE "VICTREX" POLYARYLETHERKETONE PEEK

JOHN B. ROSE
Formerly ICI Plastics Division, Welwyn Garden City, now Chemistry Department, University of Surrey, Guildford, Surrey GU2 5XH, U.K.

INTRODUCTION

Polyaryletherketones were reported independently by ICI and du Pont [1,2] in the patent literature as novel materials at about the same time as the corresponding polyethersulphones in the early 1960s. It is likely therefore that these discoveries arose from scientific and technological considerations similar to those detailed in the previous chapter for the polyethersulphones except that the polyketone inventors were probably looking for fibre forming polymers as Goodman et. al. [1] were working in the Fibres Division of ICI and Bonner [2] reported drawing fibres from a melt of one of his polymers. Union Carbide was also interested in polyetherketones at this time and exemplified a polyaryletherketone-sulphone and a polyaryletherketone containing a bis-phenol. A residue in the repeat unit in the British Patent [3] describing their polyether synthesis and concerned mainly with polyarylethersulphones.

The main synthetic routes to polyaryletherketones, reactions (1) and (2), are analogous to those used for the corresponding polysulphones, polyaroylations, (1), replacing the polysulphonylation processes described in the previous chapter (see p.169), while the halogen atoms displaced in the polyether syntheses, (2), are activated by carbonyl rather than by sulphonyl groups:

(1a) H-Ar-H + ClCO-Ar'-COCl → -Ar-CO-Ar'-CO- + 2HCl

(1b) H-Ar-COCl → -Ar-CO- + HCl

(2a) Hal-Ar''-Hal + MO-Ar'''-OM → -Ar''-O-Ar'''-O + 2MHal

(2b) Hal-Ar''-OM -Ar''-O- + MHal

There is, however, an important practical difference between the reaction conditions required for successful operation of the routes to polyetherketones as compared with those for the polysulphones and this arises from a crucial difference between the two classes of polymers. Polyethersulphones are amorphous or only slightly crystalline and dissolve readily in polar organic solvents such as nitrobenzene or dimethyl sulphoxide at room temperature, but many polyetherketones develop considerable crystallinity and dissolve only at temperatures close to their melting points in this type of solvent. The insolubility of these polymers presented a major synthetic problem as it limits the molecular weights that could be obtained before the growing chains crystallised out from the polycondensation system and this is the main reason why commercial development of the polyetherketones lagged behind that of the polyethersulphones. Solutions to this problem for the polyaroylation reactions, (1), were found first by du Pont and then by Raychem, while for the polyether syntheses the problem was solved by ICI. Raychem manufactured a polymer of structure I, named Stilan, between 1972 and 1976, and manufacture of the ICI polymer, II, trade name "Victrex PEEK", started in 1982.

Copyright 1986 by Elsevier Science Publishing Co., Inc.
High Performance Polymers: Their Origin and Development
R.B. Seymour and G.S. Kirshenbaum, Editors

$-C_6H_4-CO-C_6H_4-O-$ I

$-C_6H_4-CO-C_6H_4-O-C_6H_4-O-$ II

DISCOVERY OF THE POLYARYLETHERKETONES

The first preparation of completely aromatic polyetherketones was reported by Bonner of du Pont [1], who obtained polymers of low molecular weight (Inherent. Viscosity, IV, 0.13-0.18) by the polyaroylation reactions (3) using both isophthaloyl and terephthaloyl chlorides as monomers with

(3) $C_6H_5-O-C_6H_5$ + $ClCO-C_6H_4-COCl$ ⟶ $-C_6H_4-O-C_6H_4-CO-C_6H_4-CO-$ + 2HCl

nitrobenzene as solvent and aluminium chloride as the catalyst. In ICI, Goodman et al. [2] performed similar experiments and were the first to report a polymer of structure I which they made by reaction (4) using aluminium chloride as catalyst and methylene chloride as solvent. This polymer appeared not to melt at 350°C and had a Reduced Viscosity, RV, of 0.57 (equivalent to IV = 0.50). A polymer of the same structure (IV = 0.5) was prepared via reaction (5) by Iwakura [4] who noted that the polymer was insoluble in all the usual organic solvents and used polyphosphoric acid as the solvent/catalyst for his reactions.

(4) $C_6H_5-O-C_6H_4-COCl$ ⟶

(5) $C_6H_5-O-C_6H_4-CO_2H$ ⟶

$-C_6H_4-O-C_6H_4-CO-$ I

A much more effective acidic system was devised by Marks [5], working for du Pont, who found that liquid hydrogen fluoride was a good solvent for polyaryletherketones (these solutions are bright yellow indicating dissolution due to protonation of the carbonyl groups by the acid) and that in this solvent molar quantities of BF_3 catalysed polycondensations such as (3) or (4) at room temperature. Use of the HF/BF_3 system solved the problem of obtaining polymers of high molecular weight by polyaroylation, and a polymer of structure I with IV = 1.33 was prepared, using reaction (4), which could be moulded at 400°C to a strong film with an ultimate elongation of 7%. It appears that du Pont did not take this work further but the process has seen considerable development by Dahl and his colleagues at Raychem [6] who pointed out the importance of molecular weight, showing that for polymers such as I an IV $\geqslant$ 0.8 (equivalent to RV $\geqslant$ 1.02) is required to develop a useful degree of toughness in partially crystalline specimens. This work led to the manufacture of Stilan, structure I, by Raychem for "in-house" use as extruded coating for high temperature performance electrical wiring.

DEVELOPMENT OF THE POLYETHER SYNTHESIS TO GIVE PEEK

The polyether syntheses (2) provide an alternative route to polyaryletherketones, but the problem of finding a solvent for the polymers is more acute here as protic solvents cannot be used. Thus,

Farnham et al., working at Union Carbide, failed to obtain high molecular weight samples of I from reaction (6) due to premature crystallisation of the polymer from their preferred solvent, Sulpholane [7], and using the corresponding single monomer reaction, (7), in the same solvent [8] obtained only polymer with RV = 0.27.

(6) F–C₆H₄–CO–C₆H₄–F + KO–C₆H₄–CO–C₆H₄–OK → –C₆H₄–CO–C₆H₄–O–

(7) F–C₆H₄–CO–C₆H₄–OK → –C₆H₄–CO–C₆H₄–O–

In ICI the preliminary work at Fibres Division aimed at preparing polyetherketones such as I by polyaroylation was not taken beyond the preliminary stage and interest in these polymers transferred to the Plastics Division where they were regarded as crystalline analogues of the polyethersulphones. After the decision in 1965 to concentrate on polyether synthesis as the route to PES (see last chapter) an exploratory programme was started at Welwyn to examine application of this developing technology to the synthesis of other aromatic polymers and the preparation of high molecular weight polyetherketones became an important objective of this programme. The incentive here was to add a crystalline polymer to the range of amorphous aromatic polymers being developed under the PES programme.

A major step forward came in 1972 when it was shown [9] that crystalline copolyetherketonesulphones of high molecular weight could be obtained by conducting polycondensations such as (8) in phenyl sulphone as solvent at temperatures close to the polymers melting points.

(8)
0.5 KO–C₆H₄–CO–C₆H₄–OK
0.3 Cl–C₆H₄–CO–C₆H₄–Cl
0.2 Cl–C₆H₄–SO_2–C₆H₄–Cl
→
$(–C_6H_4–CO–C_6H_4–O–)_{0.8}$ I
$(–C_6H_4–SO_2–C_6H_4–O–)_{0.2}$ III

Phenyl sulphone and certain of its homologues have high thermal stability and do not appear to react significantly with the phenoxide functional groups during polycondensations at the high temperatures required to hold the polymers in solution. The importance of using aryl sulphones in place of the more usual dipolar aprotic solvents is indicated by the data in Table I for reaction (8). With all the solvents other than the diaryl sulphones the reaction gave polymers which were either insoluble in sulphuric acid or have RVs < 0.4 in this solvent and the solutions were often deeply coloured. Most of the solvents used were unstable under the polycondensation conditions giving the decomposition products shown in the table. With the diaryl sulphones decomposition products were not evolved and the polymers obtained were of high molecular weight (RV ≥ 1.3) giving compression moulded specimens which were tough.

TABLE I. Synthesis of 80-20 copolyetherketone-sulphone (reaction 8) in various solvents.

Solvent	Polymn. temperatures(°C)	Polymer RV*	Volatile products
Hexamethyl phosphoramide	238	insoluble	NMe_3
Dimethyl sulphone	250	0.35+	MeSH, Me_2S
Sulpholane	280	0.28+	SO_2, C_4 olefine
N,N-dimethyl acetamide	290	insoluble	CH_3COCH_3
Ditolyl sulphone	290	0.18+	SO_2, PhMe
Methyl phenyl sulphone	290	0.17+	SO_2, PhH
Benzophenone	290	0.14	-
Diphenyl sulphone	290	2.57	-
4-Phenylsulphonyl biphenyl	290	1.35	-
Dibenzothiophene dioxide	290	1.49	-

* For 1% solution in H_2SO_4 which had been filtered if necessary to remove solids

\+ Solution deeply coloured

A series of copolymers comprised of repeat units I and III were prepared by altering the proportions of _bis_-chlorophenyl ketone and _bis_-chlorophenyl sulphone in reaction (8) and the crystallisation characteristics of the copolymers obtained were determined (Table II) [10]. It was found that to obtain adequate crystallinity in melt fabricated samples the copolymers should not contain more than 10 mol% of PES repeat units, III. Unfortunately, reducing the proportion of _bis_-chlorophenyl sulphone employed reduced the average halogen reactivity in the system and increased the temperature required to keep the more crystalline copolymers in solution. The effect of both these factors was to make the preparation of linear, high molecular weight polymers more

TABLE II. T_m and crystallizability for copolyetherketone-sulphones comprising repeat units I and III

Mol % III	T_m	Crystallizability from the melt*
0	367°	Well crystalline
5	358	Well crystalline
10	348	Fairly well crystalline
15	333	-
20	318	Poorly crystalline
25	308	Amorphous+
25	302	Amorphous+
40	-	Amorphous+
50	-	Amorphous
100	-	Amorphous

* Cooling at 63°C min^{-1}; degree of crystallinity as indicated by the height of the d.s.c. crystallization exotherm

\+ Crystalline, as indicated by X-ray diffraction, as made

difficult, and although a copolymer containing 10 mol% PES repeats was established as a compromise in the laboratory, attempts to scale up this work failed.

Although the work with bis-4-chlorophenyl ketone was not successful, the corresponding difluoride is considerably more reactive and when reaction (6) was carried out at 335°C using diphenyl sulphone as solvent high molecular weight samples of the polyetherketone, I, were rapidly obtained. The molecular weight of these polymers could be controlled by employing the difluoride in small excess and in this way it was shown that tough moulded films were obtained from polymers with solution viscosities greater than or equal to the lower level for toughness given in Dahl's patent [6] for polymers made by polyaroylation in liquid HF [10]. Thus, polyetherification was shown capable of producing polymers of structure I comparable in properties to those made by the best polyaroylation procedure.

Further work showed that this improved ether synthesis was of wide application and a series of different polyaryletherketones was prepared as indicated in Table III. During this work it was found that difficulties

TABLE III. Polyaryletherketones from bis-4-fluorophenyl ketone (FPK) or 1,4-bis(4-fluorobenzoyl)benzene (DBB) and bis-phenols [10].

FAK = F-Ø-CO-Ø-F; DBB = F-Ø-CO-Ø-CO-Ø-F: Ø = -⌬-

Dihalide	Bisphenol	Polymer repeat unit	T_m(°C)	T_g(°C)
FPK	HO-Ø-OH	-O-Ø-O-Ø-CO-Ø-	335*	144
FPK	HO-Ø-OH	-O-Ø-O-Ø-CO-Ø-		
	+	+	345	154
	HO-O-CO-O-OH	--O-Ø-CO-Ø--		
FPK	HO-Ø-CO-Ø-OH	-O-Ø-CO-Ø-	367+	154
DBB	HO-Ø-OH	-O-Ø-O-Ø-CO-Ø-CO-Ø-	358	154
FPK	HO-Ø-CO-Ø-CO-Ø-OH	-O-Ø-CO-Ø-CO-Ø-O-Ø-CO-Ø-	383	-
DBB	HO-Ø-CO-Ø-CO-Ø-OH	-O-Ø-CO-Ø-CO-Ø-	384	-
FPK	HO-Ø-Ø-OH	-O-Ø-O-Ø-Ø-CO-Ø-	416	167
FPK	HO-Ø-OH	-O-Ø-O-Ø-CO-Ø-		
	+	+	341	160
	HO-Ø-Ø-OH	-O-Ø-Ø-O-Ø-CO-Ø-		

* Farnham et al. report T_m = 350°C   + Dahl reports T_m = 365°C

associated with prior preparation of the di-potassium derivative of the <u>bis</u>-phenol could be avoided by employing a slight excess of potassium carbonate as the base and adding this to the mixture of <u>bis</u>-phenol and difluoride in phenyl sulphone. This technique increased the effectiveness of the polyether synthesis considerably and was used in 1977-8 to obtain high molecular weight samples of the polymer II for the first time by reaction (9) [11].

(9) F-⌬-CO-⌬-F + HO-⌬-OH + K_2CO_3 ⟶ -O-⌬-O-⌬-CO-⌬- (II) + 2KF + CO_2 + H_2O

Polymer II, now known as PEEK, was selected by ICI for commercial development in preference to the alternative structure, I, (Raychem's Stilan otherwise known as PEK) mainly because the required bis-phenol, hydroquinone, was available commercially as a photographic chemical but there was no commercial source of the bis-4-hydroxyphenyl ketone required to make I. There was little difference in properties between the two polymers and the lower melting point of PEEK, 343°C, was still well above the expected upper continuous use temperature of both polymers which was set by their oxidative stability.

PROCESS DEVELOPMENT

A decision to develop PEEK so as to extend ICI's "VICTREX" range of aromatic polymers with a crystalline material was taken in 1979 and work started to scale up the laboratory process using semi-technical equipment previously assembled for the unsuccessful development of the ketone-sulphone co-polymer. The process so developed was established in production at ICI's Hillhouse factory by 1982 so that the new material could be offered for sale in Europe and the USA. Since then development has occurred rapidly, construction of a 1-200 tpa plant at Atlas Point in the USA was announced in 1983 and in 1985 ICI decided to build a new PEEK plant at Hillhouse.

PRODUCT DEVELOPMENT

PEEK was envisaged as the crystalline analogue of PES, the crystallinity and high melting point conferring additional performance at high temperatures and greater resistance to solvents. These expectations have been confirmed and an Underwriters Laboratories temperature rating close to 250° C is now expected while detailed testing shows the material to have outstanding resistance to aqueous acids, alkalis and organic solvents. Other key properties include good toughness and abrasion resistance at high temperatures, low flamability together with low emission of smoke and toxic gases in fires and excellent resistance to nuclear radiation and to water at temperatures up to 260°C.

The first application area for the polymer was as extruded insulation for high performance wires and cables, but there is now a wide spread of applications including injection moulded parts, chemically resistant surface coatings, monofiliament for industrial belts and filters and as the matrix in carbon fibre composites for aerospace components. A range of grades to match these applications is now available and there is increasing emphasis on filled material three of the eight grades listed in ICI's product brochure for 1984 being,filled with glass or carbon fibre; special formulations have also been developed for lining bearings (for further details see ref [12]. Three new products containing PEEK have been announced recently by ICI, an extruded film called "Stabar K" (1984), a yarn called Zyex (1985) and a continuous carbon fibre pre-preg called APC (1983) for fabrication into composite parts for aerospace applications [13].

CONCLUSIONS

The research programme which led to the discovery of PEEK was started to extend the scope of the polyether synthesis adopted in 1965 for development into a production process for PES. This work was continued in spite of early technical disappointments and its eventual success illustrates the importance of persisting with long term exploratory research alongside a major development programme to extend the range of products from that programme. Inevitably, there was some interchange of staff between the two programmes so that some people appearing in the composite PEEK management and project team photograph, shown below, also appear in that for the PES project.

PEEK Research and Development Team

REFERENCES

1. I. Goodman, J.E. McIntyre and W. Russell, British Patent 971,227 (1964).
2. W.H. Bonner, U.S. Patent 3,065,205 (1962).
3. A.G. Farnham and R.N. Johnson, British Patent 1,078,234 (1963).
4. Y. Iwakura et. al. J. Polym. Sci., A1, 6, 3345 (1968).
5. B.M. Marks, U.S. Patent 3,442,857 (1969).
6. K.J. Dahl, British Patent 1,387,303 (1975).
7. A.G. Farnham et. al., J. Polym. Sci., A1, 5, 2375 (1967).
8. R.A. Clendinning, British Patent 1,177,183 (1976).
9. J.B. Rose, British Patents 1,414,421 and 1,414,422 (1975).
10. J.B. Rose, et. al. Polymer, 22, 1096 (1981).
11. J.B. Rose and P.A. Staniland, U.S. Patent 4,320,224 (1982).
12. C.P. Smith, Swiss Plastics, 3, 37 (1981).
13. G.R. Belbin, Proc. Instn. Mech. Engs., 198C, 1 (1984).

DISCOVERY AND DEVELOPMENT OF POLYETHERIMIDES

General Electric Company
Plastics Business Group
Joseph G. Wirth
Vice President & General Manager
Plastics Technology Department

Good morning. It is a pleasure for me to be here and participate in this symposium. I would like to express my thanks to the organizers for the opportunity to talk about an area of polymer chemistry and technology in which I have been involved for a period of over fifteen years . . . . initially as a synthetic chemist in GE's Corporate Research and Development Center in Schenectady, New York, and more recently in a managerial role in GE's Plastics Business in Pittsfield, Massachusetts, where this work attained commercial status in 1982.

During that year ULTEM® polyetherimides were added to General Electric Plastics' family of polymers which, at the time, included polycarbonates, modified polyphenylene oxides and polybutylene terephthalate. This addition was noteworthy because it occurred at a time when bringing totally new polymers from the research laboratory to the marketplace had become a rare event.

From GE's Research and Development Center in Schenectady, New York, where basic development was done, the work was continued on both a laboratory and a pilot plant scale at GE Plastics facilities in Pittsfield, Massachusetts, and the new material was ultimately produced on a commercial scale in a new manufacturing plant located at our site in Mt. Vernon, Indiana.

Polyetherimides were developed as a result of research interests in aromatic nucleophilic displacement chemistry combined with a perceived marketplace need for high performance polymers which could be readily fabricated by standard plastics extrusion and injection molding processes.

Published 1986 by Elsevier Science Publishing Co., Inc.
High Performance Polymers: Their Origin and Development
R.B. Seymour and G.S. Kirshenbaum, Editors

A note by Gorvin in Chemistry and Industry in 1967 describing methoxide displacement of aromatic nitro groups in 2- and 4-nitrobenzophenones provided the initial impetus for this work. The paper described reactions in which nitro in high yields under mild conditions as, for example, in groups were displaced 4,4'-dinitrobenzophenone. Under the same conditions 4,4'-dichlorobenzophenone showed only minimal replacement of halogen.

In our laboratories we reacted 4,4'- dinitrobenzophenone with sodium phenoxide in DMSO and obtained essentially quantitative yields of 4,4'- diphenoxybenzophenone. We quickly followed this experiment by reaction of 4,4' - dinitrobenzophenone with bisphenol a disodium salt in DMSO, and obtained a polyetherketone of molecular weights typical for engineering polymers. This same result was reported by Radlman in 1969.

At this point it was clear that we had a polymer-forming reaction and that the essential feature was a nitro group activated through ortho or para substitution to a strong electron withdrawing group in an aromatic system. We decided to explore the scope of the reaction to see where it might lead us.

Reaction of excess sodium phenoxide with 2, 4-dinitrobenzoic acid gave no evidence of nitro displacement, obviously due to carboxylate ion information. However, aliphatic esters of 2,4- dinitrobenzoic acid did undergo nitro displacement by phenoxide but not in the yields required to produce high molecular weight polymers. Aliphatic esters of 3- and 4- nitrophthalic acid reacted readily with phenoxide to give the respective phenoxy derivatives. Nitrophthalonitriles react even more rapidly to give phenoxyphthalonitriles or with bisphenoxides the corresponding tetranitriles. Either 2, 4-dinitrobenzonitrile or 2-, 6- dinitrobenzonitrile reacts with bisphenoxides to give essentially quantitative yields of the expected polymeric products.

However, the benzonitrile arylether polymers had no outstanding properties which would render them commercially interesting.

Subsequently, we realized that N-substituted nitrophthalimides should undergo the nitro displacement reaction.

We were able to demonstrate that phenoxides react easily with both 3- and 4-nitro-N-phenylphthalimides, producing the corresponding phenyletherimides... in excellent yields. As synthetic organic chemists, we were both surprised . . . and pleased . . . with the excellent yields and ease of these reactions.

The potential synthetic utility of this reaction for preparing novel polymers was obvious at this stage of our work, but the driving force at the onset of our interest was the novel organic chemistry of these displacement reactions. What followed was:

- An exploration of the ease of halogen - versus nitro-displacement
- An exploration of the relative reactivity of different phenolate anions and
- An exploration of the effectiveness of the phthalimide ring at activating nitro- or halogen-displacement

In our hands, the order of leaving group reactivity in these systems was observed as nitro > fluoro > chloro, with 3-substituted phthalimide isomers being more reactive toward displacement than 4-substituted isomers. Looking back through the literature, when one can find appropriate data, the usual order is F > NO_2 > Cl - reversing nitro and fluoro. We also found that electron-withdrawing groups on the imide nitrogen additionally enhanced the rate of nitro-displacement. Compare, for example, p-chlorophenyl versus phenyl.

We found large numbers of phenoxide derivatives that reacted cleanly - often quantitatively, in a matter of minutes - to give the corresponding ether imides. Note yields of 94-98 percent. One would have expected electron-releasing groups to increase reaction rates;... and that is what was observed. And it is not at all surprising to find that thiophenoxide derivatives are much better nucleophiles than phenoxides themselves - the commonly observed increase in nucleophilic character as one moves down a column of elements in the periodic table... in this case by a factor of more than 100 in reaction rate.

Having synthesized a number of etherimides, we found they were subject to ether exchange reactions with other phenoxides, yielding equilibrium mixtures. The position of the equilibrium was related to the nucleophilicities of the phenoxides involved. Note the enhanced reactivity of a second row element .. S- versus -O ... the equilibrium lies almost completely toward the thioether.

POLYMERIZATION

At this point in our work we successfully demonstrated the potential synthetic and economic accessibility of an entire family of nitro-substituted phthalic derivatives; and as just discussed their propensity for undergoing facile displacement of the nitro group by aryloxyanions.

We were now in a position to address the problem of polymer synthesis. The generalized polyetherimide structure, suggests two possible approaches, each incorporating a nitro-displacement reaction, but at different stages in the overall synthetic procedure. One could imagine a difunctional imide monomer forming an ether linkage during polymerization with a difunctional aryloxy anion; or one could incorporate the ether linkage into a phthalic monomer and react it with an appropriate diamine, forming an imide linkage during the polymerization. We examined both approaches.

To form the ether linkage by nitro displacement polymerization we first prepared bis-nitro-bis-imide monomers from the appropriate 3- and 4-nitrophthalic anhydrides and diamines in refluxing acetic acid. Here are just a few of the nitro-phthalimides.

The bis-phenoxide salts were prepared in anhydrous form by azeotropic drying, starting with the bis-phenol and an alkali hydroxide. They were polymerized by displacement reaction with the bis-imide in dipolar aprotic solvents, or mixtures of these with other non-interfering solvents such as toluene or chlorobenzene. The nitro-displacement reactions are very rapid, close to quantitative - 95 to 100 percent yields -, and produce high molecular weight polymers.

As you can see, it is an extremely versatile reaction. Note especially where the Tg values generally fall - mostly above 200°C.

End-groups are always a concern in polymerization reactions and products. It is worth noting that polyetherimides formed by nitro-displacement polymerization will contain phenolic and aromatic nitro end-groups ... which unfortunately will adversely affect thermo-oxidative stability of the polymer.

And there is yet another critical constraint on this synthetic approach... essentially anhydrous conditions. Any water at all rapidly hydrolyzes nitro-imides to nitro-amic acid salts ... which will not undergo the necessary nitro-displacement reaction, limiting chain growth and hampering one's ability to reproduce a given molecular composition on demand.

FORMING THE IMIDE LINKAGE

In an alternate approach to polyetherimide synthesis, bis-ether dianhydride monomers were prepared in a series of steps in which we again made use of the nitro-displacement reaction.

Unfortunately, the nitro-phthalic anhydrides cannot be used directly for the displacement reactions since aryloxyanions preferentially attack at the carbonyl center, leading to formation of ester acid salts via ring-opening. For example, reacting the 4-isomer with sodium phenoxide at room temperature produced ring-opening mixtures of ester acids. This not only resulted in consumption of aryloxide anions but simultaneously generated materials which were not susceptible to nitro-displacement.

Our solution to this problem was to use a "protected" nitro-phthalic anhydride .. a nitro-N-alkyl phthalimide. As we have shown, when treated with bis-phenoxide salts, the nitro group is smoothly and selectively displaced, producing bis-ether-bis-imides. These intermediates can be highly purified to remove impurities which the alternate method of polymerization built into the polymer as unstable end groups.

Purified bis-imides can then be converted to bis-ether dianhydrides by hydrolysis to the tetra acids, followed by ring-closure to the dianhydrides.

Treatment of the dianhydrides with various diamines finally gives polyetherimides in very high yields. Again, we could list a very large number of polyetherimides prepared via imidization polymerization. Suffice it to say the most important one at the moment is that derived from bis-phenol-A, 4-nitro phthalimide, and meta-phenylenediamine. It's glass transition temperature is 218°C.

One general rule that emerges is this. For a given bis-phenol/diamine pair, Tg values for polymers containing the 3-substituted phthalate ring are always greater than those containing the 4-isomer. The data clearly suggest that by judicious choice of bis-phenol, diamine, and phthalate ring isomer, polyetherimides can be tailored to one's own choice of Tg requirements.

Furthermore, with this polymerization scheme, end groups can be readily converted to thermo-oxidatively stable imides. For example:

- anhydrides, by treatment with aniline, yielding N-phenyletherimide end groups
- amines, by treatment with phthalic anhydride yield N-aryl phthalimide end groups

The presence of small amounts of moisture is inconsequential in imidization polymerization since it is simply removed along with that formed as a by-product.

As you should expect, there are commercial refinements and improvements for a number of the steps in the imidization polymerization route. These are the kinds of changes which make large scale synthesis feasibile and commercially viable.

ULTEM PROPERTIES

Of all the polyetherimides, the prime candidate from the standpoint of physical properties, environmental resistance, processability - especially injection molding - and the commercial attractiveness of the process for manufacturing ... was the product prepared from bis-phenol A, 4-nitrophthalimide and m-phenylenediamine. It is called ULTEM® 1000.

The repeating aromatic imide units in the molecule are connected by aromatic ether units, yielding a polymer with flexible linkages for good heat flow characteristics, while retaining sufficient stiffness for good engineering properties. In addition, designing the polymer backbone with highly aromatic character combined with ether and imide connecting units lends itself to good thermal stability, and flame retardancy. Let's examine the properties in a little more detail to understand why the material was commercialized.

MECHANICAL PROPERTIES

First some characteristic mechanical properties. The polymer is exceptionally strong, with tensile strength of 15,000 psi and flexural strength of 21,000 psi. The flexural modulus of the material, at 480,000 psi, is outstanding for an unfilled amorphous thermoplastic resin.

THERMAL PROPERTIES

Looking at the thermal properties, the polymer has a heat distortion temperature of 217°C at 264 psi. The small difference between softening point and heat distortion temperature suggests the polymer has very good property retention at elevated temperatures. This is confirmed by measurements of the flexural modulus of the material versus temperature, which shows that even at 180°C, the flexural modulus is over 300,000 psi. Add to that the well-known thermal-oxidative stability of polyimides (in general) and it should not be at all surprising that the continuous use temperature is 170°C or higher.

ELECTRICAL PROPERTIES

The electrical properties of the polymer are also exceptional - in many respects comparable to those of typical high performance engineering materials at room temperature, but in addition, changing little over a wide range of elevated temperatures. This is an important property, and of considerable interest when combined with high temperature capability. All of which has led to rapid acceptance of this polyetherimide in both electrical and electronic applications.

PROCESSABILITY

For a long time the mechanical, thermal and electrical properties of polyimides excited technical people in end-use communities. However, these polymers and their properties were of limited utility - they lacked melt processability - the capability of being conveniently converted into useful parts or shapes.

Back in the mid-70's when the polyetherimide chemistry and physical properties were under study, considerable skepticism and prejudice existed. Previous failures to melt process high temperature polyimides were at the roots of our concerns.

The prejudice was just that. The developing body of thermal and rheological data suggested that polyetherimides could not only be melt processed - but that they would process extremely well by injection molding and extrusion.

The first and key requirement for successful melt processing is melt stability at processing temperatures. The melt stability of polyetherimide was demonstrated utilizing thermogravemetric studies to 600°C. Data clearly showed no weight loss up to and beyond 400°C, well above the anticipated processing range. From a chemical point of view, these tests confirmed the stability of the ether linkage, produced by the new nitro-displacement chemistry. Stability of the imide linkage was reconfirmed.

The second critical requirement for successful melt processing is resistance to mechanical shear. This was demonstrated for polyetherimides by measuring stability under mechanical deformation in a torque rheometer at high temperatures. There was no change in torque over a 90 minute period indicating that there was no degradation .. lowering of torque, polymer unzipping or increase in torque, no crosslinking. The data clearly demonstrated the high melt stability of the ether-imide structure under either mechanical or thermal melt processing stresses.

A third, and equally critical requirement, is rheological properties ... viscosity vs. shear. The first, range finding measurements, established the viscosity and shear stress behavior in the temperature range required for melt processing. These data clearly showed non-Newtonian behavior ... for example considerable lowering of viscosity at high shear rates. This same phenomenon could also be demonstrated by measurement of viscosity vs. temperature at constant shear rates. Calculations using Arrhenius plot data yielded an activation energy for flow of 20.7 kcal/mol for polyetherimides. This value is in line with other engineering plastics.

To put all of this rheological data into perspective, comparisons were made with other melt processable materials, such as high impact polystyrene, polycarbonates, and modified polyphenylenethers. The comparisons were made at temperatures required for processing of each polymer.

At these temperatures and at very low shear rates of 10^1 to 10^2 reciprocal seconds, typical of extrusion processes, the viscosity of ULTEM® was, surprisingly, found to be lower than that of high impact polystyrene and judged to be suitable for low shear processing. However, the injection molding process requires higher workable viscosities at higher shear rates of 10^3 reciprocal seconds or greater. The data showed that polyetherimide again performed like polymers that could be injection molded.

Assessments were made on a variety of polyetherimides and it was found that the thermal and rheological behavior were quite similar for this entire polymer family.

The experts with negative views on processing were wrong ... the thermal and rheological studies led polyetherimides from the laboratory to the marketplace.

FLAMMABILITY

The inherent flammability characteristics of ULTEM® polyetherimide are also noteworthy - especially because of their significance in a number of important commercial application areas ... aircraft and construction industries.

In a wide variety of flammability tests designed to establish a fire performance profile for materials, the polyetherimides have displayed outstanding behavior ... especially for a polymer containing no additives to enhance flame retardancy.

A brief comparison of three critical thermal and flammability characteristics of polyimides vs. commodity plastics, engineering plastics and wood a common material in the construction industry tells the whole story.

Thermal Stability - There are several mechanisms by which polymers degrade, an important one in fire is thermal decomposition into gaseous fragments that burn. Commodity Plastics generally start to decompose as measured by TGA at 200 to 300°C, Engineering Plastics at 300 to 400°C, and ULTEM® exhibits stability above 400°C - a reluctant fuel source.

Flash Ignition - Is the point at which sufficient gaseous fragments are present to ignite in the presence of a flame. Predictably, the ether-imides have an extremely high value of 521°C when compared not only to the other plastic materials, but also to wood products.

Oxygen Index - Is an excellent measure by which to characterize materials with regard to the propensity to burn or extinguish. The numerical values assigned to a material .. oxygen index ... represents the minimum percentage of oxygen in a nitrogen mixture to sustain burning in the absence of external heat or flame. The O.I. value of polyetherimide is a remarkable 47, excellent when compared to other materials. For example, the value for wood products is 22 and engineering plastics are in general 25-30. However, it must be pointed out that for almost all engineering plastics to demonstrate values of 30, they require additives. The polyetherimide value of 47 is attained without additives. The excellent flammability performance of polyetherimide translates directly to acceptance in markets requiring the best with regard to flame resistance.

In today's markets, the use of thermoplastic products is regulated from a standpoint of fire safety. Minimum performance in dozens of fire tests is spelled out by standards, regulations, and building codes that regulate over 90% of the products specified and used. In tests for transportation, electrical and construction applications, polyetherimides show excellent performance.

Let me conclude by observing that the creative contributions of many individuals, as shown on the slides, are necessary to take a new polymer from a laboratory discovery to the marketplace, and then to make it commercially successful.

A tremendous dollar investment in research and development is required to bring the technology to the manufacturing stage - some 25 to 50 million dollars - and 8 to 10 years to complete. Strong financial support and a great deal of patience are as essential as unique chemistry and an outstanding balance of polymer properties.

I'm pleased to tell you that aromatic polyetherimides have a bright future. Hard evidence in support of that viewpoint is the fact that General Electric Plastics has under construction a $74MM ULTEM® capacity expansion at our site in Mt. Vernon, Indiana. We obviously believe strongly in the materials resulting from nucleophilic displacement of aromatic nitro groups.

Blends and Alloys

DISCOVERY AND COMMERCIALIZATION OF NORYL^{R} RESINS

ALLAN S. HAY, General Electric Company, Research & Development Center,
PO Box 8, Schenectady, NY

ABSTRACT

The oxidation of 2,6-dimethylphenol in solution, at room temperature, with oxygen in the presence of an amine complex of a copper salt as catalyst yields a high molecular weight linear polyphenylene oxide as product.

When the substituents are bulky groups, e.g., t-butyl, the only product is the diphenoquinone. PPO^{TM} resin, the polymer obtained when the substituents are methyl groups, can be molded or cast into tough products however the polymer is subject to oxidative degradation at elevated temperatures.

PPO is completely miscible with polystyrene and blends of the two products have glass temperatures between those of the two polymers. This result led to the commercialization of NORYL resins, one of the class of engineering thermoplastics. The discoveries which led to the commercial product as well as the scope of the oxidative coupling reaction are the subject of this paper.

In 1956 we became interested in a very simple synthesis of azobenzene which involved the catalyst oxidation of aniline that had recently been described by Terent'ev and Mogilyanskii in a Russian journal[1].

$$C_6H_5NH_2 \xrightarrow[\text{CuCl, pyridine}]{O_2} C_6H_5N{=}NC_6H_5$$

ANILINE AZOBENZENE

The reaction took place at room temperature simply by blowing oxygen through a pyridine solution of aniline in the presence of cuprous chloride as catalyst. We first became interested in the possibility of making polymers by oxidizing diamines such as benzidine to polyazo compounds.

$$n\ H_2N{-}C_6H_4{-}C_6H_4{-}NH_2 \xrightarrow[\text{CuCl, pyridine}]{O_2} \left[C_6H_4{-}N{=}N{-}C_6H_4 \right]_n$$

BENZIDINE POLYAZOBIPHENYLENE

The products we obtained were insoluble and infusible solids which we decided not to pursue further. Instead, we turned our attention to the oxidation of phenols. On the assumption that the first step in the oxidation of aniline was

$$C_6H_5NH_2 \longrightarrow C_6H_5NH\cdot$$

Copyright 1986 by Elsevier Science Publishing Co., Inc.
High Performance Polymers: Their Origin and Development
R.B. Seymour and G.S. Kirshenbaum, Editors

it seemed probable that phenol would oxidize readily in this system also and give as the first intermediate, the phenoxy radical.

PHENOL PHENOXY RADICAL

When we carried out the experiment, the product obtained from the oxidation of phenol was a dark brown tar! In retrospect this was not too surprising because the phenol molecule has four active sites at which a reaction can take place and after the initial reaction takes place any possible product you can imagine would oxidize further and easier than the starting material.

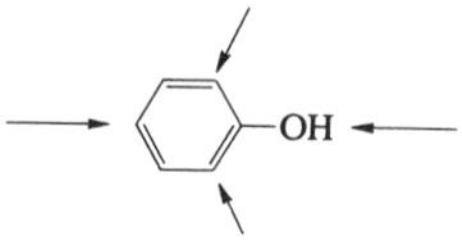

To limit the number of possibilities we began to study the oxidation of phenols with various numbers of substituents blocking the active sites. When we oxidized 2,6-dimethylphenol we were surprised to find that a very exothermic reaction took place at room temperature by simply bubbling oxygen through the reaction mixture and that in a very short time the reaction mixture became too viscous to stir. We soon found out that a highly specific reaction had taken place which had resulted in oxidative removal of two hydrogen atoms to give a linear, high molecular weight polyphenylene oxide as product which was soon christened PPO[R] resin.

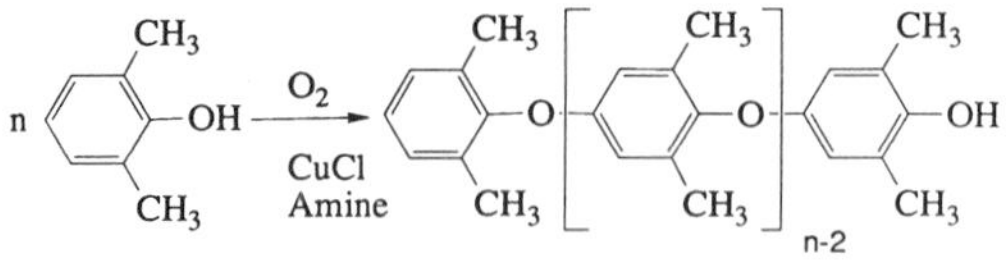

2,6-DIMETHYLPHENOL PPO

We had discovered a new method for synthesizing high molecular weight polymers which we called polymerization by oxidative coupling[2].

We began immediately to make somewhat large quantities of PPO resin in order to learn more about its properties. The polymer has a very high glass temperature (208 deg. C) and when molded or cast it is a tough, ductile material. It was recognized that a polymer of this structure would probably not be useable for extended periods of time at temperatures near its glass temperature because of oxidation of the methyl groups. But the very high glass temperature would be expected to lead to exceptional dimensional stability at lower temperatures and from the structure we could predict that it would have excellent hydrolytic stability. The monomer for this polymer, 2,6-dimethylphenol, was only available in small quantities in an impure form as a coal tar byproduct. Our Chemical department was fully occupied with the commercial development of Lexan polycarbonate which was discovered by Dr. Daniel W. Fox and is being described at this symposium. There was, therefore, little or no interest in pursuing at that time the development of another new material.

We then began a study of this new polymer forming reaction and it was several years before we had a reasonable understanding of the mechanism. PPO resin is only one member of a family of polymers so we also began a study of the oxidation of other 2,6-disubstituted phenols. We found that

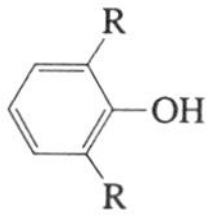

2,6-DISUBSTITUTED PHENOL

polymers, each with different properties, were readily formed from a variety of other monomers (e.g. ethyl, propyl, etc.). However, none of these monomers were readily available either. If the substituents were large, bulky groups (e.g. t-butyl) the molecules were not able to join together to give the polyethers by a carbon-oxygen coupling reaction, but instead they joined together by a carbon-carbon coupling reaction to give a diphenoquinone.

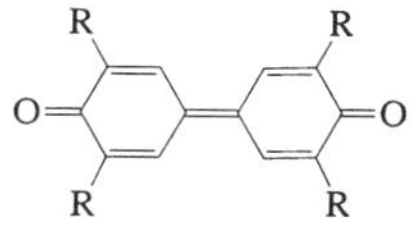

DIPHENOQUINONE

The parent unsubstituted molecule, phenol, and monosubstituted phenols such as o-cresol are readily available and inexpensive commodity materials and the polymers would be expected to have attractive properties. Our initial attempts to polymerize these materials had failed. As our understanding of the mechanism of this reaction increased it became apparent that we were dealing with a reaction that was very similar to enzymatic oxidations. The evidence indicated that in the first step of the reaction the copper catalyst was directly coordinated to the phenol molecule. Therefore it appeared that by appropriate choice of the ligands of the copper catalyst we might physically block out the unwanted reactions at the ortho positions and get increased selectivity. This is shown schematically below.

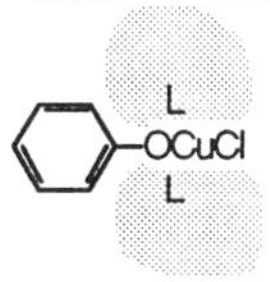

INTERMEDIATE COMPLEX

Using this approach we were eventually able to make high molecular weight polymers from phenol and o-cresol however we were never able to completely exclude reactions at the ortho positions and the resulting polymers were always highly branched materials that did not have useful properties.

We were, however, at a later date able to synthesize the completely aromatic polyphenylene oxide from 2,6-diphenylphenol which we called P_3O. In contrast to PPO which crystallizes only to a small extent (melting point approximately 260 deg. C.) and with great difficulty, P_3O is highly crystalline and as expected very thermally and oxidatively stable and melts at 480 deg. C. With such a high melting point the material could not be used as a molding compound but was of interest as a fiber or film. The polymer reached a semicommercial stage, under license to AKZO in the Netherlands, as a fiber (TENAX^R).

n 2,6-diphenylphenol $\xrightarrow[\text{CuCl, Amine}]{O_2}$ P_3O

2,6-DIPHENYLPHENOL P$_3$O

In 1960 the commercialization of LEXAN polycarbonate was well underway and our Chemical Development Operation then began looking for the next candidate for commercial development. They settled on PPO which by this time was fairly well characterized and had an attractive property profile. Around this time also a fresh Ph.D. chemical engineer from the University of Illinois, Jack Welch, was hired and soon became a key player in the development. A miniplant was soon in operation. A major hurdle to commercialization of PPO was the unavailability of the monomer, therefore an effort was mounted to find an attractive synthesis. Dr. Stephen Hamilton made the key discovery of a "hot tube" reaction which produced the monomer in high yield by simply passing phenol and methanol over a magnesium oxide catalyst at 400 deg. C. With a monomer process in hand a separate business

Phenol $+ 2\ CH_3OH \xrightarrow[\text{heat}]{MgO}$ 2,6-dimethylphenol

PHENOL METHANOL 2,6- DIMETHYLPHENOL

unit, named Polymer Products Operation, was established and commercialization proceeded apace. A commercial plant was announced in 1965.

When PPO was introduced into the marketplace problems arose which were traced to the very high temperatures necessary for processing because of the very high glass temperatures of PPO. At the elevated processing temperatures attained during extrusion and molding, degradation occurred due to oxidation of the polymer by oxygen. The problems were serious enough to put the fledgling business in jeopardy. But fortunately it had been found that PPO, rather remarkably, was completely miscible with polystyrene, a large volume commodity resin. Therefore by blending the two polymers one could obtain a new composition that had a glass temperature in between that of PPO (208 deg.C.) and polystyrene (100 deg. C.). This meant that we had

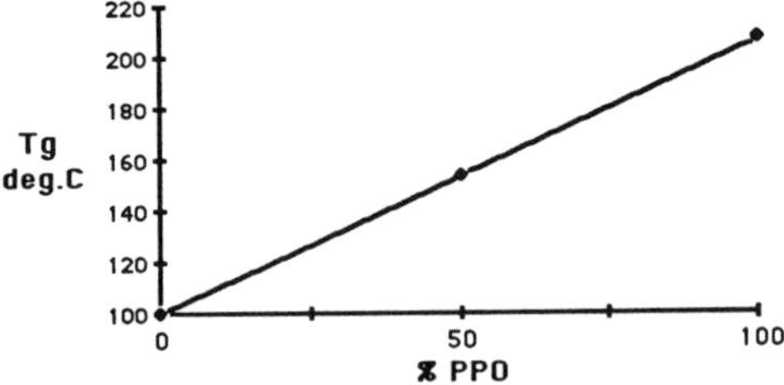

the capability of maintaining many of the useful properties of PPO-excellent hydrolytic stability, dimensional stability, excellent electrical properties-while lowering the heat distortion temperature but still maintaining it high enough for most of the applications being pursued. In addition we obtained dramatically improved processability and substantially lower cost.

At this point in time a business decision was made by Dr. Jack Welch, the introduction of modified PPO resins which were given the name NORYL[R] resins. The resins were projected to fill a market niche between the engineering plastics known at that time and the commodity resins. Today Noryl resin is one of the largest volume engineering thermoplastics.

REFERENCES

1. A. Terent'ev and R.D. Mogilyanski, Doklady, 103, 91 (1955).
2. Allan S. Hay, Polymer Science and Engineering, 16, 1 (1976).

XENOY® AND NORYL® GTX ENGINEERING THERMOPLASTICS BLENDS

DR. JEAN M. HEUSCHEN
General Electric Plastics Europe
PB 117, 4600 AC Bergen op Zoom, The Netherlands

INTRODUCTION

Thanks to their unusual miscibility polymer blends of polyphenylene oxide and polystyrene were the first commercially successful amorphous engineering thermoplastics blends, introduced back in 1968. Pete Juliano of GE Corporate Research Laboratories presented a comprehensive review at the last IUPAC Meeting of The Hague(1) where he showed how the evolution of science and technology of blends (2,3) based on polyphenylene oxide, bisphenol A polycarbonate, polybutylene teraphthalate, polyamides and polyacetals, created many more opportunities for the development of engineering thermoplastics with attractive combination of attributes.

POLYMER BLENDS OF AMORPHOUS AND SEMICRYSTALLINE POLYMERS.

While high glass transition amorphous polymers such as non modified bisphenol A polycarbonate or polyphenylene oxide exhibit excessive gasoline and stress cracking sensitivity, nonmodified semicrystalline thermoplastics, such as polyamides and polybutylene terephthalate lack intrinsic dimensional stability and ductility. Blending them together to overcome those shortcomings seemed evident.

Both systems take advantage of the high glass transition, high ductility of the amorphous polymers and the chemical resistance, high melting point, easy processing of the semi-crystalline polymers. They, however, are fundamentally different in their morphology:

- Xenoy® PC/PBT blends are mainly characterized by the partial miscibility of bisphenol A polycarbonate and polybutylene terephthalate, stabilization and impact modification technology being applied to upgrade properties to the need of the applications.
- Noryl® GTX PPO®/PA blends get their unique properties thanks to the true incompatibility of PPO® and polyamide. In this case morphology, rheology and properties are mainly controlled through compatibilization and impact modification technology.

DEVELOPMENTS OF XENOY® PC/PBT BLENDS.

The history of Xenoy® PC/PBT commercial engineering thermoplastics blends is still rather recent: in the late sixties, early seventies many activities in blending polycarbonate with polyesters took place, leading to over 120 patents and many academic studies, but very few products.

It is only in 1978 that the ductility of elastomer impact modified polycarbonate was associated to car bumpers, but it failed to meet the basic gasoline resistance requirements. This led to the polycarbonate-polyethylene terephthalate, non-miscible amorphous blends developments in 1979. Those blends showed improved resistance to accidental contact with gasoline in the absence of stress.

Copyright 1986 by Elsevier Science Publishing Co., Inc.
High Performance Polymers: Their Origin and Development
R.B. Seymour and G.S. Kirshenbaum, Editors

The gasoline resistance required for unpainted, selfcolored bumper application was, however, only achieved with elastomer modified PC/PBT blends, referred to as Xenoy® CL 100 introduced late 1980. The more stringent low temperature, 5 mph. impact requirements of painted U.S. bumper were met in 1984 with Xenoy® CX1101. The property profiles of Xenoy® modified PC/PBT blends can be defined as a medium heat of 125°C, 20°C below polycarbonate, high ductility and dimensional stability of amorphous polymers, stress cracking and gasoline resistance of semi-crystalline polymers (Table 1). Prototyping and processing studies of the largest injection molded, unpainted engineering thermoplastic automotive parts could be completed.

This resulted in the mass production of the vibration welded box bumper of the European Ford Sierra in 1981, meeting 4 kph. impact standards.

TECHNOLOGY AND SCIENCE OF POLYCARBONATE-POLYESTER BLENDS.

Simultaneously to those developments the technology of polycarbonate-polyester blends has been extensively studied by various laboratories, confirming the early technical developments and further contributing to polymer science in this field:

- polycarbonate-polyester blends miscibility by D.R. Paul(4),
- ester interchange chemistry by J.P. Mercier (5)
- PBT crystallization studies by R.S. Stein (6) and S.Y. Hobbs (7).

The measurement of the dynamic mechanical properties as a function of temperature is one of the most powerful tools to assess polymer blends

TABLE I. Xenoy® Bumper Grades Properties.

			CL 100 (1980)	CX 1101 (1984)
Tensile Strength Yield	ASTM D 638	N/mm^2 (Psi)	51 (7,400)	47 (6,800)
Break		N/mm^2 (Psi)	55 (8,000)	50 (7,250)
Elongation Yield		%	5	5
Break		%	145	145
Tensile Modulus		N/mm^2 (Psi)	2075 (301,000)	1900 (276,000)
Flexural Strength	ASTM D 790	N/mm^2 (Psi)	90 (13,000)	80 (11,600)
Flexural Modulus		N/mm^2 (Psi)	2050 (297,000)	1900 (276,000)
Charpy Notched Impact	DIN 53453	KJ/mm^2 ($ft.lb/in^2$)		
23 °C			38 (18.1)	45 (21.4)
0 °C			26 (12.4)	45 (21.4)
- 20 °C			8 (3.8)	40 (19.0)
- 40 °C			6 (2.9)	20 (9.5)
Izod Notched Impact	ASTM D 256	J/m (ft.lb/in.)		
23 °C			790 (14.2)	750 (14.1)
0 °C			700 (13.1)	700 (13.1)
- 20 °C			250 (4.7)	650 (12.2)
- 40 °C			160 (3.0)	450 (8.4)
HDT 1.82 N/mm^2	ASTM D 648	°C (F)	96 (205)	95 (203)
Vicat VST B/120 (5 kgs)	DIN 53460	°C (F)	127 (261)	124 (255)
Mould Shrinkage	ASTM D 1299	%	0.8	0.8
Specific Gravity	ASTM D 792	grs/cm^3	1.23	1.21
Petrol Resistance	FORD TEST	-	passes	passes

morphology and in particular polycarbonate/polyester miscibility. Fig. 1a shows a typical scan of bisphenol A polycarbonate with a sharp glass transition at 150 °C. Similar trace for PBT (fig 1b) shows the typical behavior of a semi-crystalline polymer with a sharp melting point at 225 °C and broad glass transition associated to the spectrum of relaxation times of the crystal-amorphous polymer interphase.

Blending both resins in various ratios results in a complex morphology exhibiting partial miscibility of polycarbonate and PBT as indicated by the depressed glass transition of polycarbonate, shifting from 150 °C to 136 °C, in an 80/20 PBT/PC blend (fig. 2a) or to 141 °C in a 55/45 PC/PBT blend (fig. 2b). In both cases Differential Scanning Calorimetry studies indicate a crystallization and melting behaviour of PBT hardly affected even under abusive processing conditions with a sharp melting point of 225 °C (fig. 3a).

Such a behaviour is, however, only achieved in the case those blends, as for Xenoy®, are properly stabilized against transesterification. In a series of publications J.P. Mercier (5) reported the chemistry of carbonate-ester interchange taking place at processing temperature (fig. 4). The interchain reaction leads to the change from a physical blend to block copolymer and finally single phase random copolymer as the number of reactions increases within the same chain. This behaviour is illustrated

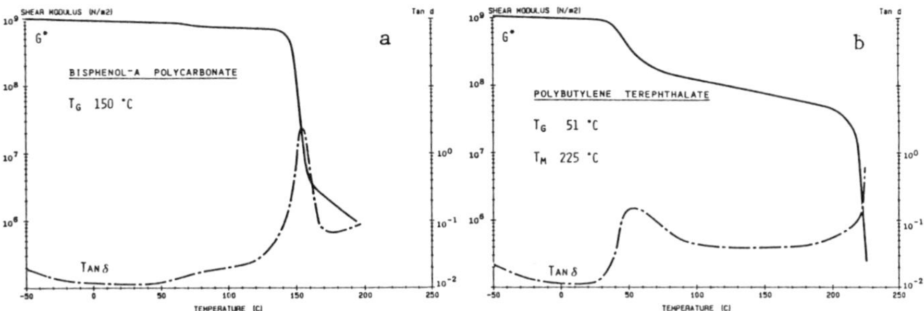

FIG. 1. Dynamic Mechanical Properties of Polycarbonate and Polybutylene Terephthalate.

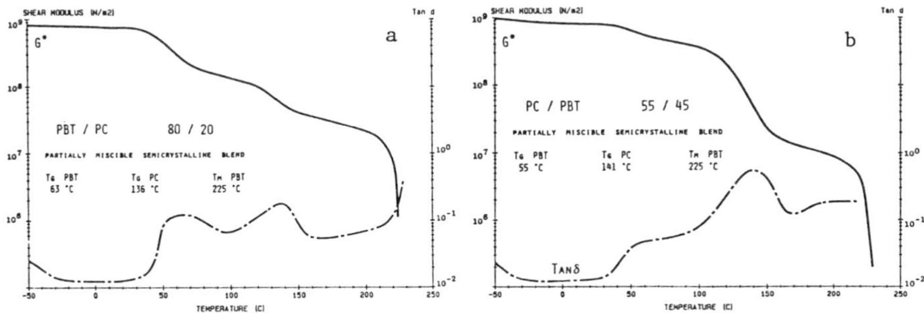

FIG. 2. Dynamic Mechanical Properties of Polycarbonate/Polybutylene Terephthalate Blends.

by DSC where an unstabilized PC/PBT blend held for 6 minutes at 285 °C undergoes a major morphology change (fig.3b). Tentatively isothermally crystallized for 10 minutes at 200 °C this sample remained totally amorphous with a sharp glass transition of fully miscible polycarbonate and polybutylene terephthalate structures at 82 °C. The exotherm of crystallization at 160 °C being equal to the endotherm of melting at 208 °C confirms the presence of PBT segments crystallizing upon heat scanning but with a different kinetics and morphology than virgin PBT.

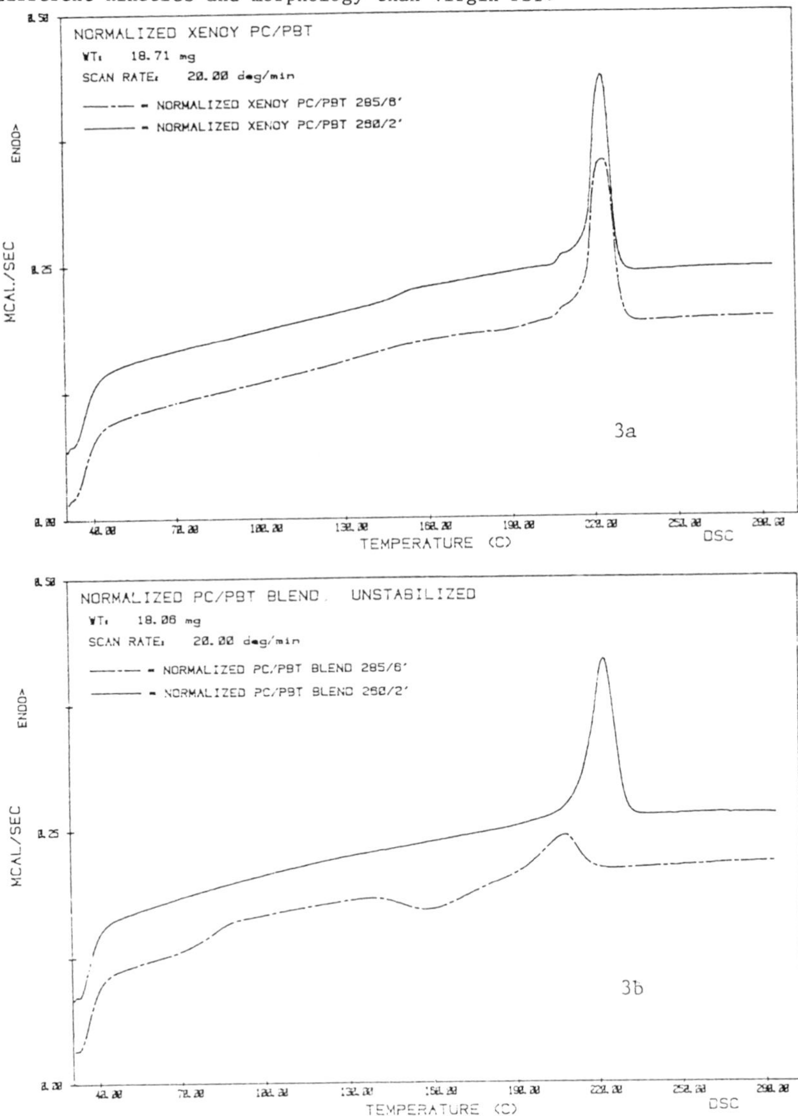

FIG. 3. Differential Scanning Calorimetry of 55/45 PC/PBT Blends.

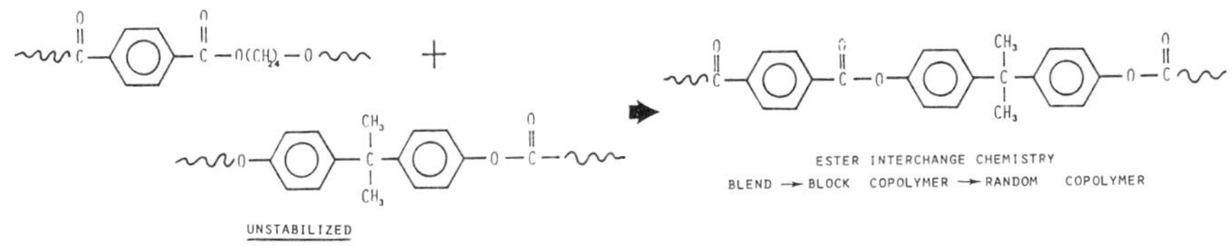

UNSTABILIZED

ESTER INTERCHANGE CHEMISTRY

BLEND → BLOCK COPOLYMER → RANDOM COPOLYMER

FIG. 4. Polybutylene Terephthalate - Polycarbonate Blends Ester Interchange.

POLYCARBONATE/POLYESTER BLENDS PRODUCT LINES.

The PC/Polyester product line and its technology are, however, not limited to those PC/PBT blends, but also includes impact modified amorphous PC/PET nonmiscible blends as indicated by the DMS scan (fig. 5a) with two sharp glass transitions at 81 °C for PET and slightly broadened damping characteristics of polycarbonate at 148 °C.

Blends of polycarbonate with copolyester of hexane dimethanol with isoterephthalic acids known as Kodar A 150 are fully miscible as indicated by the sharp tan delta peak at 120 °C (fig. 5b). These single glass transition, transparent blends published by R. Avakian and R. Allen in 1984 (12) find their applications in the medical field thanks to their transparency, good stress cracking resistance and excellent gamma radiation sterilization stability.

The evolution of technology of these impact modified blends has led to the current Xenoy® PC/PBT blends with izod notched impact strength, high ductility down to minus 50 °C, mainly used for off-line painted bumper applications (fig. 6).

The self colored bumper grades are evolving towards improved processability as indicated by the viscosity curves as a function of shear rate (fig. 6b). New grades with much improved flow for thin wall large automotive front ends still exhibit similar practical toughness (fig. 6 (3)). Those grades meet the color UV weathering requirements of car body paints in a broad variety of colors. Accelerated weathering by carbon arc leads to no obvious visual color change between original, 500 or 1600 hrs exposed samples equivalent to 12 months Florida outdoor ageing.

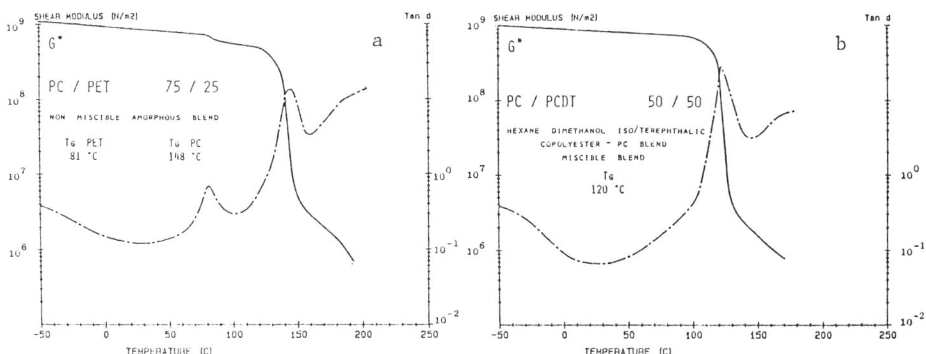

FIG. 5. Dynamic Mechanical properties of Polycarbonate/Polyesters Blends.

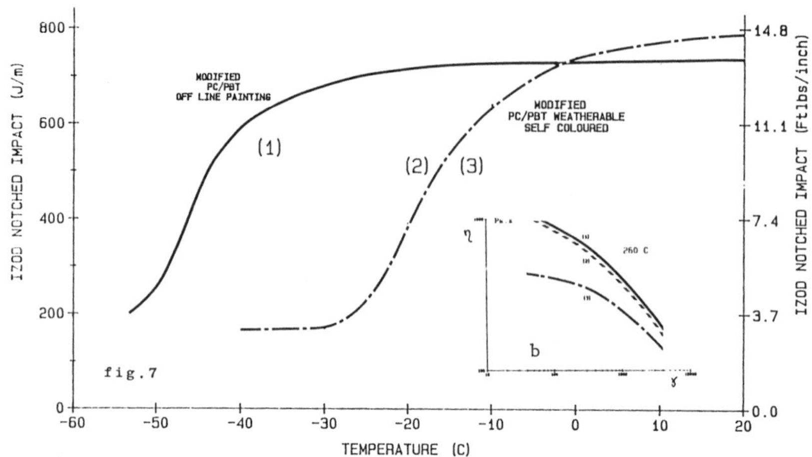

FIG. 6. Impact Strength and Rheology of Impact Modified PC/PBT Blends.

The proliferation of European car models carrying either unpainted or painted modified PC/PBT blend thermoplastic bumpers has considerably changed the market share of non-metal passenger car bumpers. While in 1979 of the 30 % nonmetal bumpers, modified polycarbonate did only account for about one percent, (next to major polypropylene covers, SMC and RIM) (fig.7), in 1986 modified PC/PBT blends are enjoying 25 % market share of now 94 % plastic bumper applications. (Painted versus unpainted being about 50/50).

This successful trend of engineered thermoplastic bumpers is more and more evolving towards integrated front ends, not only in Europe, but also in the U.S. as evidenced by the new Ford Taurus and Mercury Sable advanced design carrying an off-line painted Xenoy® bumper, which meet the 5 mph bumper impact requirement at - 20 °F. This industry trend leads obviously to the second part of this presentation dealing with the development of Noryl® GTX PPO®-polyamide blends for on-line painted automotive applications aimed at replacing metal as body parts.

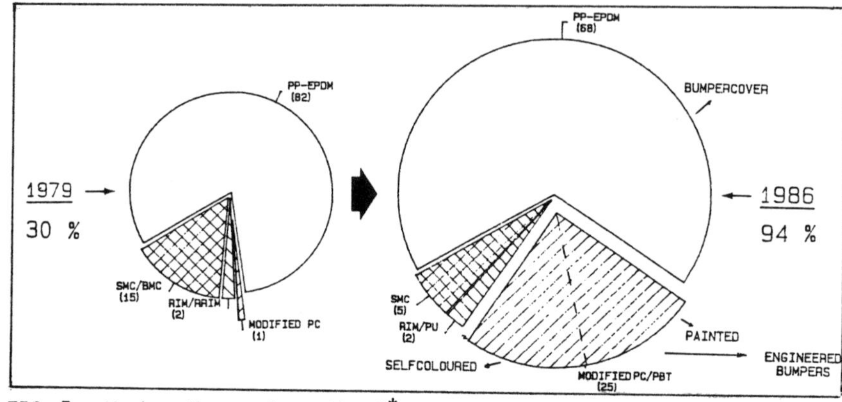

FIG. 7. Market Share of non Metal* European Passenger Car Bumpers.

NORYL® GTX POLYPHENYLENE OXIDE - POLYAMIDE BLENDS DEVELOPMENT.

Developing well defined blends of truly incompatible polymers requires the use of compatibilization technology, which has been pioneered in the laboratories of Professors Paul (8), Riess (9), Teyssie (10), and Heikens (11), who have demonstrated and illustrated how block and graft copolymers when judiciously selected can act as polymeric emulsifiers stabilizing the dispersion of truly incompatible polymers (fig. 8). Enhanced properties are then obtained through morphology control.

Applying this technology to PPO®-polyamide blends resulted in a new thermoplastic concept based on well defined PPO® dispersion in a continuous polyamide matrix with good interphase adhesion.

This unique morphology has led to defining Noryl® GTX as an amorphous PPO®-polymer reinforcement of a semi-crystalline polyamide matrix, taking advantage of polyamide chemical resistance, processability and paintability, while adding heat resistance, dimensional stability and low temperture thoughness through PPO®, without sacrifying ductility and density.

The heat resistance advantage of Noryl® GTX is clearly described in the shear modulus curve as a function of temperature (fig. 9). Thanks to the non-miscibility of PPO® and polyamide, the modulus is maintained almost constant up to the plasticization of the modified PPO® phase at 205 °C with only a minor drop around 50 °C due to the glass transition of the amorphous polyamide phase. The continuous polyamide phase having a melting point at 265 °C together with the specific morphology allows the processing of this high heat resistant thermoplastic at temperatures as low as 280 °C, just 75 °C above the PPO® phase Tg. The stiffness advantage over straight polyamide is even more pronounced in humid conditions: a major drop of modulus is known for polyamide at room temperature when glass transition of the amorphous phase drops from 60 °C down to sub zero temperatures upon water absorption and plasticization.

While the water absorbtion of PPO®-PA blend is a straight function of the polyamide content (fig. 10), the net effect on the mechanical properties shows major synergism as indicated in the tensile stress-strain diagram (fig. 11) where the dry as molded and 50 % relative humidity conditioned PPO®/PA blends maintains very close properties. A major drop in tensile strength is observed for its parent polyamide 66.

This unique balance of properties (Table 2) and economics of a less than two years old material have already led to the mass production of on-line painted large automotive parts such as the valence panels of the new Peugeot 505, the Peugeot 205 GTI, the BMW 300 serie and the specification by various car manufacturers around the world of Noryl® GTX as the first injection molded thermoplastic fender grade. Noryl® GTX passes the stringent heat requirements of electrophoreses paint line referred to as E-

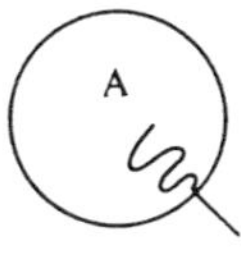

FIG. 8. Polymer Blend Compatibilization

coat up to 190 °C without part distortion. More automotive body parts such as hatchback rear doors or complete integrated front ends are under study using advanced design concepts and assembly techniques as developed in the General Electric Plastics "SCOPE" car concept.

ENGINEERING THERMOPLASTIC BLENDS AS EXTERIOR AUTOMOTIVE BODY MATERIALS.

Figure 12 positions these new engineering thermoplastics as a function of paint line temperature resistance and practicle low temperature toughness of finished parts. Before PPO®-polyamide blends appeared only filled

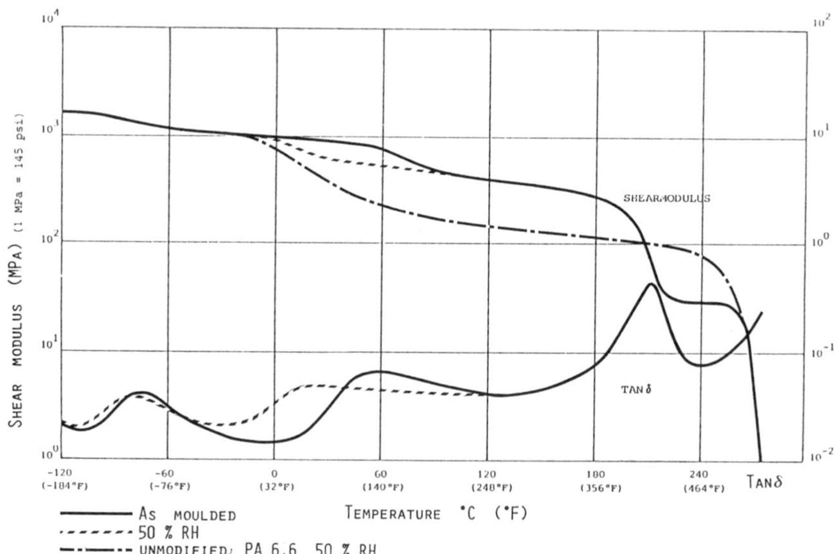

FIG. 9. Dynamic Mechanical Properties of PPO®/PA 6.6 Noryl® GTX 910 and PA 6.6.

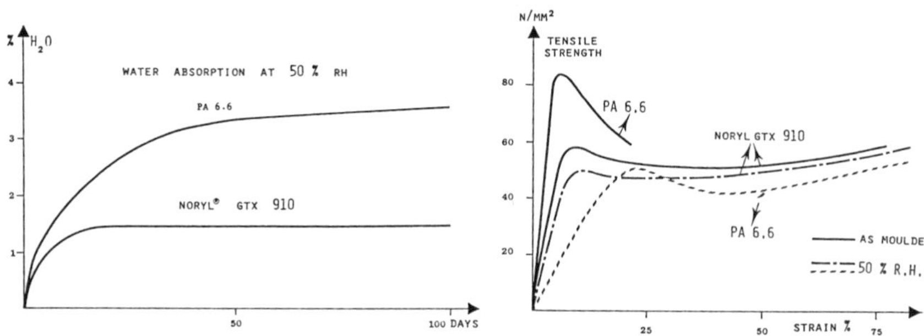

FIG. 10. Moisture Sensitivity of Noryl® GTX PPO® - Polyamide Blend.

FIG. 11. Tensile Stress - Strain Curve at 23 °C.

TABLE II. Noryl® GTX and Polyamides Properties.

PROPERTIES	ASTM/ UNITS	PA 6.6 DRY	PA 6.6 50% RH	MOD. PA 6.6 DRY	MOD. PA 6.6 50% RH	UNFILLED GTX DRY	UNFILLED GTX 50% RH	40% MIN. PA DRY	40% MIN. PA 50% RH	30% GF PA DRY	30% GF PA 50% RH
Tensile Strength	D 638/										
at Yield	N/mm²	83	59	61	52	60	55	-	-	-	-
at Break	N/mm²	83	77	62	52	57	50	90	64	185	125
Elongation	D 638/										
at Break	%	60	300	80	270	100	120	14	44	3	4
Flexural Modulus	D 790/ N/mm²	2,800	1,200	1,950	1,100	2,100	1,750	5,500	2,100	9,000	6,200
Izod Notched Impact Strength	D 256/ J/m	53	112	187	240	240	255	64	123	95	140
Density	D 792	1.14		1.09		1.10		1.48		1.38	

semi crystalline thermoplastics or SMC could meet the electrophorese or E-coat high heat requirements. Those materials, however, either lack ductility, surface finish or ease of processing, whereas Noryl® GTX meets all those requirements in once. It is also pointed out that the Noryl® GTX technology is not limited to pure PPO®/polyamide 66 blends but commercial grades based on polyamide 6 and PPO®/polystyrene/polyamide ternary blends have already been introduced to meet specific application requirements following the trend of further enhanced ductility.

PC/PBT blends have their fit either in the unpainted grades meeting the retouch oven paint temperatures or in off-line painted applications. PBT/PC blends occasionally will meet the top coat online painted requirement when those do not exceed 140 °C. In the years to come one expects that this chart will get more and more crowded especially in the high ductility, upto 160 °C heat area which will be sufficient as car industries develop the all plastic body car no longer requiring the E-coat treatment. More new materials based on the here described polymer blends concept are obviously under development.

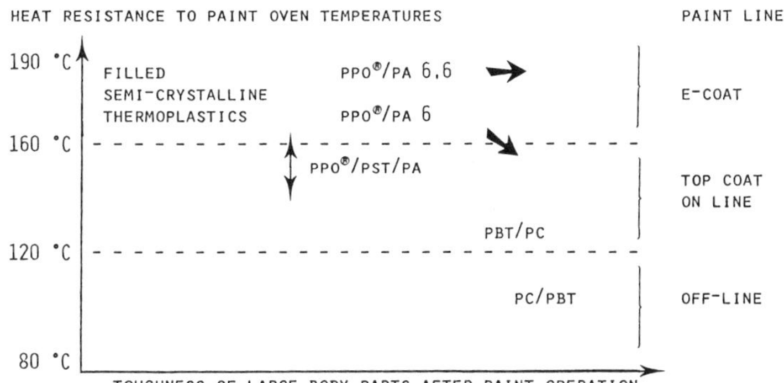

FIG. 12. Modified PC/PBT Blends and PPO®/PA Blends as Exterior Automotive Thermoplastic Materials.

SUMMARY.

In summary, the fast development of those new engineering thermoplastic blends has not only opened new commercial application areas, but also has and hopefully will continue to contribute to the science and technology of new polymer based materials.

ACKNOWLEDGEMENT.

The author wishes to associate to this work the specific contribution of Drs.J.Verhoeven (NL) and Dr.R.Hepp (IND) for the Xenoy® developments, Dr.R.van der Meer (NL), Dr. B. Gallucci (MASS) and R.Avakian (NL) for the Noryl® GTX developments as well as the scientific support from GE Central Research & Development Polymer Engineering & Physics Laboratories from Dr.P.Juliano and Dr.S.Hobbs. General Electric Plastics acknowledges the spearheading scientific contributions of professors D.Heikens, J.P. Mercier, D.R.Paul and P.Teyssie and their laboratories to these specific fields of new polymer blends.

REFERENCES.

(1) P.C. Juliano, IUPAC Meeting, The Hage, August 1985.
(2) J.A. Manson & L.H. Sperling, "Polymer Blends & Composites", Plenum Press N.Y. (1976).
(3) O. Olabisi, L.M. Robeson & M.T. Show, "Polymer-Polymer Miscibility" Academic Press, N.Y. (1979)
(4) D.R. Paul, C.A. Cruz and J.W. Barlow, Macromolecules, 12, 726 (1979).
(5) J.P. Mercier, J. Polym. Sci., Polym. Physics Ed., 20 1875 (1982).
(6) R.S. Stein & A. Misra J. Polym. Sci., Polym. Physics Ed., 18 327 (1980).
(7) S.Y. Hobbs & C.F. Pratt, Journal of Applied Polymer Sci., 19 1701 (1975).
(8) D.R. Paul, "Polymer Blends", D.R. Paul & S. Newman, Academic Press N.Y., Vol. 2 (1978)
(9) G. Riess, J. Kohler & A. Banderet, Eur. Polym. Journal 4 187 (1968).
(10) P. Teyssie & All, ACS Symposium Series, 59 (1977).
(11) D. Heikens, N. Hoen, W. Barentsen, W. Piet & H.J. Ladan, J. Polym. Sci (Polym. Symp.) 62, 309 (1978).
(12) R.W. Avakian & R.B. Allen, Polym. Eng. & Sci., 25, 462 (1985).

History and Development of Interpenetrating Polymer Networks

L. H. Sperling
Polymer Science and Engineering Program
Department of Chemical Engineering
Materials Research Center #32
Lehigh University
Bethlehem, Pa.

INTRODUCTION

An interpenetrating polymer network, IPN, is defined as a combination of two polymers in network form, at least one of which is synthesized and/or crosslinked in the immediate presence of the other. While closely related to other multicomponent polymer materials such as the polymer blends, grafts, and blocks, the IPN's can be distinguished from them in two ways: (1) An IPN swells, but does not dissolve in solvents, and (2) creep and flow are suppressed. The IPN topology is compared with that of other multicomponent polymer structures in Figure 1.

Over the years, people have discovered many different ways of synthesizing IPN's. Figure 2 illustrates the sequential IPN synthesis, top, and the simultaneous interpenetrating network, SIN, synthesis, bottom. In the sequential synthesis, polymer network I is swollen with monomer II plus crosslinker and activator, and polymerized *in situ*. The SIN synthesis begins with a mutual solution of both monomers or prepolymers and their respective crosslinkers, which are then polymerized simultaneously by noninterfering modes, such as stepwise and chain polymerizations. These methods have been used in the bulk, suspension, and latex states. Each will yield a distinguishable composition, even for the same polymer pair.

THE FIRST IPN

IPN's were invented and reinvented over and over again. It is interesting to consider some of these. In the late 1960's, IPN's were invented independently by the Frisch team(1), and by Sperling(2). The Frisch team envisaged a macrocyclic catenane structure, Figure 3. Of course, the catenanes are interlocking rings which are not bound chemically together. For carbon-based rings, 30 or more backbone atoms are required. Since most simple networks have about 100 atoms in their "rings," they fulfill this requirement easily.

When Sperling first came to Lehigh University in 1967, he was interested in studying polymer blends, but had no equipment to do so. At that time, as part of a senior research project, a student by the name of Dave Friedman constructed a UV polymerization box for Sperling. Sperling concieved the idea that a finely divided blend might be made by first polymerizing a network in the UV box, and then swelling in a second monomer and crosslinker, and continuing the polymerization. Sperling reasoned that the pre-existing crosslinks might cause the second phase to be finely divided, this simulating a blend by inexpensive means. One of the first compositions was poly[(*cross*-ethyl acrylate)]-*inter*-poly(*cross*-styrene). Sperling and Friedman noted that a very tough leathery product was formed, which withstood repeated folding. This was the beginning of Sperling's program.

Copyright 1986 by Elsevier Science Publishing Co., Inc.
High Performance Polymers: Their Origin and Development
R.B. Seymour and G.S. Kirshenbaum, Editors

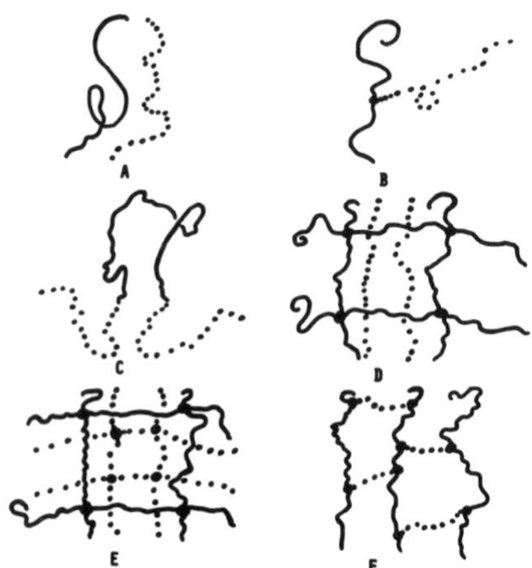

Figure 1. Six basic ways to organize two chains in space. A, With no bond linking the two chains, a blend is formed. B, a graft copolymer; C, a block copolymer; D, a semi-IPN; E, a full IPN; F, conterminously grafted copolymer.

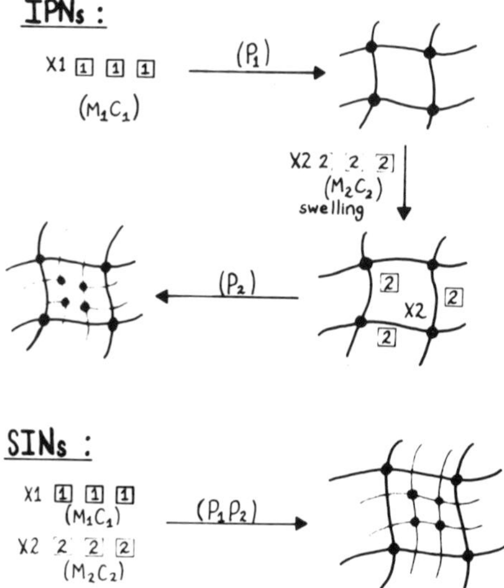

Figure 2. Schematic of the synthesis of sequential IPN's and SIN's. On identical polymer pairs, the chemical structure may be the same, but the morphology and mechanical behavior will differ.

However, neither Sperling nor Frisch were the first to discover IPN's. Shortly after Sperling started work on IPN's, Dr. J. A. Manson suggested that he look in the Journal of the Chemical Society, 1960. There he found an article by John Millar(3), using the exact same nomenclature, interpenetrating polymer networks. It turned out that he was the first to coin that term. However, subsequent discussions with Millar revealed that he was not the first to synthesize an IPN. Indeed, Millar said that IPN's were used commerically in the 1950's as ion exchange resins(4). In that case, one network was charged anionically, while the other network was charged cationically, providing a very efficient way of doing business.

J. J. Staudinger, H. Staudinger's son, had a still earlier patent, with an application date of 1941(5), which was issued ten years later. This patent was on the manufacture of smooth transparent plastics. The idea was that single networks polymerized in sheet form had wavy surfaces, with poor esthetic properties. When the network was swelled with new monomer, the surface was stretched and hence smoothed. This was the first known case of what are now designated homo-IPN's, that is, both networks have the same composition.

However, well before J. J. Staudinger's time, another IPN structure had been invented by Aylsworth(6), in 1914. According to Aylsworth, "My invention...comprises the incorporation with masses of india rubber or cautchouc, or of gutta percha compounded with sulfur or other vulcanizing agent, of a powdered infusible condensation product of phenol or its homologues or derivatives and formaldehyde or other substance containing the methylene radical CH2, which will react with the phenol on application of sufficient heat to form a final condensation product." This is a type of simultaneous interpenetrating network, and turns out to be the earliest known example of an IPN.

Aylsworth's IPN was actually a modified Bakelite product. Just a few years before, Leo Baekeland had patented the first truly synthetic polymer, the condensation product of phenol and formaldehyde(7). The other component, natural rubber plus sulfur, of course, was discovered much earlier by Charles Goodyear(8). Interestingly, although Aylsworth's language is quite modern his invention preceeded the enunciation of the macromolecular hypothesis by Staudinger in 1920(9), and hence Aylsworth's patent makes no mention of any polymeric structure, long chains, networks, IPN's, etc.

(Before proceeding, perhaps it will be of interest to the reader to note that in the years immediately following Baekeland's discovery of Bakelite, chemists tried to mix all available materials with the new polymer. In one case, for example, Ludwig Berend(10) mixed animal glue, shellac, and dinitrocellulose with the phenol-formaldehyde two-part formulation.)

Among other applications for his new material, Alysworth mentions floor tiling. In fact, it is easy to imagine that mid-range compositions of the type he was talking about would make a fine, tough linoleum. Alysworth also understood that his IPN's were multiphased materials. "The result of the process is the formation of a composition in which an infinite number of particles which may be fairly coarse like sea sand, or ultra fine like a precipitate or a ground mineral pigment, are all intimately cemented to their surrounding neighbors by a tenacious elastic bond." To date, there have been about ten Ph. D. theses and a number of M. S. theses devoted to obtaining an understanding of IPN morphology!

It is also interesting to note that prior to his discovery of IPN´s, Aylsworth had worked in Thomas Edison´s laboratory in Menlo Park, where he was chief chemist(11). It seems that Aylsworth was first hired in 1889 by Edison as a consultant, to make a compound suitable for Edison´s new phonograph record(12-14). The new material was made out of wax, gum, stearites, resin, and paste. For this Edison paid him $35,000, a princely sum in those days. Later, working at Menlo Park for Edison, he continued his work on recording materials, and also worked on batteries. At the time that he invented IPN´s, Aylsworth had formed his own company, The Condensite Company of America, in East Orange, N. J.

TRIGLYCERIDE OIL BASED IPN´S

While a significant fraction of the early activity on IPN´s at Lehigh University has already been recounted(15), the history of the "Oils Program" has not yet been recorded. The story begins when Dr. L. A. Wenzel, then head of the Department of Chemical Engineering at Lehigh University went on sabbatical to the Universidad Industrial de Santander, Bucaramanga, Colombia, to help set up a graduate program there. A young faculty member by the name of Alfonso Conde, interested in castor oil, did his M.S. thesis under Dr. Wenzel at that time(16).

A short time later, Conde came to Lehigh to study for his Ph. D. degree, and did his research under Dr. J. A. Manson at Lehigh. As part of his Ph. D. general examination requirements, Conde wrote a proposal suggesting that castor oil could be polymerized and made into an IPN, thus creating a new type of impact resistant plastic. He proposed that such a product could create a new industry for his native country, Colombia. It must be pointed out that at that time Conde was "rubbing elbows" with Sperling´s students in the Polymer Laboratory at Lehigh. Because of his long interest in castor oil, he knew that castor beans grew wild in Colombia. (Castor beans are grown commercially in Brazil, Colombia´s back door neighbor.)

On Wenzel´s recommendation, a proposal based on Conde´s ideas was submitted to the International Division of the National Science Foundation in the fall of 1973. Six weeks later, the first oil crisis struck, with cars lined up for blocks to get half a tank of gas. It must be emphasized that castor oil does not come from an oil well. This proposal resulted in a research program of several year´s standing at Lehigh.

It must be pointed out that castor oil consists mainly of the triglyceride of ricinoleic acid (90%); the structural formula may be written:

$$
\begin{array}{l}
CH_2-O-\overset{O}{\overset{\|}{C}}-(CH_2)_7-CH=CH-CH_2-\overset{OH}{\overset{|}{C}}H-(CH_2)_5-CH_3 \\
| \\
CH-O-\overset{O}{\overset{\|}{C}}-(CH_2)_7-CH=CH-CH_2-\overset{OH}{\overset{|}{C}}H-(CH_2)_5-CH_3 \\
| \\
CH_2-O-\overset{O}{\overset{\|}{C}}-(CH_2)_7-CH=CH-CH_2-\overset{OH}{\overset{|}{C}}H-(CH_2)_5-CH_3
\end{array} \qquad (1)
$$

The original synthesis as suggested by Conde involved crosslinking

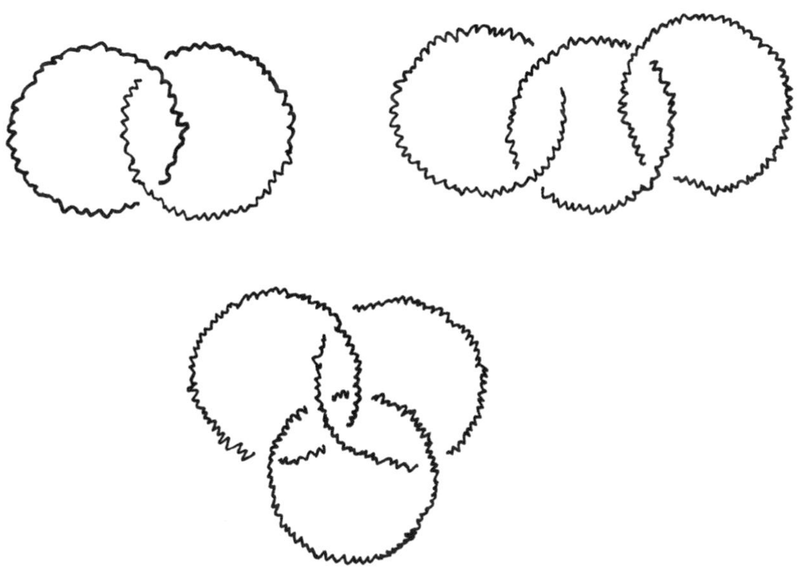

Figure 3. Small molecule catenanes. Dr. Frisch considdered the IPN's to be their macrocylic analogs.

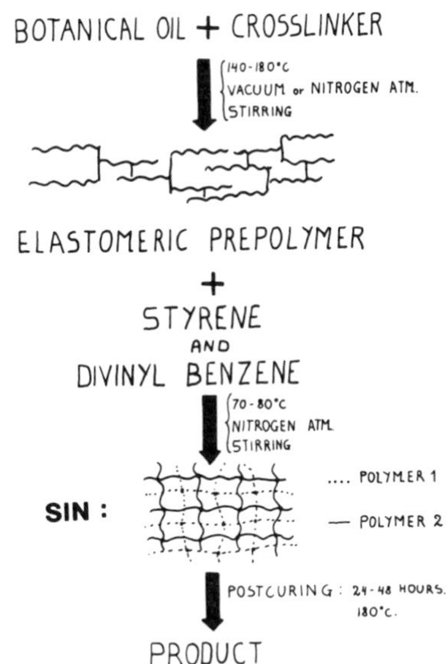

Figure 4. A schematic of the SIN synthesis using botanical oil and a plastic-forming monomer such as styrene.

the structure through its double bonds via sulfur((17). Sulfur crosslinking of unsaturated triglyceride oils reacts to produce a product known as factice. This material is a soft, brittle product widely used as lead pencil erasers. The first Ph. D. candidate working on the program was Godfrey Yenwo. The outside member of his doctoral committee was Mr. Frank Naughton of the Baker Castor Oil Company. He suggested that Yenwo try to make the urethane derivative through the hydroxyl groups. He pointed out that the urethane derivatives of castor oil are widely used in such products as urethane foam rubber. Indeed, the synthesis of the urethane with 2,4-tolylene diisocyanate, TDI, yielded a tougher elastomer, and subsequently a new class of tough, impact resistant IPN plastics. While Yenwo developed the concept of castor oil-based IPN´s using sequential IPN technology, the next Ph. D. candidate in the program, Napoleon Devia, studied the SIN route. Figure 4 illustrates the synthesis of an SIN starting from a triglyceride oil(18). A number of papers were produced in this work, which were reviewed(19). A patent was also issued(20).

Subsequent to work on castor oil-based IPN´s, other oils were investigated. Two of the most important were Vernonia and Lesquerella. Vernonia oil contains epoxy groups, one for nearly every acid residue. In this case, sebacic acid could be used as the crosslinker(21,22). Of course, sebacic acid is itself produced commercially from castor oil. Both Vernonia oil and castor oil produce soft elastomers with glass transition temperatures in the range of -50 to -60 ^{o}C. Vernonia oil grows wild in Kenya, Africa. Presently, experimental plantations of Vernonia are being grown in Zimbobwe, Africa(23). Besides possible IPN use, it can be used as a plasticizer in the manufacture of poly(vinyl chloride), and the oil is suitable for baked films and coatings. Besides the naturally grown epoxidized Vernonia, of course any double bonded triglyceride oil can be epoxidized in the laboratory(18).

Lesquerella oil is a native of the United States, growing wild in Arizona and surrounding areas. Better known to the residents as either PopWeeds or Bladder Pods, Lesquerella is actually a desert wild flower. Lesquerella oil has a structure similar to castor oil, containing hydroxyl groups. Because it has two additional -CH_2-groups between the hydroxyl and the glyceride moiety, the glass transition temperatures of the polymerized oil tend to be about 10 degrees lower than castor oil(18). It must be noted that Marc Linne, a Ph. D. candidate in Sperling´s laboratory, won the American Oil Chemists´ Society´s student award in 1985 for his research on Lesquerella oil. Lesquerella seeds contain about 30% oil, and this plant could possibly be grown as a new crop where today virtually nothing is grown.

While no brief listing could possibly do justice to the history of IPN´s, a brief chronology is given in Table I.

Table I. History of IPN´s(15)

Event	Investigators	Year
Vulcanization	Goodyear	1844
IPN Structure	Aylsworth	1914
Macro. Hypoth.	Staudinger	1920
Graft Copolym.	Ostromislensky	1927
IPN´s	Staudinger and Hutchinson	1941
Block copolym.	Dunn and Melville	1952

HiPS and ABS	Amos et al	1954
Block cop. surf.	Lunsted	1954
Homo-IPN's	Millar	1960
Thermoplast. Elast.	Holden and Milkovich	1966
AB cross. cop.	Bamford, et al	1967
Sequential IPN	Sperling and Friedman	1969
Latex IEN	Frisch et al	1969
SIN's	Sperling and Arnts	1971
Thermopl. IPN	Fischer	1973
IPN nomencl.	Sperling	1974
IPN book	Lipatov and Sergeeva	1979
IPN book	Sperling	1981

ACKNOWLEDGEMENT

The author wishes to thank the National Science Foundation for support through Grant No. DMR-8405053, Polymers Program.

REFERENCES

(1) H. L. Frisch, D. Klempner, and K. C. Frisch, Polym. Lett. 7, 775 (1969).
(2) L. H. Sperling and D. W. Friedman, J. Polym. Sci., A-2 7, 425 (1969).
(3) J. R. Millar, J. Chem. Soc., 1311, (1960).
(4) G. S. Solt, Br. Pat. 728,508 (1955).
(5) J. J. P. Staudinger and H. M. Hutchinson, U. S. 2,539,377 (1951).
(6) J. W. Aylsworth, U. S. 1,111,284 (1914).
(7) L. H. Baekeland, U. S. 942,699 (1907).
(8) C. Goodyear, U. S. 3,633 (1844).
(9) H. Staudinger, Ber., 53, 1073 (1920).
(10) L. Berend, U. S. 952,724 (1910).
(11) R. B. Seymour, in "History of Polymer Science and Technology," R. B. Seymour, Ed., Marcel Dekker, N. Y., 1982.
(12) R. Conot, "A Streak of Luck," Seaview Books, N. Y., 1979.
(13) M. Josephson, "Edison, a Biography," McGraw-Hill, New York, 1959.
(14) C. J. Brown, Private communication, January, 1986.
(15) L. H. Sperling, in "A Century of Chemical Engineering," W. F. Furter, Ed., Plenum Press, N. Y., 1982.
(16) A. Conde Cotes and L. A. Wenzel, Rev. Latinoamer. Ing. Quim. Appl., 4, 125 (1974).
(17) G. M. Yenwo, J. A. Manson, J. Pulido, L. H. Sperling, A. Conde, and N. Devia, J. Appl. Polym. Sci., 21, 1531 (1977).
(18) M. A. Linne, L. H. Sperling, A. M. Fernandez, Shahid Qureshi,and J. A. Manson, in "Rubber-Modified Thermoset Resins," ACS Adv. in Chem. Series No. 208, American Chemical Society, Washington, D. C. 1984.
(19) L. H. Sperling, J. A. Manson, S. Qureshi, and A. M. Fernandez, Ind. Eng. Chem. Prod. Res. Dev. 20, 163 (1981).
(20) L. H. Sperling, J. A. Manson, and N. Devia-Manjarres, U. S. 4,254,002 (1981).
(21) A. M. Fernandez, C. J. Murphy, M. T. DeCrosta, J. A. Manson, and L. H. Sperling, in "Polymer Applications of Renewable-Resource Materials," C. E. Carraher and L. H. Sperling, Eds. Plenum Press, 1983.
(22) A. M. Fernandez, J. A. Manson, and L. H. Sperling, Polym. Mater. Sci. Eng. Preprints, 52, 169 (1985).
(23) R. E. Perdue, private communication.

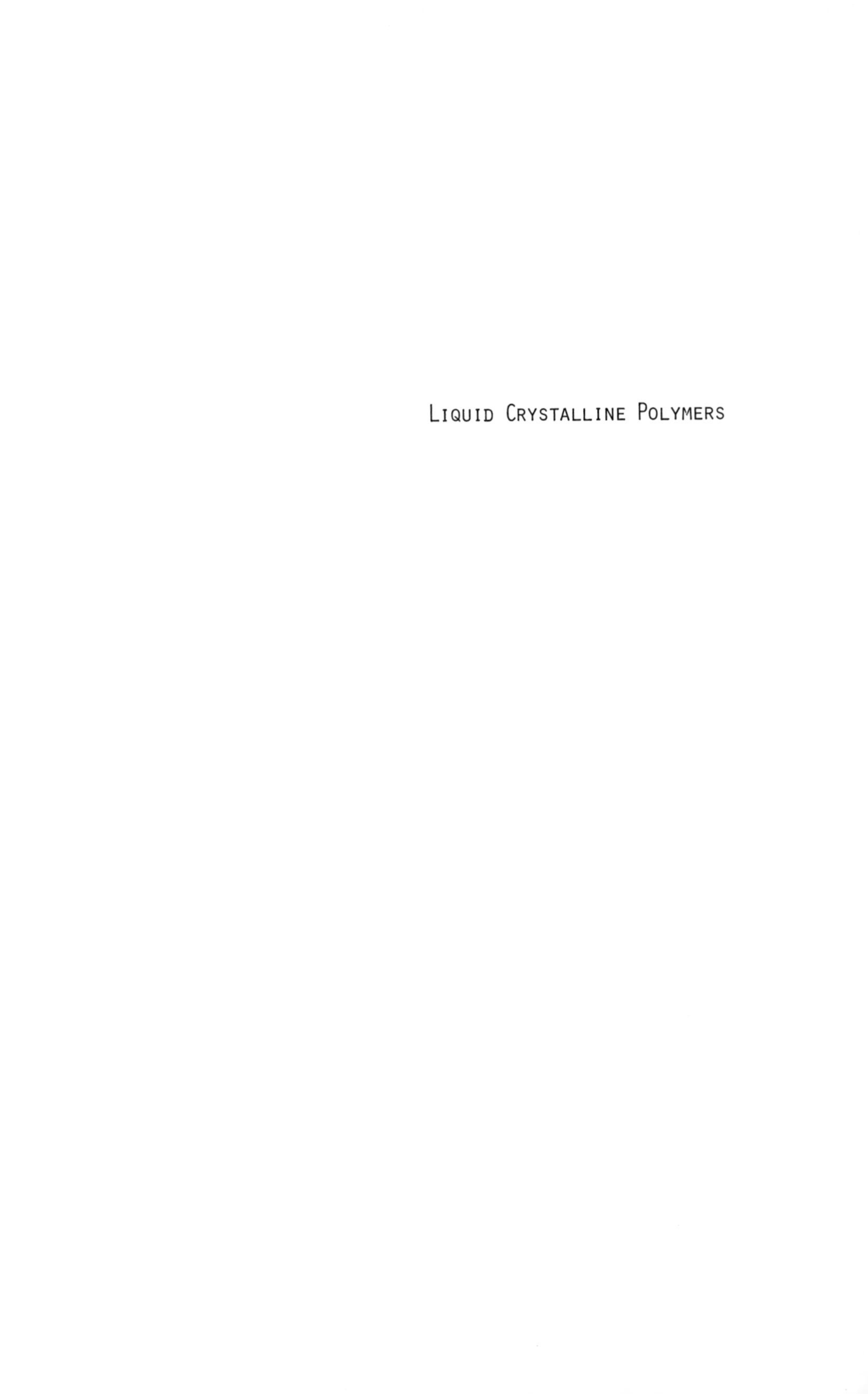

Liquid Crystalline Polymers

Industrial Development of Thermotropic Polyesters

G.W. Calundann

Celanese Research Company
86 Morris Avenue
Summit, NJ

The primary motivation for industrial participation in the development of liquid crystalline polymers was the search for high performance tensile properties, i.e., fiber tensile strength and modulus. As it later turned out, the wholly aromatic, thermotropic polyesters were found to offer a great many more useful properties than just their now well-known tensile capabilities.

By way of background, polymer anisotropy can be looked upon as a quality to be exploited for the development of very high polymer chain extension and chain orientation in a fiber extrusion process. A conventional polymer, poly(ethylene terephthalate), for example, is assumed to be in a random coil conformation in the melt. Such a conformation generally extrudes to a polyphasic structure of low chain continuity with crystalline and amorphous regions and with low chain extension. Low mechanical properties result. A liquid crystalline polymer starts with a high degree of order in the solution or melt. Extruding such solutions or melts produces fibers with a high degree of molecular order "frozen" in. One develops high chain continuity, highly extended chain structures and high mechanical properties, especially a high tensile modulus. This is believed to be the type of structure one obtains from certain aromatic polyamide solutions and from certain thermotropic polyester melts.

As a result of their anisotropic nature and consequent processability to near extended chain solid state structures, Kevlar® (poly(p-phenylene terephthalamide) and the aromatic thermotropic polyesters approach their theoretical tensile moduli rather closely. Their fibers calculate to about 1500 g/d and can be formed experimentally in a range of 1000-1400 g/d.

A history of the industrial development of thermotropic polymers would not be complete without a brief review of preceding technology, that is, the discoveries and developments made in lyotropic polymers. Thus, the timeline of milestones in liquid crystalline polymers proceeds from the initial observation of small molecule liquid crystallinity to the discovery of lyotropic and thermotropic high performance polymers and on through to the recent commercialization of thermotropic polyesters with the introduction of the Vectra® (Celanese Corporation) and Xydar® (Dartco Manufacturing) families of engineering resins.

Shown in the table below, the earliest recognition of liquid crystalline behavior, observed on a thermotropic cholesteric small molecule, is attributed to Reinitzer in 1888. In the Forties and Fifties, polymeric liquid crystallinity was noted and widely researched. The systems studied were mostly natural and synthetic lyotropic biopolymers such as

Copyright 1986 by Elsevier Science Publishing Co., Inc.
High Performance Polymers: Their Origin and Development
R.B. Seymour and G.S. Kirshenbaum, Editors

INDUSTRIAL DEVELOPMENT OF THERMOTROPIC POLYESTERS

MILESTONES

1988	LIQUID CRYSTAL RECOGNITION
1940 - 1956	LYOTROPIC BIOPOLYMERS - TOBACCO MOSAIC VIRUS - COLLAGEN - POLY γ -BENZYL-L-GLUTAMATE)
1956	THEORY - P. J. FLORY - "RIGID ROD" —— ANISOTROPIC SOLUTION
1965	LYOTROPIC AROMATIC POLYAMIDES - KEVLAR
1972	MELT PROCESSABLE WHOLLY AROMATIC POLYESTER - EKKCEL I-2000 (CARBORUNDUM)
1974	THERMOTROPIC POLYESTER - AROMATIC/ALIPHATIC COPOLYMERS-X7G (EASTMAN)
1975 - 1976	WHOLLY AROMATIC POLYESTERS, POLYAZOMETHINES
1975	INTENSE PATENT ACTIVITY - INDUSTRIAL AND ACADEMIC INTEREST INCREASING
1982 - PRESENT	TEIJIN, RHONE-POULENC, others.......
1982 - PRESENT	MONSANTO TEST MARKETING THERMOTROPIC POLYARYLATE (BPA)
1983 - PRESENT	ICI TEST MARKETING CELANESE LCPs
1984	DARTCO MFG. (DART-KRAFT) COMMERCIAL WITH XYDAR
1985	CELANESE COMMERCIAL WITH VECTRA
1985	SUMITOMO TO BE COMMERCIAL WITH "E KONOL" FIBER
1985	MITSUBISHI CHEM., UNITIKA TEST MARKETING "X-7G" VARIANTS

tobacco mosaic virus, collagen and poly(γ-benzyl-l-glutamate). Paul Flory proposed in 1956, in essence, that a solution of a rod-like synthetic polymer of suitable rigidity at the appropriate concentration would be a lyotropic system. This theoretical prediction was well demonstrated when, in 1965, Stephanie Kwolek of the DuPont Company discovered that certain wholly aromatic polyamides gave anisotropic solutions in alkylamide and alkylurea solvents, an observation that led ultimately to the development of Kevlar aramid fiber (1). This was a key milestone in that it raised the question in the fibers industry - could a similar phenomenon occur in the melt? Were thermotropic polymers possible? They were possible, and indeed the number reported to date by academic and industrial researchers is nearly beyond count.

The situation in wholly aromatic polyesters in 1974 was fairly straightforward with the known polymers falling into two groups. The first consisted of the highly crystalline, very high melting and essentially intractable materials such as poly(p-oxybenzoate) or poly(p-phenylene terephthalate). These polymers are not readily melt spinnable or injection moldable. An exception within this category is the copolyester of p-hydroxybenzoic acid, terephthalic and 4,4'-biphenol, with the trade name Ekkcel®I-2000, reported in 1972 by the Carborundum Company (2). This polymer is injection moldable albeit at temperatures in the vicinity of 400°C, a temperature not compatible with common melt spinning equipment. The rate of thermal degradation in such a polyester at 400°C makes stable fiber production particularly difficult. Moreover, most conventional injection molding equipment requires modification to higher temperature operating capability to ensure reasonable processing of this polymer.

The second category of well known polyarylates contained polymers of a much softer, amorphous nature. These polymers form clear, glassy solids from isotropic melts; the polyesters tend to have high levels of relatively asymmetric monomers such as resorcinol or bisphenol-A. The glassy polyarylates have, not surprisingly, unexceptional physical properties.

The 1972 patent (2) to Carborundum showed data comparing the melting point/composition curves of two wholly aromatic copolyesters: the copolymer of p-hydroxybenzoic acid (HBA), terephthalic acid (TA) and hydroquinone (HQ) and the copolyester made from HBA, TA and 4,4'-biphenol (BP), the Ekkcel I-2000 polymer (Figure 1). The minimum melting temperature of the former copolymer is about 420°C while the biphenol based injection moldable copolyester has a minimum melting temperature of approximately 395°C. In 1976, in another patent to Carborundum (3), an HBA/TA/BP copolymer system, modified and softened with isophthalic acid (IA), was reported to be melt spinnable to produce fiber with as-spun properties of 5 g/d tenacity and 390 g/d modulus. These researchers did not report thermotropic behavior which in fact does manifest at least for some compositions of this terpolyester. (Part of this patent property was subsequently sold to Dart-Kraft and has become the basic technology supporting Dartco Manufacturing's engineering resin, Xydar.)

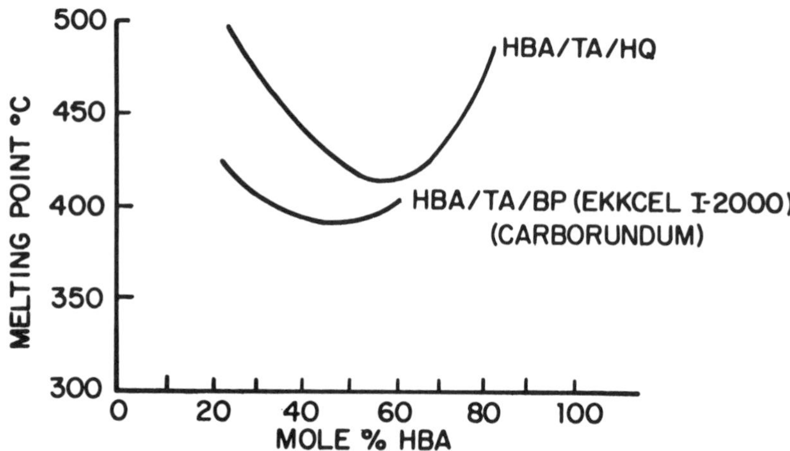

Effect of Composition on Polymer Tm, Ekkcel I-2000

Figure 1

Figure 2 (4,5) summarizes some key points on what was the first reported and well-characterized thermotropic polymer. The Eastman workers melt copolymerized p-acetoxybenzoic acid with poly(ethylene terephthalate) (PET) to form a series of thermotropic aliphatic-aromatic polyesters (X7G®polymers). Some of these compositions showed the characteristic melt optical anisotropy typical of nematic small molecules, that is, the melt between crossed polaroids on a hot stage microscope transmits light. This nematic order was persistent. Of course, shear birefringence is also shown by isotropic polymer melts but light transmission decay is quite rapid.

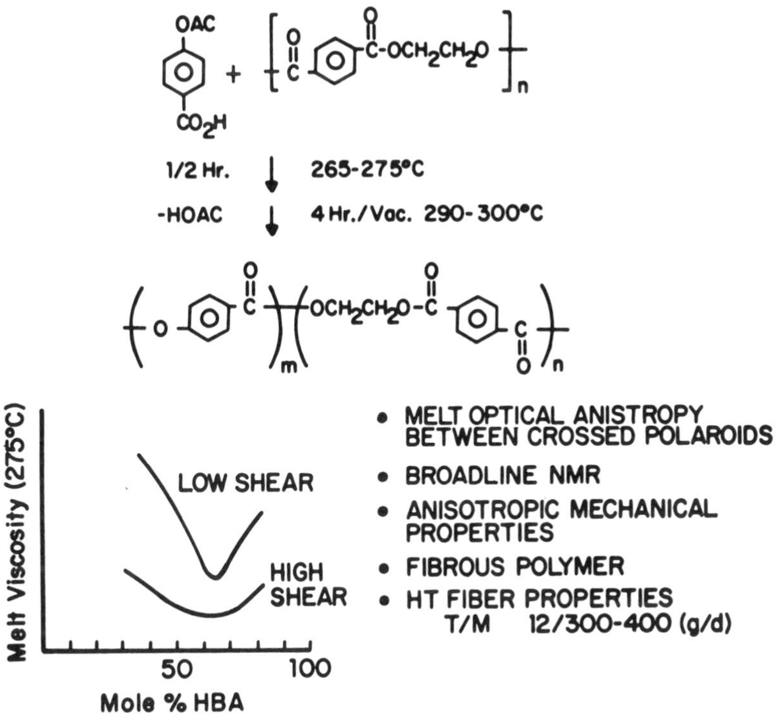

Synthesis and Characteristics of X7G Polyesters

Figure 2

Broadline nmr, relating melt order with polymer composition and temperature, also provided evidence supporting the thermotropic nature of these materials. The physical properties of molded parts of certain compositions showed a definite mechanical anisotropy. Strength and modulus values in the machine direction were considerably higher than those values transverse to the machine direction, that is, perpendicular to the direction of flow. Again, facile and high molecular orientation was noted for these polymers in the flow direction and this became more pronounced as part thickness was reduced. Further, the melt viscosities of the X7G copolyesters, at about equivalent molecular weights, were measured as a function of polymer oxybenzoyl content under low and high shear conditions (Figure 2). Lower melt viscosities were found for polymers containing from 40 to perhaps 70 mole % p-oxybenzoyl. The lower melt viscosities noted, with a minimum at about 60-70 mole % HBA content, was additional evidence supporting nematic melt structure for the copolyesters within these compositional limits. Increasing oxybenzoyl blockiness, causing incomplete polymer fusion at 275°C , the measurement temperature, would explain the rapid increase in melt viscosity for polymers of greater than 70 mole % p-HBA. The fiber properties reported for Eastman's oxybenzoyl-PET copolyesters are not particularly impressive with tensile moduli in the range of 300-400 g/d and maximum tensile strengths rarely exceeding 12 g/d.

Perhaps the most obvious characteristic of thermotropic polyesters is their bulk appearance. These polyesters, whether aliphatic-aromatic as the X7G type or the wholly aromatic structures, all have a fibrous, almost wood-like texture. The fracture surface of a strand shows the typical highly fibrillar morphology. If such a strand is put through a conventional cutting process, a fuzzy and very fibrillar chip often results. Another plainly visible characteristic of thermotropic polyesters is the opalescent, almost metallic-like surface sheen.

Paul Morgan, in 1977 (6), reported high strength, high modulus fiber spun from anisotropic melts of aromatic polyazomethines. These Schiff's base polymers were made, in the usual case, by solution condensation of an aryl diamine with an aryl dialdehyde or diketone. After an extruder molecular weight advancement, an example of one of these anisotropic polymers, that from methyl-p-phenylene diamine and terephthalaldehyde, was melt spun to fiber with properties of tenacity (T)/% elongation to break (E)/modulus (M) = 7.3 g/d/1.1%/916 g/d. This fiber was heat treated to rather extraordinary properties: strength increased to 38 g/d, elongation to 4.4% and tensile modulus to 1012 g/d. Also reported was a single filament break of 44 g/d. This was a good demonstration of the potential of thermotropic, wholly aromatic polymers.

There are, however, some drawbacks to this kind of structure. As a poly Schiff's base, one would expect hydrolytic instability and that is the case; polyazomethine fiber, for the most part, under hot-wet conditions, loses physical properties quite rapidly. Secondly, most of these structures have a glass transition temperature (Tg) at about 100°C; at or near this temperature, those fibers show a dramatic loss of tensile modulus. A third concern involves the difficulty in controlling polyazomethine molecular weight during polymerization and melt processing. If melt residence time is not carefully monitored, polymer solidification in the apparatus can become a problem.

Figure 3 summarizes the molecular architecture used to promote polyester melt anisotropy at reasonable temperatures. The Celanese, DuPont and Eastman laboratories have outlined most of the key approaches to lower polymer melting point via crystalline order disruption. This is accomplished while maintaining sufficient molecular symmetry to preserve the melt anisotropy inherent in linear aromatic polyesters, that is, polyesters derived typically from symmetric monomers as p-hydroxybenzoic acid, terephthalic acid, hydroquinone, 4,4'-biphenol and the like.

Eastman's pioneering work has already been reviewed; their approach involved introduction of aromatic monomers into aliphatic-aromatic polyesters such as PET. The increased symmetric ring content imparts melt anisotropy to the copolyester but, as mentioned, the aliphatic-aromatic structure has not given the mechanical property levels noted for the wholly aromatic polyesters.

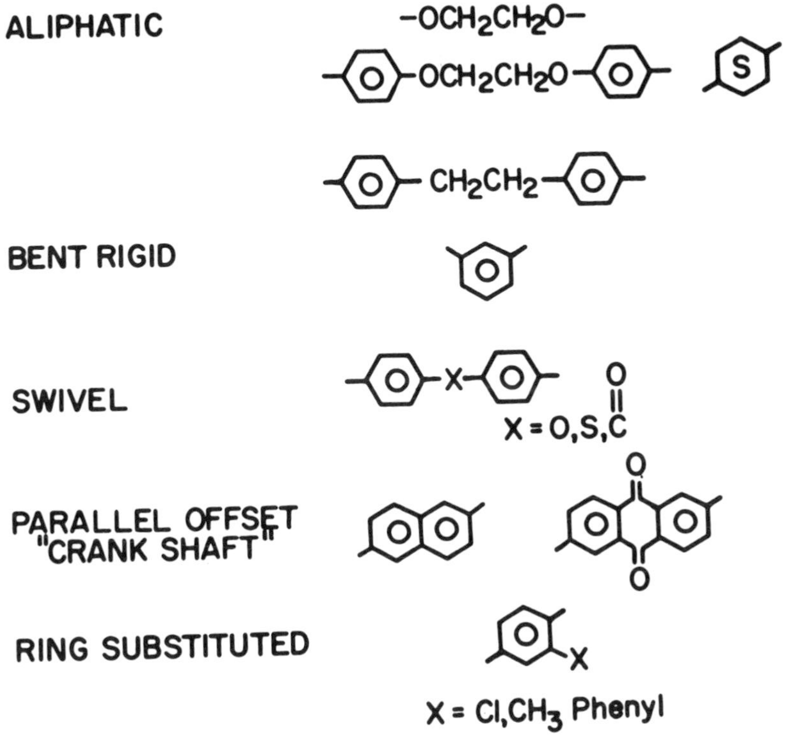

ALIPHATIC

$-OCH_2CH_2O-$

BENT RIGID

SWIVEL

X = O,S,C

PARALLEL OFFSET "CRANK SHAFT"

RING SUBSTITUTED

X = Cl, CH_3 Phenyl

Structures Providing Tractability in Aromatic Thermotropic Polyesters

Figure 3

Introduction of bent rigid units derivable, for example, from isophthalic acid or resorcinol, is an obvious route towards increasing polymer tractability. However, there are problems associated with the meta linkage. While in many systems, incorporation of low levels of isophthalic acid provides an increase in polymer tractability, increasing isophthaloyl content beyond a certain point tends to offset this gain by reducing polymer melt anisotropy with consequent negative impact on polymer rheology and ultimately on fiber (film, molding, etc.) properties. An example of this effect will be shown later. Resorcinol, even at rather low levels, generally results in amorphous polyesters with isotropic melts.

DuPont workers have described (7) a wide variety of tractabilizing molecules and the thermotropic polyesters derived therefrom. Their research has focused for the most part on ring substituted monomers such as chloro, methyl or phenyl substituted hydroquinone and "swivel" or linked ring molecules, examples of which are 3,4' or 4,4' functionally (hydroxy or carboxy) disubstituted diphenyl ether, sulfide or ketone monomers. Use of the parallel offset or "crankshaft" geometry provided by 2,6 functionally disubstituted naphthalene monomers has been the major thrust of the Celanese Corporation development of easily processed, high performance wholly aromatic thermotropic polyesters.

Specifically, Celanese has defined families of thermotropic polyesters derived from 2,6-naphthalene dicarboxylic acid (NDA), 2,6-dihydroxynaphthalene (DHN) or 6-hydroxy-2-naphthoic acid (HNA). Indeed, the Vectra LCP engineering resins of Celanese Corporation are based on these monomers.

To illustrate the dramatic effect on polymer melting point depression of the 2,6-naphthylene nucleus, Figure 4 compares the T_m/copolyester composition curves of NDA, DHN and HNA based polymers with those of the copolyester, HBA/TA/HQ (8). Thus, replacement of TA with 2,6-naphthalene di-acid in the polymerization charge gives a series of copolyesters with a melting point minimum at about 325°C, near 60 mole % HBA; the 2,6-naphthalene diol, replacing hydroquinone, produces a series with a polymer T_m low near 280°C(~50 mole % HBA), and much of the composition range of the two component polyester of p-hydroxybenzoic acid and 6-hydroxy-2-naphthoic acid (HNA/HBA) falls within the industrially convenient melt temperature zone of 250-310°C. A great many compositions of the three groups of copolyesters shown were investigated in detail; all formed anisotropic melts and all were melt spun to high strength, high modulus fibers.

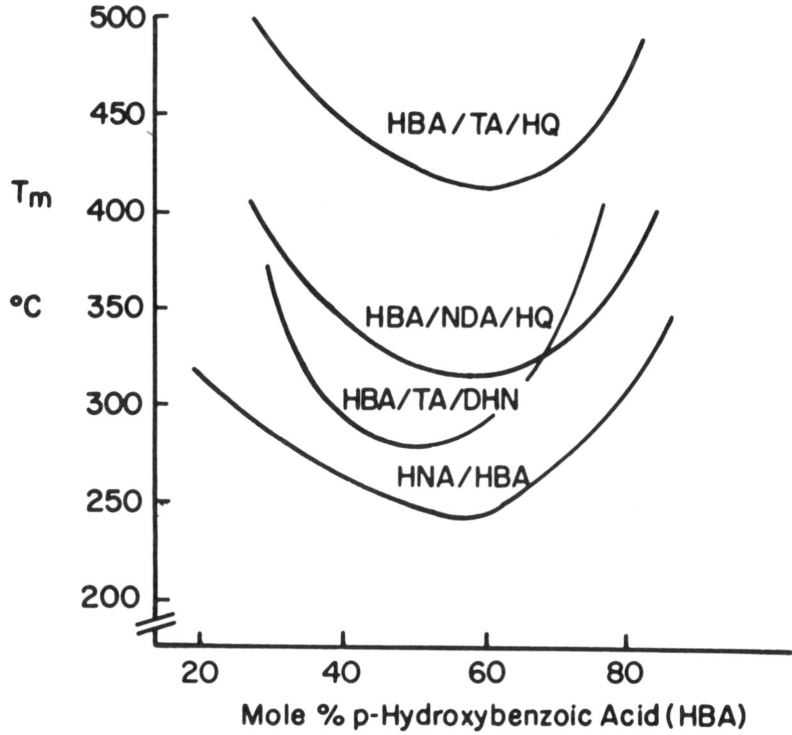

Composition/Tm Relationships of Naphthalene Derived Thermotropic Polyesters

Figure 4

Replacement of the p-oxybenzoyl fraction with p-phenylene terephthloyl units in the two component polyester resulted in a more symmetric, higher melting copolyester series, shown as polymer II, HNA/TA/HQ, in Figure 5. Similarly, all of the compositions of II that were examined were thermotropic and ultimately gave high strength and modulus fibers.

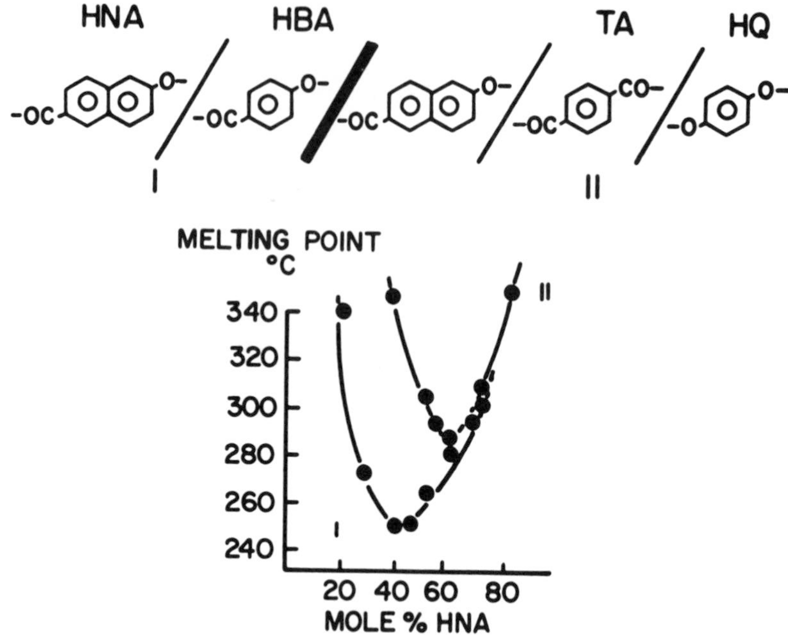

Composition/Tm Relationship of HNA Based Copolyesters

Figure 5

As an example of a more complex system, isophthalic acid was interpolymerized with HBA, NDA and HQ to form a four component terpolyester, HBA/NDA/IA/HQ (9) with increased tractability over the parent copolymer, HBA/NDA/HQ.

Figure 6 the ternary T_m/composition diagram of this terpolyester, shows contour plots representing groups of terpolyester compositions with the same crystalline melting point; note the small inner contour with compositions melting near the minimum for this series at about 255°C. Melt rheology, spinning behavior and fiber properties were by no means similar across this diagram. The shaded area encloses those terpolyesters with close to the optimum processability-fiber property profile, that is, systems melting at 300°C or less, with equal to or less than 15 mole % isophthaloyl units and from 55-75 mole % p-oxybenzoyl units. Above 300°C, stable fiber spinning becomes increasingly difficult. Lower melting terpolyesters of increased isophthaloyl content tend to have less melt anisotropy and thus less favorable rheology and generally become more difficult to process. Further, fiber from higher isophthaloyl terpolyesters show a lower level of mechanical properties and considerably less chemical stability.

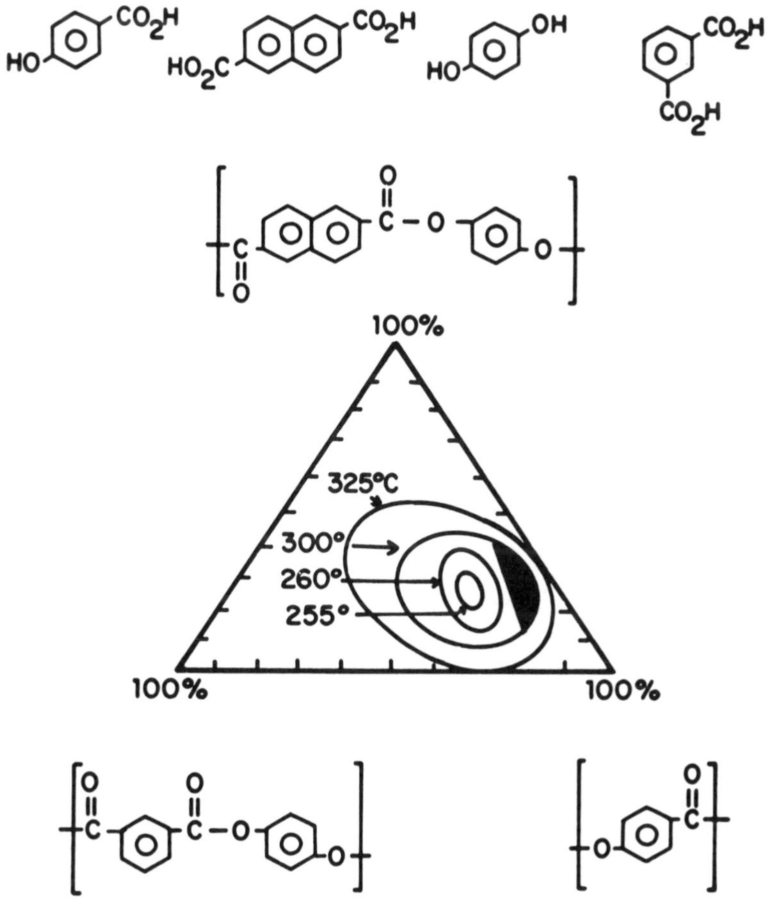

Composition/Tm Relationship of NDA Based Terpolyesters

Figure 6

Work with NDA, DHN and HNA monomers has also led to the development of thermotropic aromatic poly(ester-amides). The question arose as to whether the introduction of hydrogen bonding capability into the chain might result in some useful property improvements. The greater interchain interaction could result in increased fiber fatigue resistance, higher filament shear modulus (and tensile modulus) and improved fiber surface adhesion. Early data are promising, but considerable research remains before the potential of the ester-amide variants can be fully defined.

The introduction of a mixed linkage does not appear to offer any gain in system tractability, specifically, an increase in the range of compositions melting below 320°C. Figure 7 compares the T_m/composition curves of the HNA/TA/HQ copolyester and an analogous poly(ester-amide) in which p-acetoxyacetanilide replaced hydroquinone diacetate in the initial polymerization charge. In fact, amide linkage incorporation resulted in a reduced tractable zone although minimum melting points for both systems were nearly the same at about 275-280°C (~60 mole % HBA). The maximum amide linkage, tolerable from a tractability standpoint, was about 25 mole % (10).

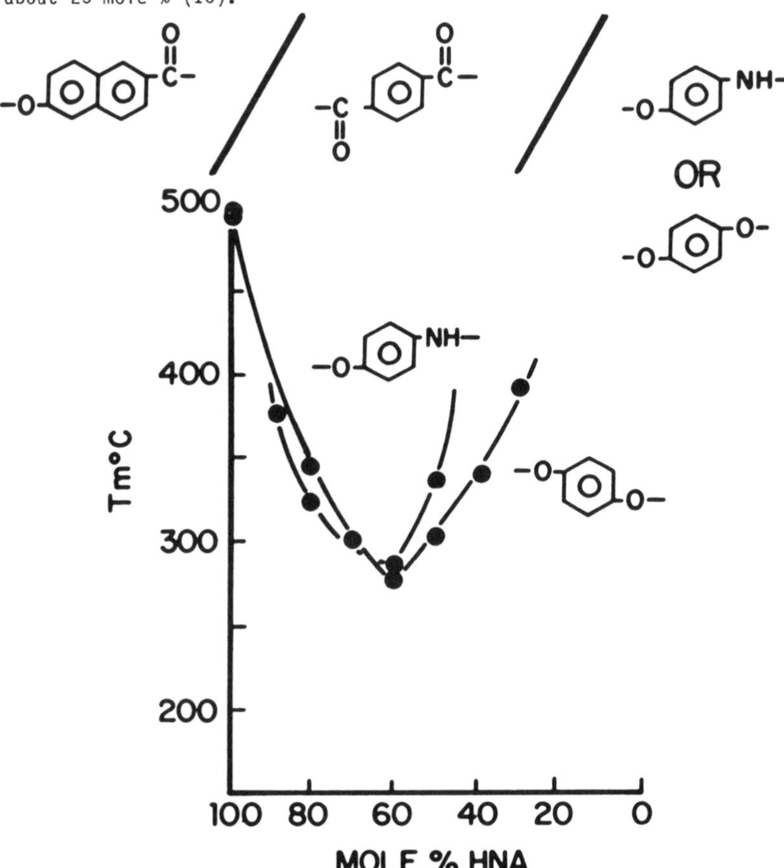

Composition/Tm Relationship of HNA Based Poly(ester-amide)

Figure 7

Thus, the factors critical in the molecular design of a commercially viable thermotropic polymer for fiber/resin applications include: (1) processability -- ideally the polymer should process in the range of 250-350°C; (2) melt anisotropy -- a careful balance of molecular symmetry is required; (3) end-use properties -- considerable structure-property experimentation needed to optimize; and, of course, (4) minimum monomer cost.

A fifth factor is certainly ease of preparation and in this characteristic the melt prepared thermotropic polymers are particularly favored. All of the polymers described thus far may be made in a conventional melt acidolysis process starting with the acetoxy derivatives of the hydroxyl containing monomers used. A typical polymerization scheme is shown in Figure 8, the preparation of the two component polyester derived from the acetylated hydroxybenzoic and hydroxynaphthoic acids. The polymerization may be carried out with or without added catalysts. The poly(ester-amides) commented on here, and the more recently reported aromatic, thermotropic poly(ester-carbonates) and poly(ester-imides), may all be synthesized in a similar manner.

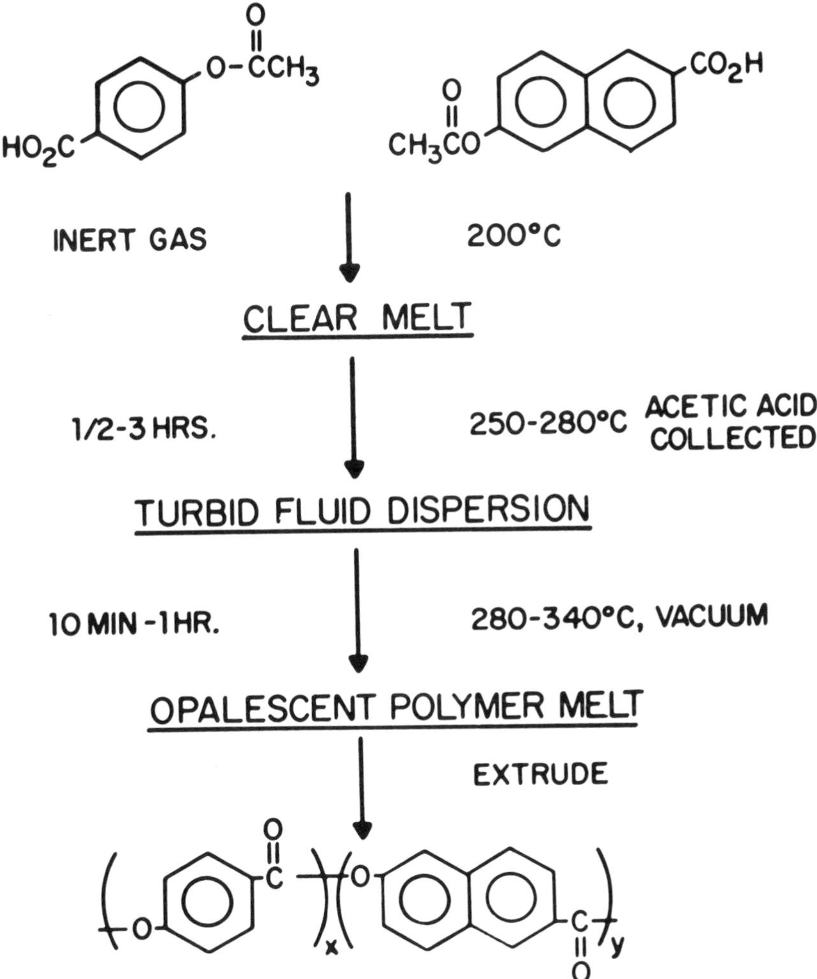

Figure 8 Typical Thermotropic Polyester Polymerization Scheme

Reference was made earlier to useful polymer properties other than high tensile strength and modulus. There are many. For example, the Vectra resins, even highly filled variants, show remarkable ease of processing derived from the polymer's low melt viscosity and high melt strength. In an injection molding process, this is combined with rapid machine cycle time, low mold shrinkage and excellent mold detail reproduction. Impressive electrical properties, impact strength, thermal stability, i.e., high continuous use temperature, controllable part dimensional stability, high heat deflection temperatures and extraordinary chemical and solvent resistance, even under stress, are further characteristics of aromatic thermotropic polyesters (11). These materials also show low flammability and smoke generation and resistance to gamma and microwave radiation as well as sunlight. In fact, test data show that the hydrolytic stability of the two component polyester of Figure 8 compares favorably with other engineering resins, and is without peer in the polyester class.

In conclusion, the industrial development of thermotropic polyesters was traced through its theoretical origins, the identification of useful compositions, and ultimately to full commercialization of Xydar and Vectra. Commercialization of other thermotropic resins will, in all likelihood, be announced soon, as will high performance fibers made from variants of polymers described herein.

The industrial challenge ahead will be to define and develop applications which take advantage of the unique properties of aromatic thermotropic polyesters. Already, commercial demand is growing in the fiber optic, chemical process, electrical/electronic, automotive and houseware markets, and new areas of potential application continue to emerge.

References

(1) See for example, S.L. Kwolek, (DuPont), U.S. Patent 3,600,350 (1971).

(2) S.G. Cottis, J. Economy and B.E. Nowak (Carborundum), U.S. Patent 3,637,595 (1972).

(3) S.G. Cottis, J. Economy and L.C. Wohrer (Carborundum), U.S. Patent 3,975,487 (1976).

(4) F.E. McFarlane, lecture, Gordon Research Conference on Polymers (1974); H.F. Kuhfuss and W.J. Jackson, Jr. (Eastman Kodak), U.S. Patent 3,778,410 (1973).

(5) W.J. Jackson, Jr. and H.F. Kuhfuss, J. Polym. Sci., Polym. Chem. Ed., 14,2043 (1976).

(6) P.W. Morgan (DuPont), U.S. Patent 4,048,148 (1977).

(7) J.J. Kleinschuster (DuPont), U.S. Patent 3,991,014 (1976); T.C. Pletcher (DuPont), U.S. Patent 3,991,013 (1976); J.J. Kleinschuster and T.C. Pletcher (DuPont), U.S. Patent 4,066,620 (1978); J.R. Schaefgen (DuPont), U.S. Patents 4,075,262 and 4,118,372 (1978); C.R. Payet (DuPont), U.S. Patent 4,159,365 (1979); R.S. Irwin (DuPont), U.S. Patents 4,232,143 and 4,232,144 (1980) and 4,245,082 (1981).

(8) G.W. Calundann (Celanese), U.S. Patents 4,067,852 (1978), 4,185,996 (1980), 4,161,470 (1979), and 4,256,624 (1981).

(9) G.W. Calundann, H.L. Davis, F.J. Gorman and R.M. Mininni (Celanese), U.S. Patent 4,083,829 (1978).

(10) A.J. East, L.F. Charbonneau and G.W. Calundann (Celanese), U.S. Patent 4,330,457 (1982).

(11) G.W. Calundann and M. Jaffe, in Proceedings of The Robert A. Welch Conferences on Chemical Research XXVI. Synthetic Polymers (Houston, TX, Nov. 15-17, 1982) p. 247.

EARLY WORK ON THERMOTROPIC LIQUID CRYSTALLINE POLYMERS HAVING A RIGID-FLEXIBLE REGULARLY ALTERNATING STRUCTURE IN THE MAIN CHAIN

ANSELM C. GRIFFIN
Departments of Chemistry and Polymer Science
University of Southern Mississippi
Hattiesburg, Mississippi

ABSTRACT

Thermotropic liquid crystalline polymers having a regularly alternating rigid-flexible structure along the main chain were first reported in 1975 by Roviello and Sirigu. Early work in the field was dominated by Roviello and Sirigu although other scientists made significant contributions to this emerging field of polymer science.

INTRODUCTION

The connection between liquid crystallinity (mesomorphism) and polymers can arguably be dated from 1950 with Elliott and Ambrose's[1] description of birefringent chloroform solutions of the synthetic polypeptide poly(γ-benzyl-L-glutamate). Robinson's reports, begun in 1956[2], on careful studies of such solutions, firmly established the phenomenon of lyotropic (solvent induced) liquid crystallinity in polymers. In the 1960's industrial research into what later proved to be thermotropic liquid crystalline polymers - linear para-substituted aromatic polyesters - was actively pursued.[3] The now famous discovery[4] in 1965 by Kwolek at du Pont of lyotropic liquid crystallinity in aromatic polyamides led to the highly successful Kevlar® materials. These (and other) landmarks related to the phenomenon of polymeric liquid crystallinity set the stage for the discovery in 1975 of thermotropic polymers having liquid crystallinity derived from mesogenic structural elements in the polymer main chain. As described by Dobb and McIntyre in reference 3, three independent reports of thermotropic main chain liquid crystalline polymers appeared in 1975. Kleinschuster, Pletcher and Schaefgen at du Pont and also Jackson and Kuhfuss at Tennessee Eastman reported, in 1975, thermotropic liquid crystalline main chain polymers. (These types of polymers are of commercial importance and are treated historically by Calundann elsewhere in this volume.) This review will describe early work on thermotropic liquid crystalline polymers also commencing in 1975, but on a different type of main chain configuration, one with regularly alternating rigid and flexible segments along the polymer main chain. In that year Augusto Sirigu along with Antonio Roviello at the University of Naples described in a historical paper[5] the first liquid crystalline polymer with these structural features. In order to fully appreciate the context of this contribution to the field of liquid crystal science, a brief description of liquid crystals follows.

The "typical" liquid crystalline molecule is composed of a rigid-usually aromatic - central core and flexible tails - alkyl or alkoxy. A representative small molecule liquid crystal with these features, 4-pentyloxyphenyl 4'-pentyloxybenzoate, is shown on the following page. This molecular structure can be pictorally represented by the figure below it. Back in the 1920's the prolific German synthetic

Copyright 1986 by Elsevier Science Publishing Co., Inc.
High Performance Polymers: Their Origin and Development
R.B. Seymour and G.S. Kirshenbaum, Editors

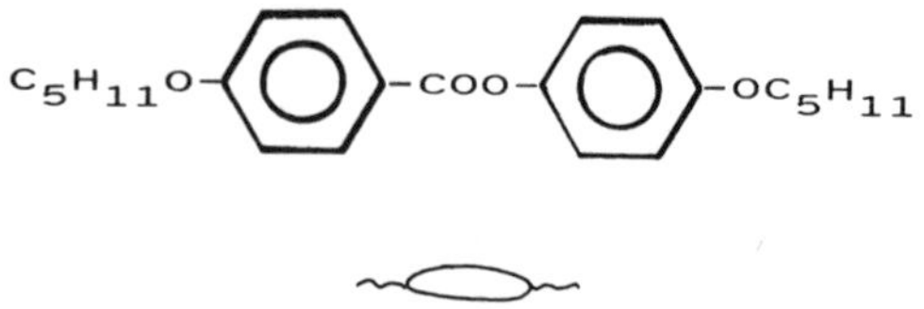

$C_5H_{11}O-C_6H_4-COO-C_6H_4-OC_5H_{11}$

chemist Daniel Vorlander synthesized a number of mesogens having two rigid cores separated by a flexible spacer group. These molecules can be considered as the conceptual Twin dimer formed by joining one of the flexible tails from each of two molecules of a typical "monomeric" small molecule liquid crystal. The structure of one of Vorlander's[6] typical Twin dimer molecules is shown below with its pictorial representation underneath. A straightforward extrapolation from monomer to dimer to

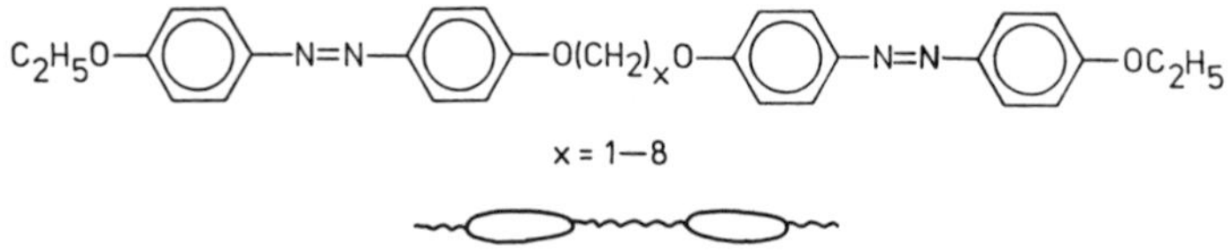

$C_2H_5O-C_6H_4-N{=}N-C_6H_4-O(CH_2)_xO-C_6H_4-N{=}N-C_6H_4-OC_2H_5$

x = 1—8

polymeric mesogen is today an obvious extension, i.e., the coupling of many rigid cores via flexible spacer groups.[7] However, in Vorlander's time Staudinger's macromolecular hypothesis was still struggling for acceptance. It does appear, however, that Vorlander[8] did give some thought to the possibility of polymeric liquid crystals even prior to Staudinger's classic work. There does not appear to be any evidence of synthetic efforts toward polymeric mesogens by Vorlander even though his flexible-rigid-flexible-rigid-flexible dimers certainly are suggestive of them. These Twin dimers have recently received attention as monodisperse models for polymeric mesogens of similar constitution.[9]

Thermotropic mesomorphism in linear polymers had been considered theoretically prior to 1975 (see reference 3 for leading references). However, in 1975, de Gennes[10] explicitly considered linear polymers having alternating rigid mesogenic groups and flexible spacers in the polymeric main chain. He predicted possible nematic behavior for such polymers.

DISCUSSION

Augusto Sirigu was experienced in crystallography and came to the areas of liquid crystallinity and polymers from a structural perspective. In his now classic 1975 paper (with A. Roviello) it is clearly evident that their interest was in studying mesophase morphology in the polymeric solid state. They also presented an analysis of what type of mesophase would be expected from an ... ABAB ... sequence depending on the extent of interaction of AA, BB and AB units in neighboring chains. Their ideas could only be tested by synthesizing such a polymer which they did. The structure on the following page describes these polymers.

Interfacial polycondensation was used to prepare the polyesters. Differential scanning calorimetry, optical microscopy, x-ray diffraction and solution viscosity measurements were reported on them. All three

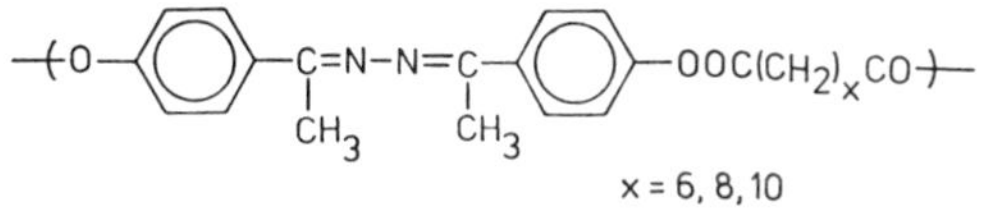

$$-\!\!\left(O-C_6H_4-\underset{CH_3}{\underset{|}{C}}=N-N=\underset{CH_3}{\underset{|}{C}}-C_6H_4-OOC(CH_2)_xCO\right)\!\!-$$

x = 6, 8, 10

polymers were liquid crystalline. Much of what is now known about regularly repeating rigid-flexible mesomorphic polymers was uncovered by Sirigu in this first paper. Double endotherms for the melting transition, numerous discontinuities in the optical textures, x-ray diffraction maxima corresponding to the length of the polymeric repeat unit - all these phenomena were described.

Recounting the factors leading to his seminal 1975 work, Sirigu writes, "I have been interested since the beginning of my scientific career in the structural (molecular and packing) organization of the matter. Around 1973 I developed a strong interest on the structure of non crystalline materials and particularly on those systems whose structure might be either spontaneously ordered or induced to be so by physical means.

"At that time the discussion about the structure of the amorphous phase of ordinary polymers was still alive, although not at its highest point, and my feeling was that one had to look for totally unusual molecular structures in order to escape from the ambiguity and evanescence of what was at that time quoted as "local order in amorphous polymers".

"The decisive (and a posteriori trivial) idea came to me from the concepts of liquid crystallinity and from the examination of the most recurrent and general structural features of the low molecular weight compounds known to exhibit liquid crystal behaviour. The basic assumptions were the following:

1 - The nematic phase of small molecules of the type is dominated by the rigid group.

2 - The flexible tails, keeping a very high conformational mobility, have to a large extent the function of lowering the melting temperature accomodating themselves without particularly severe sterical requirements and

3 - Provided the flexible tails are not too short no particularly relevant effects should follow by connecting them to form a continuous (polymeric) linear structure.

These were the hypotheses and, as you know, they are not entirely correct!

"The next step was that of selecting a specific compound which could be prepared as easily as possible. We were lucky enough because the first trial was successful."[11]

In 1976 the year following the appearance of his pioneering paper, Sirigu published on the liquid crystal behavior of the exact small molecule analogues of his dimethylbenzalazine polyesters (shown on the following page). This study[12], derived directly from the polymer series, enables one to analyze the effect on liquid crystallinity of coupling (entropic restriction) the "monomeric" liquid crystals to form the related polymer.

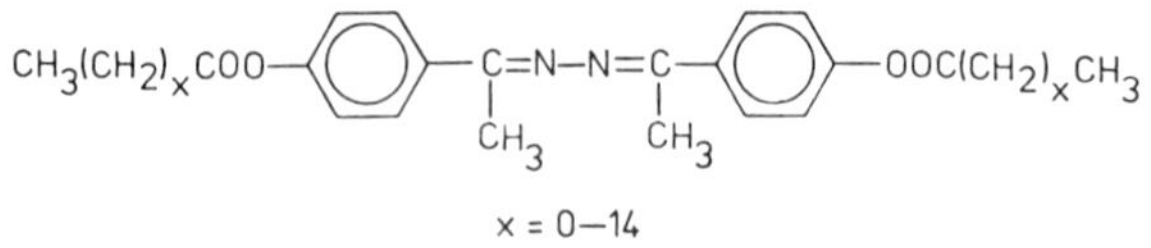

$$CH_3(CH_2)_xCOO-C_6H_4-C(CH_3)=N-N=C(CH_3)-C_6H_4-OOC(CH_2)_xCH_3$$

x = 0–14

Three years after Roviello and Sirigu's initial paper two reports appeared on regularly alternating rigid-flexible liquid crystalline polymers. One of these was from Guillon and Skoulios[13] at Strasbourg. Their polymer, a polyimine ether, is shown below. It contained a four-ring

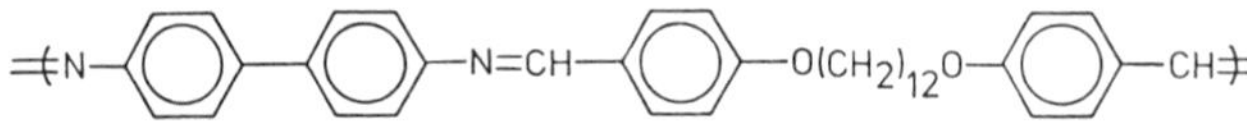

$$\text{≠}\!\!+\!N-C_6H_4-C_6H_4-N=CH-C_6H_4-O(CH_2)_{12}O-C_6H_4-CH\!+\!\!\text{≠}$$

rigid core and a twelve methylene diether spacer. Based on x-ray evidence the polymer was described as possessing a smectic B mesophase. Also in 1978, Samulski and Wiercinski[14] at the University of Connecticut used polymers having a polyimine ether structure (below) to probe the effects

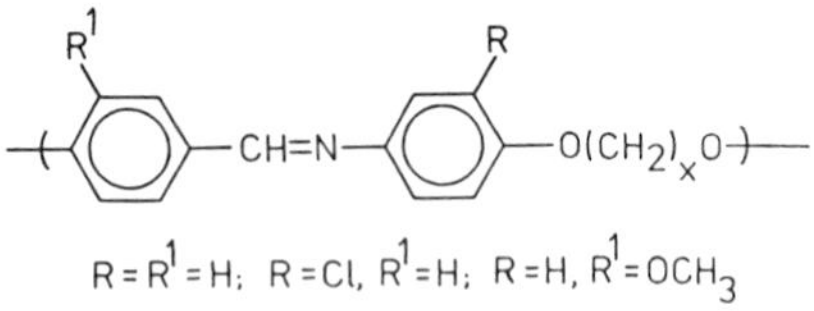

$$-\!\!\left(C_6H_3(R^1)-CH=N-C_6H_3(R)-O(CH_2)_xO\right)\!\!-$$

$R = R^1 = H$; $R = Cl$, $R^1 = H$; $R = H$, $R^1 = OCH_3$

of lateral substitution and spacer length on mesophase properties. It was found that the mesophase-isotropic transition temperature decreases with increasing substituent size and also that there is a pronounced even:odd alternation in both solid-mesophase and mesophase-isotropic transition temperatures as a function of spacer length (x).[15]

By 1979 other scientists were reporting on thermotropic liquid crystalline polymers having a regular rigid-flexible alternating structure in the polymer main chain. Among these were Fayolle, Noel and Billard[16] in France who studied an interesting polyester (below) having a

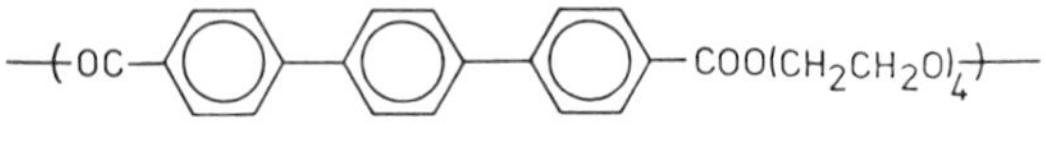

$$-\!\!\left(OC-C_6H_4-C_6H_4-C_6H_4-COO(CH_2CH_2O)_4\right)\!\!-$$

para-terphenyl rigid core and an oxyethylene spacer. This polymer was reported to be smectic C based on isomorphy with known smectic C phases. From the University of Lowell the Blumsteins and coworkers[17] reported on various rigid groups in polymeric liquid crystals. An example is the polyester on the following page which has a _trans_-stilbene rigid core. Influence of flexibility of the spacer on location and nature of the solid-liquid crystalline transition was also examined. A preliminary account of this work appeared in a 1978 ACS Polymer Preprint.[18]

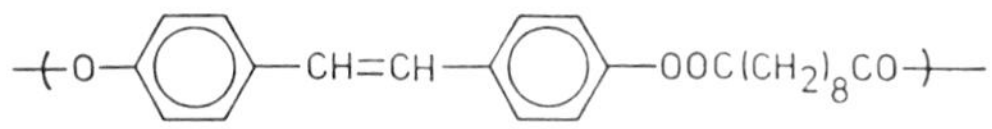

$-\!\!(O-C_6H_4-CH{=}CH-C_6H_4-OOC(CH_2)_8CO)\!\!-$

A doubly homologous series of polyesters (below) was described by Griffin and Havens[19] at the University of Southern Mississippi. Their

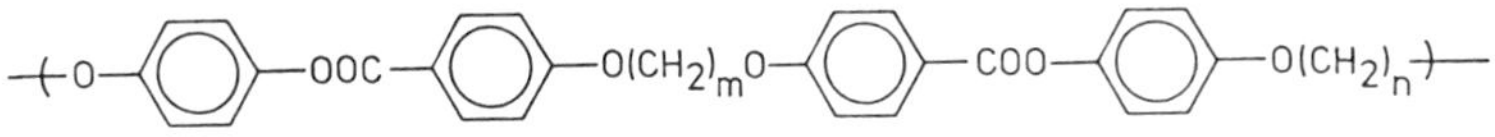

$-\!\!(O-C_6H_4-OOC-C_6H_4-O(CH_2)_mO-C_6H_4-COO-C_6H_4-O(CH_2)_n)\!\!-$

m = 2—10,12
n = 6,8,10

polymers were patterned after the stable, well-characterized small molecule liquid crystalline family of 4-alkoxyphenyl 4'-alkoxybenzoates. All thirty of these polyesters are liquid crystalline. The first case of monotropic (metastable) mesomorphism in a polymeric liquid crystal was reported by Roviello and Sirigu[20] for the α-methylstilbene polyester below. Although common in small molecule liquid crystals, the phenomenon was hitherto unknown in polymers.

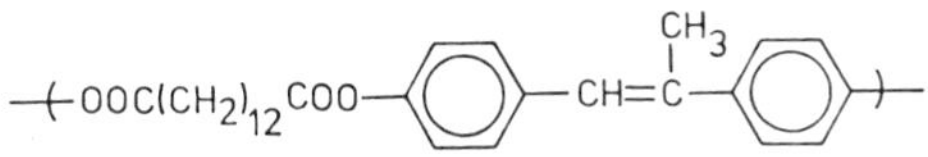

$-\!\!(OOC(CH_2)_{12}COO-C_6H_4-CH{=}C(CH_3)-C_6H_4)\!\!-$

Roviello and Sirigu reported also in 1979 on two other aspects of polymeric mesomorphism. One was on carbonate as a linking group to connect the rigid core unit to the spacer. These polycarbonates[21] (below) are

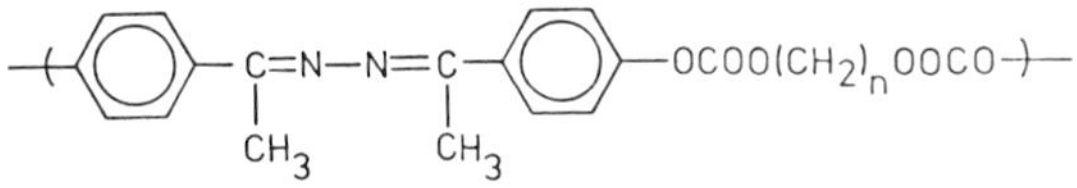

$-\!\!(C_6H_4-C(CH_3){=}N-N{=}C(CH_3)-C_6H_4-OCOO(CH_2)_nOOCO)\!\!-$

n = 6, 8, 10, 12

chemically analogous to their 1975 polyesters and have lower melting points and a wider mesophase range than the corresponding polyesters. Roviello and Sirigu also described the first copolymers[22] analogous to their 1975 polyesters. Using a fixed rigid core, copolymers were synthesized using two different spacer lengths in the diacid component. They found a depression of the melting temperature in the copolymers below that of either homopolymer. However, the mesophase-isotropic transition was linear with composition. Millard, Thierry, Strazielle and Skoulios[23] reported on a terephthaldehyde based imine ether polymer (on the following page) having a variable number of methylene groups in the spacer. They found a nematic mesophase for n = 2 and 6 and both nematic and smectic phases for the n = 12 homologue.

$$-\!\left(\!-C_6H_4-N{=}CH-C_6H_4-CH{=}N-C_6H_4-O(CH_2)_nO-\right)\!-$$

n = 2, 6, 12

Since 1979 a large volume of work on these types of polymeric mesogens has appeared. This review, however, sought to cover only the first four years, the beginning years, of research into this intriguing aspect of polymer science.

ACKNOWLEDGEMENT

This work was supported in part by a grant from the National Science Foundation (DMR-8417834).

REFERENCES

This paper is part 8 in a series on Mesogenic Polymers.

1. A. Elliott and E. J. Ambrose, Disc. Farad. Soc., 9, 246 (1950).

2. C. Robinson, Trans. Farad. Soc., 52, 571 (1956).

3. See M. G. Dobb and J. E. McIntyre in "Advances in Polymer Science 60/61: Liquid Crystal Polymers II/III", M. Gordon, editor; N. A. Platé, guest editor, Springer-Verlag, Berlin Heidelberg, 1984, p. 61 for discussion and references.

4. See S. L. Kwolek, Chemist (Washington, DC), 57, 9 (1980).

5. A. Roviello and A. Sirigu, J. Polym. Sci., Polym. Letters Ed., 13, 455 (1975).

6. D. Vorlander, Z. Phys. Chem., A126, 449 (1927).

7. See E. T. Samulski, Faraday Discuss. Chem. Soc., 79, 7 (1985) for a discussion of this topic.

8. D. Vorlander, Z. Physik. Chem., 105, 211 (1923).

9. A. C. Griffin and T. R. Britt, J. Am. Chem. Soc., 103, 4957 (1981).

10. P. G. de Gennes, C. R. Acad. Sci. Paris, Series B281, 101 (1975).

11. A. Sirigu, private communication to ACG, September, 1985.

12. A. Roviello and A. Sirigu, Mol. Cryst. Liq. Cryst., 33, 19 (1976).

13. D. Guillon and A. Skoulios, Mol. Cryst. Liq. Cryst. Letters, 49, 119 (1978).

14. E. T. Samulski and R. W. Wiercinski, Bull. Am. Phys. Soc., [2] 23, 296 (1978); R. W. Wiercinski, Ph.D. thesis, University of Connecticut, (1978).

15. For a fuller explanation of these phenomena see E. T. Samulski and D. B. DuPre in "Advances in Liquid Crystals, Vol. 4", G. H. Brown, editor, Academic Press, New York, NY, 1979, p. 121.

16. B. Fayolle, C. Noel and J. Billard, J. de Physique, 40(C3), Suppl. No. 4, C3-485 (1979).

17. A. Blumstein, K. N. Sivaramakrishnan, S. B. Clough and R. B. Blumstein, Mol. Cryst. Liq. Cryst. Letters, 49, 255 (1979).

18. K. N. Sivaramakrishnan, A. Blumstein, S. B. Clough and R. B. Blumstein, Polymer Preprints, 19, 2, 190 (1978).

19. A. C. Griffin and S. J. Havens, Mol. Cryst. Liq. Cryst. Letters, 49, 239 (1979).

20. A. Roviello and A. Sirigu, Makromol. Chem., 180, 2543 (1979).

21. A. Roviello and A. Sirigu, European Polymer Journal, 15, 423 (1979).

22. A. Roviello and A. Sirigu, European Polymer Journal, 15, 61 (1979).

23. B. Millard, A. Thierry, C. Strazielle and A. Skoulios, Mol. Cryst. Liq. Cryst. Letters, 49, 299 (1979).

FLUOROPLASTICS

THE HISTORY OF POLYTETRAFLUOROETHYLENE: DISCOVERY AND DEVELOPMENT

Roy J. Plunkett
14113 Jackfish Ave.
Corpus Christi, TX

INTRODUCTION

Good morning ladies and gentlemen. It is a pleasure for me to be here today and participate in this review program. I am proud of my contribution; I am proud of the Du Pont Company with whom I was associated all of my active career; I am proud of the fluoropolymer industry which has developed, and I am most of all proud of the benefits to mankind that have resulted from the original discovery of polytetrafluoroethylene and the development of useful applications for it.

The discovery of polytetrafluoroethylene (PTFE) has been variously described as (1) an example of serendipity, (2) a lucky accident and (3) a flash of genius. Perhaps all three were involved. There is complete agreement, however, on the results of that discovery. It revolutionized the plastics industry and led to vigorous applications not otherwise possible.

A number of events or developments had to occur before the stage was set for the discovery of PTFE. I shall describe and relate some of these events and point out their relationships to the discovery of PTFE.

In 1851 a U.S. Patent was granted to Dr. John Gorrie for "An Appliance for The Artificial Production of Ice in Tropical Climates." That invention can be considered as the ancestor of the modern household refrigerator.

Late in the 19th Century the Belgiam chemist Swartz established the chemistry for the introduction of fluorine into organic compounds which first achieved commercial status in the fluorocarbon refrigerant industry.

$$- C - Cl + HF \xrightarrow{SbF_3} - C - F + HCl$$

$$\text{eg.} \quad CCl_4 \longrightarrow CCl_2F_2$$

Swartz - 1895

Swartz found that chlorine attached to carbon could be replaced by fluorine if the chloro compound was heated with hydrofluoric acid in the presence of antimony trifluoride. He made a number of compounds that had the general composition $CxFyCl_2$. For approximately 30 years these materials remained essentially laboratory curiosities.

By the time of the 1920's decade the mechanical refrigerator had been developed to the point that it resembled closely the machine as we know it today. There was, however, one very serious drawback. The refrigerants available for use in those machines were such chemical compounds as ethylene, ammonia and sulfur dioxide. The machines of that time frequently leaked and discharged flammable, noxious and otherwise unpleasant gases to the consternation of the unsuspecting housewife.

Copyright 1986 by Elsevier Science Publishing Co., Inc.
High Performance Polymers: Their Origin and Development
R.B. Seymour and G.S. Kirshenbaum, Editors

In 1928, Boss Kettering representing General Motors, Frigidaire Division, asked Robert McNary, Thomas Midgely and Abert Henne to make another search for a safe refrigerant--one which would be colorless, odorless, tasteless, non-toxic and non-flammable. These gentlemen reviewed the literature and made a list of all known chemicals that had boiling points within the range that might make them useful as refrigerants. An analysis of the toxicity, flammability and physical constants of these compounds in relation to the periodic table led to a postulation that fluorocarbon compound should fit the requirements. Some of the work of Swartz was repeated and the compounds were evaluated against the desired properties. The compound $CCl_2F_2$was found to be almost ideally suited to the needs.

In 1930, Midgely and Henne presented a paper at an ACS meeting in Atlanta. They pointed out that some carbon chlorofluoro compounds had unique properties which would make them useful as refrigerants.

I am sure that you will be interested to know that apparently some luck or providence guided Midgely, et. al., in their work. Prior to that time it was generally believed that all compounds of fluorine would be very toxic. However, Midgely and Henne reasoned from their comparative study of properties that the fluorocarbons and chlorofluorocarbons would have a low order of toxicity. They decided to make some CCl_2F_2. Midgely, on the occasion of his being awarded the Perkin Medal, reported that he called a chemical supply house and ordered five one-ounce bottles of antimony trifluoride (the entire Country's supply). One bottle of SbF_3 was taken at random and a few grams of CCl_2F_2 were prepared. A guinea pig was placed under a bell jar with the CCl_2F_2 and instead of gasping and dying, the animal was not even irritated. This fulfilled their predictions. However, when another bottle of SbF_3 was used, the product proved to be toxic and the guinea pig died. Further examination showed that all but one of the bottles of SbF_3 contained water of crystallization which led to phosgene formation. If they had not by chance chosen the one bottle that contained anhydrous SbF_3, they might have given up the study as a "poor hunch."

Within five years of 1930 through the medium of Kinetic Chemical Corporation, a joint venture of General Motors and Du Pont, DuPont developed processes for a total of five refrigerants and introduced them to the market.

	BP°C
CCl_3F	23.8
CCl_2F_2	-29.8
$CHClF_2$	-40.8
CCl_2FCClF_2	47.6
$CClF_2CClF_2$	3.6

These materials are nonflammable; they have little odor and a low order of toxicity. They represent the backbone of the fluorocarbon industry.

In 1932 (I am sure that some of you will recall that year as the bottom of the great depression), I graduated from Manchester College with a major in chemistry. Try as I might I was unable to find any suitable employment. In the fall of 1932, I entered the Ohio State University Graduate school and four years later I received a PhD degree in organic chemistry. Immediately following graduation, I joined the Du Pont Company as a research chemist in the Jackson Laboratory and was assigned to work in the fluorocarbon field. Thus came Plunkett to the scene.

The literature records several early attempts to produce tetrafluoroethylene (TFE) with little success. In 1933, Ruff decomposed carbon tetrafluoride (CF_4) in an electric arc and obtained some crude TFE. This material was purified by reacting it with bromine and then with zinc. The purified material was used to obtain the first reliable determination of physical properties. In 1934, Henne reported the extraction of Cl_2 from $CClF_2CClF_2$ to obtain TFE.

In 1936, I was assigned the problem of developing an alternate refrigerant to $CClF_2CClF_2$ which, because of proprietary patent rights, was not available to all who desired to use it. The compound chosen for evaluation was $CClF_2CHF_2$. From a study of the information available at the time, I reasoned that a supply of the desired compound could best be obtained through the following reactions.

$$CClF_2CClF_2 \xrightarrow[\text{Alcohol}]{\text{Zn}} CF_2 = CF_2 + ZnCl_2$$

$$CF_2 = CF_2 + HCl \longrightarrow CCLF_2CHF_2$$

As a first step, I prepared approximately 100 pounds of TFE, purified it by distillation and stored it in one- to two-pound quantities in small steel cylinders at dry-ice temperatures. Then I set about converting the TFE to $CClF_2CHF_2$ by adding HCl to it over a carbon catalyst.

In carrying out the above step, a cylinder of TFE was placed on a balance and a metered stream of the TFE gas was passed over hot charcoal along with varying ratios of HCl gas.

On the morning of April 6, 1938, Jack Rebok, my assistant, selected one of the TFE cylinders that we had been using the previous day and set up the apparatus ready to go. When he opened the valve--to let the TFE gas flow under its own pressure from the cylinder--nothing happened. Jack called me over and asked whether we had used all the TFE from that cylinder. I said, I don't think so. We both tinkered with the valve a bit, and then thinking it might be stuck or closed in some way, we disconnected the cylinder from the line and pushed a wire through the valve opening. Still no TFE came out, although the weight of the cylinder showed that there was material inside. We were in a quandary. I couldn't think of anything else to do under the circumstances, so we unscrewed the valve from the cylinder. By this time it was pretty clear that there wasn't any gas left. I carefully tipped the cylinder upside down, and out came a whitish powder down onto the lab bench. We scraped around some with the wire inside the cylinder--or maybe I tapped it--I don't remember which--to get some more of the powder. What I got out that way certainly didn't add up, so I knew there must be more inside. Finally, more out of curiosity I suppose than anything else, we decided to cut open the cylinder. When we did, we found more of the powder packed onto the bottom and lower sides of the cylinder.

My first reaction to discovering that the TFE had polymerized was now we will have to start all over; my helper felt about the same. I didn't quite realize what had taken place. It did not occur to me right away what possibilities the material might have. That had to wait until I could get some lab tests run on it. There was no doubt in my mind that the TFE had in some fashion polymerized. Within a few weeks lab tests showed that the polymer was inert to all the solvents, acids and bases available. Experiments carried out showed that polymerization of TFE occurred under various conditions as (1) in the absence of a specific catalyst, (2) in the presence of a catalyst and (3) in the presence of a catalyst and a solvent.

Before the initial polymerization occurred, I did not know that given the right chance TFE would polymerize. But looking back on the discovery, what is more significant to me is that storing a cylinder with a kilogram of TFE in it could have been the end of both Jack and me. Spontaneous polymerization of TFE can lead to explosive reaction.

While I had recognized that PTFE was an unusual material, I was somewhat at a loss as to what to do with it. However, I was fortunate to be associated with a Company that had a large interest in polymers and was supporting many scientists and engineers in the field of polymer chemistry. So we asked some of my associates in the Central Research Department to characterize and study the new polymer. In short order they developed laboratory techniques for the fabrication and use of PTFE. Engineering and economic evaluations of the costs of developing commercial production processes and fabrication procedures for making useful products were so discouraging that it was difficult to justify proceeding with the development.

In the early stages of the Manhattan project during World War II, it became evident that improved materials were necessary for use as gaskets, packings and linings in plants producing and handling UF_6 at Oak Ridge, Tennessee. PTFE seemed to be the answer and with the interest of Du Pont personnel and the urging of leaders of the Manhattan project, intensified activities were pursued at Du Pont.

1. On fundamental studies in the Central Research Department.

2. On the development of a process for producing TFE at Jackson Laboratory.

3. On the development of processes for polymerization of TFE and procedures for polymer product fabrication at Arlington.

These efforts were highly successful. Pilot plants were established to produce and purify TFE to polymerize TFE to PTFE and to fabricate the polymer into useful shapes for use.

The excellent properties of PTFE led to its adoption for military uses during World War II. The first use was as a nose cone covering the proximity fuses on artillery shells.

Because of the military and other governmental interests the Patent Office placed an order of secrecy on research and development work on most fluorochemical activities. Accordingly, most of the public did not hear of these activities until after the war was over.

I was granted a patent on PTFE as a composition of matter on February 4, 1941. The patent also covered polymerization under pressure and under pressure in the presence of a catalyst.

The Du Pont trademark "TEFLON" for PTFE resins was coined in 1944.

In 1946, after being released from secrecy, papers describing the chemistry, processes of production and fabrication of PTFE were published. About 30 patents were granted by the U.S. Patent Office.

In a paper published in JACS in 1946, Hanford and Joyce reported on laboratory work at the Du Pont Experimental Station and described:

1. The polymerization of tetrafluoroethylene.

2. The properties of polytetrafluoroethylene.

3. X-Ray diffraction of PTFE.

4. Transition point of PTFE.

5. Structure and viscosity of PTFE.

6. Cohesion of PTFE.

7. Insolubility of PTFE.

In a paper published in JIEC in 1946, Renfrew and Lewis reported on pilot plant production of PTFE and provided the first detailed description of fabricating techniques and properties of the polymer. They presented:

1. A short history of TFE.

2. A detailed discussion of the PTFE and its properties.

    a. Resistance to reagents and solvents.

    b. Thermal properties.

    c. Electrical properties.

    d. Mechanical properties.

    e. Miscellaneous properties.

3. PTFE was summarily described as follows:

    "Polytetrafluoroethylene is a new plastic being manufactured on ar experimental plant scale. It is insoluble in all solvents tried so far and is not attacked below its melting point by any common corrosive agents except molten alkali metals. It withstands temperatures up to 300°C for long periods without serious degradation and is not brittle at low temperatures. An outstanding property is its combination of low power factor and low dielectric constant. The principal current uses for polytetrafluoroethylene are as gaskets and packing in equipment for handling hot corrosive liquids and as electrical

> insulation, particularly at high frequencies and under strenuous environments. The plastic is being sold in small quantities for development purposes, in the form of simple shapes such as tape, sheets, rods and tubes, gaskets and insulated wire."

Large scale development production begain in 1946; and the first commercial plant went on stream at Parkersburg, West Virginia in 1950.

Many scientists and engineers as well as independent entrepreneurs have contributed immensely to the development of multitudes of use and applications for PTFE in the years that have followed the disclosures mentioned above.

Products made from PTFE touches people in countless ways everyday from cookware to wire insulation and automobile gaskets; from soil repellants to dome roofs; from ski suits and space suits to lifesaving body parts. PTFE with its unusual properties and applications has become an integral part of the machinery that runs today's economy, affecting the fields of architecture, medicine, energy and environment conservation, fire protection, space travel, electronics and the military, among others.

It has been exciting to see something I worked on become of such great benefit to mankind in general as well as to real people I know. Some years ago, at a social affair, I was introduced by a doctor friend to a man who had been suffering from a serious heart defect. "See that fellow dancing over there" asked my friend, "he's alive today because he is wearing a "TEFLON" aorta which I installed." Over the years I have heard of and experienced many incidents like that. It sort of makes the occasional discouragements of research a lot easier--I believe I've been more than just helpful; I've made a real contribution.

POLYTETRAFLUOROETHYLENE: HISTORY OF ITS DEVELOPMENT AND SOME RECENT ADVANCES

Carleton A. Sperati
Stocker Adjunct Professor of Chemical Engineering
Ohio University, Athens Ohio
and
Polymer Products Department
E. I. DuPont de Nemours & Co., Inc.
Wilmington, Delaware

INTRODUCTION

In the first paper on fluorocarbon polymers in this Symposium [1], Dr. Plunkett described the details about his discovery of polytetrafluoroethylene (PTFE) nearly fifty years ago [2]. It is the purpose of this paper to continue his story and to review aspects of the history that are listed on Table I. This list includes an overview of the nature of PTFE, the history of some key inventions and development of scientific understanding about the polymer, and brief mention of a few recent studies. Even in this published version of the paper from the Symposium, it is not feasible to cover the history in any sort of detail. Therefore, this paper can only outline the general background, with sufficient references to the literature so that detailed information will be readily accessible. The references, shown within [ ], are selective, rather than exhaustive, leading the reader to some of the many people who made PTFE the important material it is. Usually, a reference is made to a pioneering patent or paper rather than to many that followed.

PTFE is used so many places in our daily lives that few of us even know about many of its applications. This paper, however, is concerned with history and development of the science of PTFE, so that it will include essentially no mention of applications. Even today, nearly fifty years after Dr. Plunkett's notable discovery, we are still learning many significant things about that waxy feeling, seemingly intractable white solid he found in the cut-open cylinder.

Dr. Plunkett described the inertness of that solid; a material that is insoluble in all common solvents and did not appear to melt. It was not long, however, before PTFE succumbed to simple fabrication techniques, albeit they were, and still are, novel to normal plastics technology [3-6].

Formally, PTFE is classed as a tough thermoplastic. In practice, however, the exceptionally high melt viscosity of all forms of the resin useful for plastics applications requires processing techniques that, in many respects, are similar to those used in the metals and ceramics industries.

MAJOR STAGES IN DEVELOPMENT OF PTFE

Table II is a list of inventions, inventors and researchers, and references in the history of PTFE that I feel are especially significant. Plunkett's paper summarized the development until the early 1950's. During that period, in fact even today, much of the research work is to learn how to make the monomer and polymer safely in quantities large enough to meet the demands for use in current commercial application and for developing additional applications.

The explosion of TFE in the first semiworks is included on the list to help make two points. Dr. Plunkett mentioned that TFE is a very hazardous

Copyright 1986 by Elsevier Science Publishing Co., Inc.
High Performance Polymers: Their Origin and Development
R.B. Seymour and G.S. Kirshenbaum, Editors

gas and commented on his good fortune that the cylinder of TFE did not explode. Not only does TFE have an extremely wide explosive range when mixed with air, but it can, in addition, disproportionate under a wide range of conditions to form carbon and carbon tetrafluoride with the explosive force about equal to black powder. These hazards were not known before the time of the explosion. The polymer, polytetrafluoroethylene, on the other hand, is one of the most stable organic materials known. In my own case, my work was in the building that was severely damaged, and where two of our colleagues were mortally injured. I was fortunate that the accident occurred during the night. In spite of the accident and loss of life, however, the work continued to permit those concerned to learn what caused the explosion and to learn how to make PTFE safely.

TABLE I. Topics for discussion.

The Nature of PTFE

Stages in the Development of PTFE
- Inventions of Products
- Inventions in Processing
- Scientific Understanding

Preparation of PTFE
- Granular PTFE
- Dispersion PTFE

Processing PTFE

Recent Developments

TABLE II. Major advances in development of PTFE.

Discovery of PTFE, April 6, 1938 Roy Plunkett [2]

Invention of Process to Make Granular - M. M. Brubaker [7]

Invention of Dispersion Polymer M.M. Renfrew [8]; K.L. Berry [9]; J.F. Lontz & W.B. Happoldt [10]

Invention of Lubricated Extrusion for Coagulated Dispersion Polymer J.F. Lontz, et al. [11]

Destruction of PTFE Semiworks by Explosion of TFE. - Nov. 22, 1944

Invention of Finely Ground Granular Polymer - Thomas & Wallace [15]

Invention of Free-flowing Agglomerated Granular Powders Anderson & Roberts [16], Black [17]

Invention of High Strength, Porous PTFE - R. Gore [18]

TABLE III. Manufactures and Trademarks for PTFE

Trademark	Made by:
TEFLON	DuPont in the USA & Netherlands Mitsui-DuPont in Japan
FLUON	ICI in England & USA
ALGOFLON	Montecatini in Italy
SOREFLON	Ugine-Kuhlmann in France
POLYFLON	Daikin in Japan
HALON	Allied Corporation in the USA (Montecatini)
HOSTAFLON	Hoechst in Germany

As Dr. Plunkett mentioned, all of the initial product from the pilot plant at Arlington, New Jersey went to the Manhattan Project. After World War II, the polymer became available for critical industrial uses. A combination of extreme resistance to chemicals, excellent resistance to very high temperatures, superb electrical properties, and unique surface properties indicated a promising future for this new material. Plans for commercial production led to construction of a full scale plant at the new DuPont facilities at Parkersburg, WV where operations started in 1950. Since then there have been almost continuous expansions within DuPont and the other companies that now manufacture PTFE. Table III lists various manufactures with the trademark each uses.

Yelton had an early paper that described PTFE in general terms [19], and Dr. Plunkett has already referred to the papers by Hanford and Joyce [20] and Renfrew and Lewis [21]. It is interesting that contemporary work with the composition of matter that Renfrew and Berry discovered is the basis for some of the more exciting work on PTFE that will be mentioned later. Renfrew's research on PTFE certainly is one of the items that must be added to the many activities for which he has been honored.

PTFE had many properties that were considered "mysterious" and a "language", much of it really a form of "shop jargon", evolved to describe the processing and properties that are unique to this material. As is common with material break-through discoveries, much of the early research was empirical, yielding critical inventions along with practical ways of doing the job needed without, necessarily, giving a satisfying understanding of the science underlying the work. Of particular significance was development of a form of PTFE dispersed in water as a suspensoid. The research of Berry, Renfrew, Lontz, Happoldt, W. L. Gore, and others cited in Table II was very significant and many new fields of application were opened.

In the mid 1950's, right after our basic work on structure v. properties of polyethylene, some concerted work within what was then the DuPont Polychemicals Department, led to a reasonably good understanding of the basic nature of the polymer. Those studies were summarized and presented at a series of meetings.

1) Gordon Research Polymers Conference in 1955 [22] ;

2) Society of Plastics Engineers later in 1955 [23-25];

3) A series of five papers at the fall meeting of the American Chemical Society in 1956 [26-30]; and

4) A conference for users of PTFE held September 17-18, 1957 [31].

The ACS papers from the DuPont laboratory were very significant in advancing understanding about PTFE. Available scientific information about PTFE was summarized in a review paper in 1961 [32]. In addition, Sherratt's excellent article in 1966 [33] brought into the literature many of the results from research at Imperial Chemical Industries. Through the years there have been many additional contributions. Zisman was the first person to recognize the unusual surface properties of PTFE and his research was very important [34]. Much help was obtained from Prof. William T. Miller, inventor of high molecular weight polychlorotrifluoroethylene, in his role as a consultant to the DuPont research group. Also noteworthy are papers by Clark et al. [35], McCrum [36], Eby and Sinnott [37], Wunderlich [38], a review paper by Gangal [39], and a series of papers by Starkweather and his associates [40].

PREPARATION OF POLYTETRAFLUOROETHYLENE

Polymerization of TFE

Commercially, PTFE is available in three different forms that come from two different types of polymerization. These materials are known as granular polymer, aqueous dispersions, and coagulated dispersion products, often referred to as "fine powder". Specifications for each of these materials have been prepared by the American Society for Testing and Materials and pertinent ASTM standards are listed in Table IV. These standards are a source of much useful information about PTFE that is not readily available elsewhere.

Superficially, the free radical polymerization of TFE to form PTFE seems to be a relatively simple process. One starts with an autoclave partly full of water containing a very small amount of an initiator and, perhaps, some surface active agent. The mixture is heated at a moderate temperature, very pure TFE monomer is introduced into the kettle under pressure, and the polymer forms as white, relatively coarse particles in granular polymerization and as a milky-looking suspensoid in the other type. When the desired amount of polymer is obtained, the batch is cooled and the polymer is isolated. All of these operations must be done by remote control of the well-barricaded autoclaves.

TABLE IV. ASTM Standards on PTFE Related Materials.

D 1457	Specification for PTFE Molding and Extrusion Materials
D1675	Method of Testing Polytetrafluoroethylene Tubing
D 1710	Specification for TFE Fluorocarbon Rod
D 2686	Specification for Tetrafluoroethylene-Backed Pressure-Sensitive Electrical Insulating Tape
D 2902	Specification for Fluoropolymer Resin Heat-Shrinkable Tubing
D 3293	Specification for PTFE Resin Molded Sheet
D 3294	Specification for PTFE Resin Molded Basic Shapes
D 3295	Specification for PTFE Tubing
D 3297	Practice for Molding and Machining Tolerances for PTFE Resin Parts
D 3308	Specification for PTFE Resin Skived Tape
D 3369	Specification for TFE Fluorocarbon Resin Cast Film
D 4441	Specification for Dispersions of Polytetrafluoroethylene
D 4591	Method for Determining Temperatures and Heats of Transitions of Fluoropolymers by Differential Scanning Calorimetry
E 911	Specification for Glass Stopcocks with Polytetrafluoroethylene (PTFE) Plugs
F 423	Specification for Polytetrafluoroethylene (PTFE) Plastic-Lined Ferrous Metal Pipe and Fittings
F 754	Specification for Implantable Polytetrafluoroethylene (PTFE) Polymer Fabricated in Sheet, Tube, and Rod Shapes

The usual equations that are used to describe free radical chain polymerizations are not readily applicable to the polymerization of PTFE because the reaction does not meet the first requirement for application of the usual kinetic scheme, the steady state assumption [41, 42]. The $-CF_2$ radical has a very long lifetime, and usual termination reactions (combination and disproportionation] do not seem to occur. The need to have an extremely high molecular weight in order to have adequate physical properties requires a reaction system essentially free of any material that will chain transfer, eliminating that mode of termination. The result is that molecular weight of PTFE increases with time of polymerization, unlike most other radical-chain polymerizations.

The Nature of Granular PTFE

The first PTFE that was made and sold commercially is known as granular polymer. The product mentioned above is isolated as a relatively coarse solid and the particles are reduced in size to the level desired. Thirty years ago standard product had a size about 500 micrometers. The powder, after drying, is then pressed, at or near room temperature, in a mold approximating the final shape desired, and then this preform or green piece is heated above the melting point of the polymer, and finally cooled. The molded products made in this manner often had up to several percent of voids that had a profound, deleterious effect on some properties of the material [25]. The solution was to comminute the raw polymer to a particle size between a tenth and a twentieth of the original products [15]. Moldings from these powders were essentially free of voids and had greatly improved properties. On the other hand, the small particles tend to stick together by entanglement, or more probably, by secondary valence forces holding them together resulting in poor powder flow. To improve the powder handling characteristics, many different procedures were developed to give controlled agglomeration of the small particles to form powders of controlled hardness with a size about that of the first granular products. These powders give moldings with low void content and mechanical properties only slightly inferior to those available from the unagglomerated products. [16-17]. Almost all commercial granular powders today fall in this class.

The granular powders are, in fact, not the "simple" products that they appear to be, but they are a mixture of two quite different kinds of PTFE that have been referred to as alpha and beta PTFE [43]. The naming stems from evidence that the alpha material is formed in the polymerizing mixture before the beta product. During polymerization the polymer is formed initially as an unstable suspensoid. Depending on the details of the conditions used for polymerization, concentration of the dispersed polymer may vary from a fraction of a percent to several percent. Eventually, the suspensoid coagulates and forms a solid phase that is not wet by the reaction medium. The coagulated polymer carries with it radicals active for further polymerization. The subsequent gas-solids polymerization continues at much higher rate, to form a solid polymer with much lower specific surface area. The communition (grinding) mentioned above operates differently on the two kinds of polymer because the initial suspensoid material is ground more easily. In fact, the two kinds of polymer can be separated using modification of fluid energy grinding techniques. The differences between these materials can be seen easily by examination with a microscope at about 200 X. Photomicrographs show both kinds particles when viewed between polarizing elements, with the beta polymer showing typical Maltese cross patterns. The alpha polymer has a relatively high specific surface area of about 5-6 m^2/g, while the beta material is typically less than 1 m^2/g. Other properties show similar differences.

The Nature of Dispersion Based PTFE

There is a common misconception that these products are made by an emulsion polymerization. They are not, even though the product is often called a "latex", emulsion, dispersion, etc. For a classic emulsion polymerization to occur one must have liquid monomer dispersed in an aqueous phase. During the polymerization of TFE, as it is done in many commercial processes, the temperature is well above the critical temperature of TFE and the pressure is well below the critical pressure of the monomer. There are many different ideas as to the exact mechanism of the polymerization, but this writer at least, is not aware of any results that show convincingly exactly what happens. The dispersion is really a suspensoid, with particles varying in shape, but with dimensions of the order of 0.2-0.4 micrometers [10].

As made, the dispersion is coagulated easily, by vigorous agitation, for example, to produce the coagulated product. This material is converted into useful materials by a process that appears to be unique. It is called lubricated or paste extrusion [11-14] and is widely use in the fluoropolymers industry today. In this process, outlined on Table V, the coagulated and dried dispersion polymer is blended with about forty volume percent of a liquid such as a purified kerosene. The blended powder still is free-flowing but it has the capability of being compressed into a preform or "billet" and extruded through an orifice to form sheet, ribbons, beading, thin coatings on wire, and many other shapes.

When lubricated extrusion is used to coat wire and other applications where a very thin cross section is needed, there is a very large change in diameter between the preform and the extrudate. In other words, a very high reduction ratio. Special resins have been developed to meet the need for superior performance under such conditions [44], and such resins tend to dominate the marketplace today. Many of these resins use very low concentrations of a co-monomer to provide a modified structure.

A major discovery on how the paste extrudate can be used has lead to a new form of PTFE, high strength, porous, expanded PTFE. Table VI shows pertinent information. During expansion there is little or no change in cross-sectional area of the material being stretched. The length, however is increased many fold to produce a highly porous product in which the polymer has greatly increased strength and a structure characterized by "nodes and fibrils." This material has a melting point about 60°C higher than earlier fabricated PTFE [45]. The new, high melting PTFE was discovered while processing old type materials in a new way [18, 46]. Also it has been reported to give moldings that have strength properties, up to 20,000 psi in moldings, well beyond any reported previously for PTFE [18].

TABLE V. Lubricated extrusion of PTFE.

Dried coagulated dispersion PTFE ("Fine Powder") is blended with organic "lubricant".

A preform is made at or about room temperature.

Using a ram-type extruder, the "paste" is extruded through an orfice to makes wire coating or shapes.

The lubricant is removed by drying or extraction.

Extruded material is sintered and (usually) quenched.

TABLE VI. High Strength, Porous, Expanded PTFE.

Dried extrudate from lubricated extrusion is stretched at high rates of stretching and elevated temperatures below the melting point of PTFE.

Cross sectional area does not decrease.

The expanded material can be "amorphous locked" by heating, restrained, above the melting point of PTFE.

TABLE VII. Tools for characterizing PTFE.

"Standard Specific Gravity"
Melt Viscosity
Thermal Analysis
- Dynamic methods
- Differential Scanning Calorimetry

Infrared Spectrometry
Specific Surface Area
Microscopy
Mechanical Properties

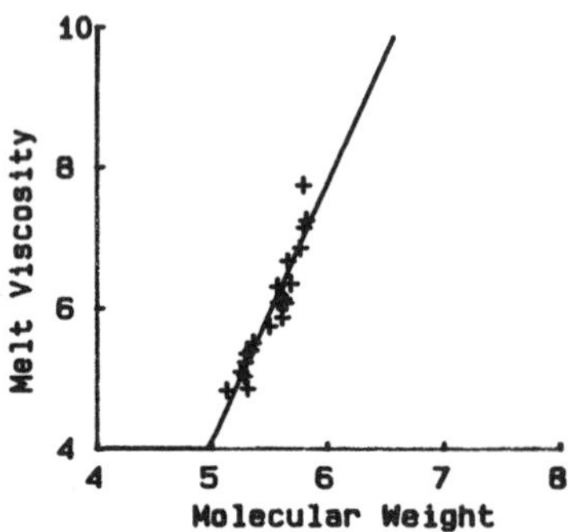

FIG. 1. Melt Viscosity vs Molecular Weight (Log scales).

The need to have base polymers that would give this new PTFE with less stringent process conditions has led to a remarkable development of coagulated dispersion resins that have much higher molecular weight and, probably a narrower distribution of molecular weights [47-49]. Extrudates from these powders can often be stretched fifty to sixty fold and have greatly enhanced the things that can be done during the manufacture of the high strength, porous PTFE. While the exact nature of these materials has not been reported, a folded structure of an extended chain is likely [50-51].

CHARACTERIZING PTFE

It can be generalized that PTFE follows the usual principles of polymer science. Nevertheless, the nature of the polymer, especially its insolubility in any available solvent, restricts use of most of the tools usually used for determining the nature of polymers. Even if general procedures can be used, usually they require modification in details. In spite of the problems, much basic information has been determined using the techniques listed on Table VII.

Molecular Weight

Molecular weight of a polymer is a fundamental property of great importance. Data from polymerizations made using radioactive sulfur in the initiator system, reported in the 1956 ACS papers are still a source of basic information. There are, however, assumptions in that work that have never been proved theoretically or experimentally. Nevertheless, the polymers made are useful in appraising new tools used to characterize PTFE [52].

The first useful procedure for assessing relative molecular weight was "standard specific gravity" (SSG) [32]. The test is incorporated into the ASTM standards (Table IV) and still is the procedure most widely used. SSG is based on the observation, common with many polymers, that the higher the molecular weight, the higher the melt viscosity and the slower the rate of crystallization under any given rate of cooling from the melt. After a "standard" thermal cycle, the specific gravity of a low molecular weight sample is higher, and the decreasing rate of crystallization gives correspondingly lower specific gravities as the molecular weight increases.

Use of differential scanning calorimetry (DSC) offers promise of much utility [53], but it is not used as widely as SSG. DSC is discussed further below. End group analysis also offers opportunity. Early studies on low molecular weight PTFE showed the nature of the end groups [54]. The technique was not available for commercial materials, however, since the very high molecular weight of the PTFE used for plastics applications requires a level of sensitivity not available until the availability of multiple scans typical of FTIR techniques. The writer suspects that there may be proprietary procedures available in industrial laboratories. The only published report of such studies known to the writer, however, are those of J. Sun [55] who found carboxyl endgroups in a samples of commercial PTFE products. There is a real opportunity for useful results in extending his work.

Melt Viscosity Relationships

The melt viscosity of PTFE is very high, up to 10^{11} poise, and usual methods for measuring MV are not applicable. Tensile creep, however, has been used successfully by Tobolsky and his students [56] and by Lontz [57]. Other useful information has come from measurements on low molecular weight PTFE made for use as lubricating powders, i. e., a "white graphite" used in a wide variety of applications. Initially these materials were made by thermal degradation of high molecular weight PTFE, and grinding the resulting high melting, brittle, wax-like product. When it was discovered that PTFE irradiated in air showed profound decrease in molecular weight, degradation by radiation became the process of choice to make the L-powders. (It is worth noting that, as part of his extensive work on radiation of PTFE [58], Wall showed that polymer irradiated in inert atmosphere is much more stable [59].)

In one study, PTFE was irradiated at increasing doses from 2 to 30 Mrads using both a gamma source and high energy electrons. Information from this study is shown on Figure 1 where the melt viscosity (at low shear rate) is plotted as a function of the molecular weight determined by combining information from infrared end group analysis with the concentration of radical ends calculated from the relationship of Siegel and Hedgepeth [60] and given in equation form in [61]. It is especially interesting that the slope of the regression line on the log-log plot is less than one standard deviation higher than the value of 3.4 characteristic of most polymers with molecular weights above the critical entanglement range.

Thermal Analysis of PTFE

Thermal analysis procedures have been powerful tools for studying PTFE. McCrum's classic work with early dynamic methods [36] clearly showed the various first and second order transitions that were elucidated further by the work of Eby, Sinnott, and others. These transitions, of course, control much of the behavior of PTFE.

Differential Scanning Calorimetry, DSC, is especially useful [62]. The heat of crystallization of a given sample is dependent on the rate of crystallization just as is the specific gravity of a molding in the SSG test discussed above. Studies at the Japanese Atomic Energy Research Institute, JAERI, led to the first paper on use of DSC to estimate relative molecular weight [53]. DSC was used effectively in other work at JAERI [63], and the work was extended using some of the original radio-sulphur samples [52]. Of particular interest is the observation that details of the DSC curves on initial melting appear to reflect differences in molecular weight and distribution of molecular weight deduced from conditions used for polymerization [47]. ASTM D4591 (see Table IV), developed especially for fluoropolymers, was approved in March 1986 and is being tested in various laboratories. Figures in references 45, 47, 52, 53, and 62 show some of the differences that can be seen using DSC.

X-ray procedures give such a diffuse amorphous halo [64] that quantitative measure of crystalline content is very imprecise. Infrared spectroscopy, on the other hand, using the procedures developed by Moynihan [27] has been a very useful tool for appraising the relative crystalline and amorphous content of PTFE.

Mechanical Properties of PTFE

One of the oldest ways of characterizing polymers, their behavior in stress-strain, seems an unlikely way to follow changes in morphological and fundamental molecular properties. In fact, with PTFE at least, the shape of a tensile strength curve can be very revealing. Initial reports on this approach are in our ACS paper from 1956 [30].

Other results are shown in Figure 2. The term "sintering" as use in PTFE "jargon" means heating above the melting point of the polymer. "Sintering" as used in powder metallurgy, however, means heating at or below the melting point of the material that is being coalesced. Figure 2 shows results from moldings made from a small particle size, high molecular weight granular powder that were preformed and heated under identical conditions except for oven temperatures of 335°C and 340°C, respectively. The tensile stress-strain curve for the specimen heated at 335°C, as identified on the

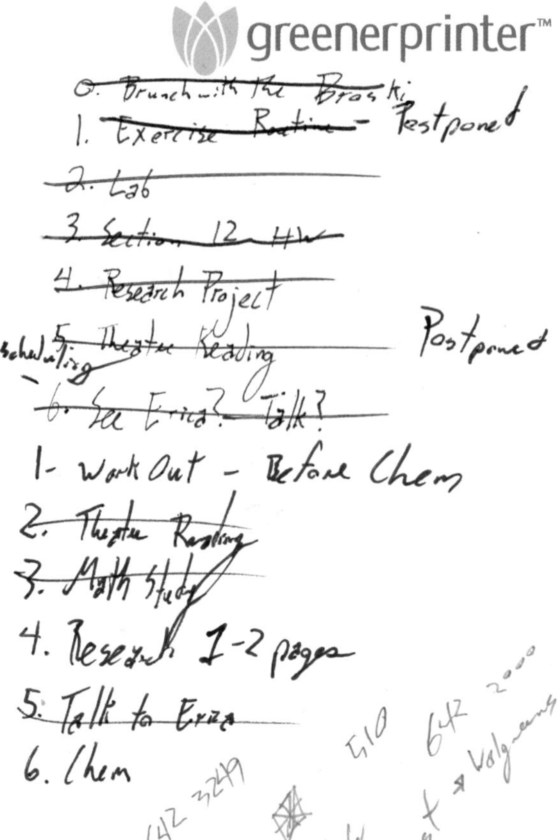
greenerprinter™
0. Brunch with the Broski
1. Exercise Routine - Postponed
2. Lab
3. Section 12 HW
4. Research Project
scheduling 5. Theater Reading Postponed
6. See Erica? Talk?
1- Work Out - Before Chem
2. Theater Reading
3. Math Study
4. Research 1-2 pages
5. Talk to Erica
6. Chem
642 3249
510 642 2000
Transfer Request Walgreens

www.greenerprinter.com • 510 898 0000

on 100% recycled content, 100% post-consumer waste, processed chlorine-free paper.

FSC SCS-C 0012

figure, shows the characteristic behavior of a high crystallinity material, i. e., a high initial modulus followed by true yield behavior and large extension with only a small increase in stress. In contrast, the curve for the material that had been melted during processing, has a lower initial modulus, no true yield point, and a rapid increase in stress up to the break point. Other properties of the moldings are summarized in Table VIII. The stress values on the figure are so-called Engineering stress in which the values shown are based on the initial cross-sectional area of the test specimen rather than the actual cross section of the specimen at the elongation shown.

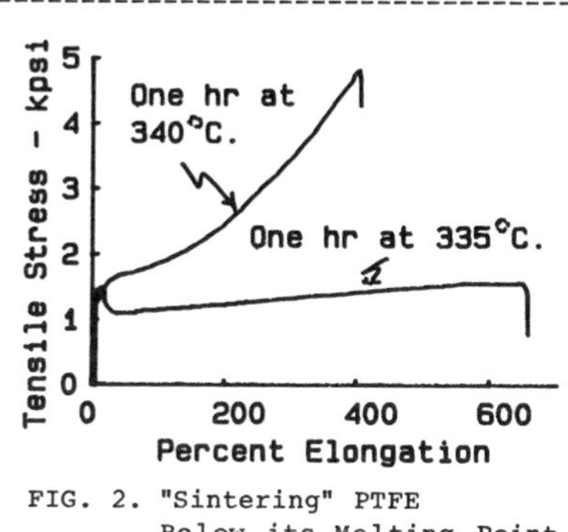

FIG. 2. "Sintering" PTFE Below its Melting Point.

TABLE VIII. "Sintering" PTFE Below its Melting Point.

Molding temp. °C.	335	340
Stress at 10% E.	1500	1474
Yield Point, psi.	1500	no yield
%E at yield.	8	----
Mod. at rupture, psi	88	1327
Break strength, psi	1530	4560
Break elongation, %.	670	390
Tensile impact strength ft lbf/in.^{2}	216	269
Specific gravity	2.25	2.17
% Increase in weight after 600 hr in boiling perchloroethylene.	0.42	1.48

BLOCK COPOLYMERS WITH PTFE

It is well known that irradiated PTFE can be used to initiate graft polymerizations on the surface of the polymer. In addition, PTFE irradiated in air was found to contain peroxy radicals with very long lifetime, at least several years [60, 65, & 66]. Recently, K. Donato [67] has studied the kinetics of the polymerization of graft block polymers starting with irradiated lubricating powder and producing block-grafts with methyl methacrylate. Some of the block-grafts have more than fifty volume percent of the graft polymer. The quantities made to date are small, but the work is continuing to give enough material for more detailed characterization.

EFFECT OF MOLECULAR STRUCTURE AND PROCESSING ON PROPERTIES

Many observations and bits and pieces of evidence over many years can be summed up to explain why PTFE behaves as it does. First is the question as to why PTFE can't be made with a melt viscosity in the range typical for thermoplastic polymers. The answer is that it can be and a lot of material is made with melt viscosity in the range of 10^2-10^4 Pa s (10^3- 10^5 poise), just right for melt extrusion and injection molding. Why then do we not use such a polymer? The answer is that it crystallizes so fast that any molding becomes highly crystalline and very brittle.

What makes the high molecular weight materials suitable for plastics applications? Theoretical studies show that the intermolecular forces between difluoromethylene groups are very low. Thus one needs either very high molecular weight of highly crystalline materials in order to have an integrated total enough for adequate strength, or one needs some way to disrupt the crystalline order. In the next paper, R. Putnam will discuss how this disruption can be done by use of comonomers. With the homopolymer, however, the best way to inhibit crystallization is to increase the viscosity of the molten material to very high values by making polymer with very high molecular weight.

Following this line of reasoning and using information such as that shown on Figure 2 and Table VIII, this writer has concluded that the main function of the "sintering" step, where a preform of the polymer is heated above its melting point, is not to effect coalescence, i. e., eliminate voids. Rather, the primary function appears to be to melt the polymer so that it can be recrystallized to a low enough crystalline content to give the sort of mechanical properties that are required for the current commercial applications of PTFE.

Until the development of the higher molecular weight coagulated dispersion powders ("fine powders") mentioned earlier, the polymer used for lubricated extrusion had a relatively low molecular weight. It was used in thin sections that could be cooled quickly from the melt to give low crystallinity products, and the toughness that goes with them. Granular products, on the other hand, are use to make moldings weighing up to 700 kg, 1500 pounds, about the largest made routinely with any thermoplastic polymer. Heat transfer requirements make it impossible to cool the center of such a molding other than slowly. To maintain a low crystalline content in the final molding, the melt viscosity, i. e., the molecular weight of the polymer, must be very high. If the molecular weight is not high to start with, or if there are reactive impurities in the polymer so that degradation occurs during the heating step, the mechanical properties of the final product are not satisfactory.

SUMMARY AND CONCLUSIONS

The history of the development of polytetrafluoroethylene was surveyed broadly. This very useful polymer is unique in its combination of properties. Most techniques for studying PTFE require procedures that are unusual for polymers. In spite of the unconventional methods needed to determine fundamental properties, the broad principles that control the polymer are those of polymer science.

PTFE homopolymer or PTFE containing very small amounts of a comonomer is the largest part of the total market for fluoropolymers. Melt processible materials, described by Dr. Putnam in the next paper of this series, are very important, but the PTFE materials are likely to hold their place as the major product for the foreseeable future.

REFERENCES

1. R. J. Plunkett, Meeting of the American Chemical Society, New York, NY, April 16, 1986.
2. R. J. Plunkett, U. S. Patent 2 230 654, Feb. 4, 1941.
3. J. Alfthan and J.L. Chynoweth, U. S. Patent 2 396 629, 3/19/46 to Du Pont.
4. J. Alfthan, U. S. Patent 2 406 127, 8/20/46 to Du Pont.
5. R.T. Fields, U. S. Patent 2 456 262, 12/14/48 to Du Pont.
6. A.J. Cheney, Jr. U. S. Patent 2 456 621, 12/21/48, to Du Pont.
7. M.M. Brubaker, U. S. Patent 2 393 967, 2/5/46 to Du Pont.
8. M.M. Renfrew, U. S. Patent 2 534 058, (Dec. 12, 1950).
9. K.L. Berry, U. S. Patent 2 559 752, July 10, 1951, to Du Pont.
10. J.F. Lontz, and W.B. Happoldt, Jr., Ind. Eng. Chem. 44, 1800-1805 (1952).
11. J.F. Lontz and L.E. Robb, U. S. Patent 2 593 582, 4/22/52 to Du Pont.
12. J.F. Lontz, J.A. Jaffe, L.E. Robb, and W. B. Happoldt, Jr., Ind. Eng. Chem. 44, 1805-1810 (1952).
13. E.E. Lewis and C.M. Winchester, Ind. Eng. Chem. 45, 1123-1127 (1953).

14. G.R. Snelling and J.F. Lontz, J. Appl. Polymer Sci. 3, 257-265 (1960).
15. P.E. Thomas and C.C. Wallace, Jr., U. S. Patent 2 936 301, 5/10/60 (to Du Pont).
16. R. Roberts and R.F. Anderson, U. S. Patent 3 766 133, 10/16/73 (to Du Pont).
17. M.B. Black, E.E. Faust, W. S. Barnhart, and R. Netsch, U. S. Patent 3 265 679. 8/9/66 (to Pennsalt).
18. R.W. Gore, U. S. Patent 3 953 566, 4/27/76; U. S. Patent 3 962 153, 6/8/76; and U. S. Patent 4 187 390, 2/5/80 to W.L. Gore & Assoc.
19. E.B. Yelton, Plastics and Resins 5, no. 5, 14-16, 36 (1946).
20. W.E. Hanford and R.M. Joyce, J. Am. Chem. Soc. 68, 2082-2085 (1946).
21. M.M. Renfrew and E.E. Lewis, Ind. Eng. Chem. 38, 870-877 (1946).
22. C.A. Sperati, Gordon Research Polymers Conference, July 7, 1955.
23. Society of Plastics Engineers Annual Technical Conference (ANTEC), Atlantic City, NJ, September 1955.
24. R.C. Doban, C.A. Sperati, and B.W. Sandt, Soc. Plastics Engrs. Journal, 11, 17-21, 24, 30 (1955).
25. P.E. Thomas, J. F. Lontz, C. A. Sperati, and J. L. McPherson, Soc. Plastics Engrs. Journal, 12 (6), 89-96 (1956).
26. J.C. Siegle, L. T. Muus, and T.P. Lin, Meeting of the American Chemical Society, Atlantic City, September 1956.
27. R.E. Moynihan, J. Am. Chem. Soc. 18, 1045-1050 (1959).
28. R.H.H. Pierce, E.S. Clark, J.F. Whitney, and W.M.D. Bryant, Atlantic City Meeting of the American Chemical Society, September, 1956.
29. R.C. Doban, A.C. Knight, J.H. Peterson, and C.A. Sperati, Meeting of the American Chemical Society, Atlantic City, September 18, 1956.
30. C.A. Sperati, and J. L. McPherson, Meeting of the American Chemical Society, Atlantic City, September 18, 1956.
31. Conference on Fluorocarbon Polymers, Wilmington, DE. September, 1957.
32. C.A. Sperati, and H.W. Starkweather, Jr., Advances in Polymer Science, 2, 465-495 (1961).
33. S. Sherratt, Kirk-Othmer Encyclopedia of Chemical Technology, 2nd ed. 9, 805-831 (1966).
34. W.A. Zisman has many papers from his research at the Naval Research Laboratory.
35. E.S. Clark, J. Macromol. Sci.-Phys., B1(4), 795-800 (1967) and other papers with L.T. Muus.
36. N.G. McCrum, B.E. Read, and G. Williams, (John Wiley & Sons, New York 1967) pp. 450-458.
37. R.K. Eby and K.M. Sinnott, J. Appl. Phys. 32, 1765-1771 (1961).
38. S-F. Lau, H. Suzuki, and B. Wunderlich, J. Polymer Sci., Polymer Phys. Ed. 22, 379-405 (1984).
39. S.V. Gangal, in Kirk-Othmer: Encyclopedia of Chemical Technology, 3rd. Ed. 11 1-24 (1980) (John Wiley & Sons, New York 1980).
40. H.W. Starkweather, Jr., R.C. Ferguson, D.B. Chase, and J.M. Minor, Macromolecules, 18 1684-1686 (1985). This reference will serve as a starting place for many other pertinent papers by HWS.
41. E. Grimaud and C. Tournut, U. S. Patent 3 870 691, March 11, 1975.
42. Many papers on the polymerization of TFE and of TFE on active radicals in PTFE have appeared in the Russian scientific literature.
43. C.H. Manwiller and C.A. Sperati, U. S. Patent 3 981 852, September 21, 1976 to Du Pont.
44. A.J. Cardinal, W.L. Edens, and J.W. Van Dyk, U. S. Patent 3 142 665, July 28, 1964, to Du Pont.
45. File history for prosecution of U. S. Patent 4 187 390, U. S. Patent office.
46. L. Rhodes, INC. 4, No. 8, 34-46 (1982).
47. S. Koizumi, S. Ichiba, T. Kadowaki, and K. Yamamoto, U. S. Patent 4 159 370, June 26, 1979; RE 31 341, Aug. 9, 1983, to Daikin Kogyo Co.
48. D.A. Holmes, U. S. Patent 4 016 345, 4/5/77 to Du Pont.
49. T. Shimizu, Private Communication, Paper given at Symposium on New

Fluorine-Containing Materials, Tokyo Institute of Technology 1985.
50. F.J. Rahl, M.A. Evanco, R.J. Fredericks, and A.C. Reimschuessel, J. Polymer Sci., A-2 10, 1337-1349 (1972).
51. T. Seguchi. T. Suwa, N. Tamura, and M. Takehisa, J. Polymer Sci., Pol. Phys. Ed. 12, 2567-2576 (1974).
52. Wan Daud, Wan Rosli, M. S. Thesis, Ohio University, March, 1981.
53. T. Suwa, M. Takehisa, and S. Machi, J. Appl. Polymer Sci. 17, 3253-3257 (1973).
54. M.I. Bro and C.A. Sperati, J. Polymer Sci. 38, 289-295 (1959).
55. J.N-P Sun, Report for M. S., Ohio University, 1951. See also J.N-P Sun, P.R. Griffiths, and C.A. Sperati, Spectrochemica Acta. 39A, 587-590 (1983).
56. A.V. Tobolsky, D. Katz, and A. Eisenberg, J. Appl. Polymer Sci., 7, 469-474 (1963). See also, Polymer Letters, 2, 129-131 (1964)
57. J.F. Lontz in: Fundamental Phenomena in the Material Sciences. Vol 1, L.J. Bonis and H.H. Hausner, eds. (Plenum Press, New York 1964) pp. 25-47.
58. L.A. Wall, Ed. Fluoropolymers, (Wiley-Interscience, New York 1972).
59. L.A. Wall and R.E. Florin, J. Appl. Polymer Sci. 2, 251 (1959).
60. S. Siegel and H. Hedgpeth, J. of Chemical Physics, 46, 3904-3912 (1967)
61. C.A. Sperati in: Polymer Handbook, 2nd. Ed., J. Brandrup & E.H. Immergut, eds., (Wiley, New York 1975) Section V, pp. 29-40.
62. C.A. Sperati, Meeting of the Fluoropolymers Division, Society of the Plastics Industry, October 15, 1982.
63. T. Suwa, T. Seguchi, M. Takehisa, and S. Machi, J. Polymer Sci., Poly. Phy. Ed., 13, 2183-2194 (1975).
64. A.L. Ryland, J. Chem. Ed. 35, 80-83 (1958).
65. D.W. Ovenall, J. Phys. and Chem. of solids 26, 81-84 (1965).
66. W.K. Fisher and J.C. Corelli, J. Polymer Sci. 19, 2465-2493 (1981).
67. K.A. Donato, M. S. Thesis, Ohio University, November, 1985.

DEVELOPMENT OF THERMOPLASTIC FLUOROPOLYMERS

Robert E. Putnam
100 Alden Avenue
Devola
Marietta, OH

INTRODUCTION

Following the discovery of TFE homopolymerization by Dr. Roy Plunkett, the commercial development of PTFE required enormous effort. Among the major obstacles that had to be overcome were the explosion hazard associated with TFE handling, its tendency to form extremely dangerous polymeric peroxides and the requirements for specialized processing techniques. This effort, coming as it did in the midst of the second World War, not only occupied all available resources, but strongly inhibited the normal inclination to search for alternative solutions to the problem of polymer intractability.

In order for PTFE to have good toughness, the highly crystalline homopolymer must be made in very high molecular weight. This need in turn leads to a melt viscosity so high that, in the temperature range between the crystalline melting point and the onset of thermal degradation, flow is a practical impossibility (Table I).

TABLE I. PTFE properties.

Molecular Weight	10^6 - 10^7
Crystalline Melting Points	327 - 345°C
Melt Viscosity (372°C)	10^{10} - 10^{11} Poises
Onset of Thermal Degradation	450°C

One theoretical possibility, of course, is to break up crystallinity just enough to permit chain entanglements to substitute for interchain forces as the mechanism for generating adequate toughness. This substitution should then permit reduction both in molecular weight and in melt viscosity to levels which would allow conventional melt processing. In the late 1940's the only olefin known that could be introduced into the PTFE structure to disrupt crystallinity without loss of the perfluorocarbon composition was hexafluoropropylene (HFP).

Fortunately for its commercial prospects, HFP could be synthesized by the same basic route as was used for TFE. That is, the pyrolysis of CF_2HCl to difluorocarbene, followed by dimerization to TFE, could be conducted in such a way as to force addition of CF_2 to TFE giving, not perfluorocyclopropane, but its rearranged isomer, HFP. While this looks simple on paper, the reader should contemplate the problems associated with removal from HFP of the dozens of impurities formed in the 800°C pyrolysis step.

$$CF_2HCl \longrightarrow [CF_2] \longrightarrow [(CF_2)_3] \longrightarrow CF_3CF = CF_2$$

Copyright 1986 by Elsevier Science Publishing Co., Inc.
High Performance Polymers: Their Origin and Development
R.B. Seymour and G.S. Kirshenbaum, Editors

"TEFLON"* FEP

At first glance it would seem that the copolymerization [1] of TFE with HFP should be a straightforward extension of the dispersion process employed to prepare TFE homopolymers. (The term dispersion is used in this case to distinguish between fluoromonomer polymerization and conventional emulsion polymerization.) And it is, if questions about end groups and their removal are ignored. A further complication results from the low reactivity of the propagating radical with HFP. This reduces both reaction rate and HFP incorporation. The process employs ammonium persulfate initiator, ammonium perfluorocarboxylate dispersing agent and careful control of TFE/HFP monomer ratio at 95°C for control of composition and molecular weight. Initiation by the sulfate ion-radical leads to carboxylic acid and ammonium carboxylate end groups. Termination is exclusively by radical coupling, though there remains much opportunity for elucidation of the exact nature of the radicals that couple. Transfer with hydrogen -

$$SO_4^{\bullet -} + CF_2 = CF_2 \longrightarrow \ominus SO_4CF_2CF_2^{\bullet} \longrightarrow \ominus SO_4CF_2CF_2\text{-}R_F -$$

$$- SO_4CF_2CF_2R_F - \; + H_2O \longrightarrow HO_2C\ CF_2R_F- \; + HSO_4^- + HF$$

containing agents can be used but is not of dramatic value in eliminating unwanted end-groups. Finally, if great care is taken to insure constancy of conditions during polymerization, this condition taken with the single termination mechanism will lead to a quite narrow molecular weight distribution. Evidence for narrow distribution can be seen in shear rate and stress plots (Figure 1) of typical "TEFLON"* FEP (fluorinated ethylene-propylene copolymer) samples [2]. Quantitative description of distribution in these polymers is still an active goal in industrial laboratories supporting commercial production of this class of polymers.

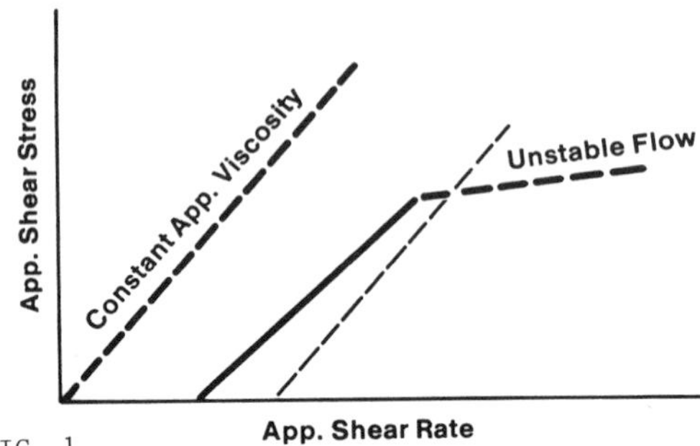

FIG. 1

In the case of TFE homopolymers, the extremely high molecular weight renders the fate and nature of end groups for all practical purposes a moot point. This is not the case with the lower molecular weight FEP polymers. Attempts to injection mold or extrude as-synthesized polymers result in excessive gas evolution. Not only this, but the gas, being a mixture of CO_2 and HF could result in serious corrosion and environmental problems.

*Du Pont's registered trademark for its fluorocarbon resin.

Once again, the solution is simple in principle[3]. It is necessary merely to heat the crude polymer to temperatures of the order of 360°C to convert the carboxylic acids and ammonium salts to a mixture of ends, primarily CF_2H, and $CF{=}CF_2$. On an industrial scale this must be carried out in the presence of air, leading to a yield loss via oxidation of the trifluorovinyl ends. This reaction in turn leads to acid fluoride ends, thus requiring that the air have a significant water content to regenerate acid ends. Under these circumstances it's a wonder that the polymer survives the experience.

$$R_FCF_2CF_2CO_2H \longrightarrow R_FCF_2CF_2H + CO_2$$

$$\longrightarrow R_FCF = CF_2 + CO_2 + HF$$

$$R_FCF_2CF = CF_2 + O_2 \longrightarrow R_FCF_2COF + COF_2$$

The final product can be had in a variety of molecular weights, depending on the degree of toughness required. In this regard there is not complete freedom of action, however. The low reactivity of HFP effectively limits the amount of HFP that can be incorporated in the product. In addition, thermal stability of the polymer is determined by the HFP content and has the same limiting effect. Finally, toughness depends steeply on both HFP content and on molecular weight (Figure 2)[4]. Thus, maximum toughness can be achieved only at the high end of the molecular weight range, a limitation forced by the limitation on HFP incorporation. Toughness and melt viscosity are thus confined to a box bounded by composition, molecular weight and thermal stability. With these constraints it is remarkable that "TEFLON"* FEP has achieved its significant commercial status. That it has is a tribute to its excellent properties (Table II) and to the great care that has been taken to insure that the polymer is made to very exacting specifications.

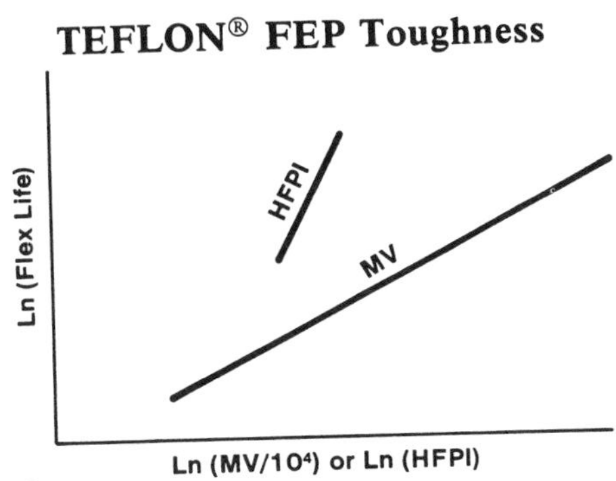

FIG. 2

*Du Pont's registered trademark for its fluorocarbon resin.

TABLE II. Properties of "TEFLON" FEP and PFA.

Property	FEP	PFA
Molecular Weight Range	250–600 K	200–450 K
Melt Viscosity (372°C)	3-45 x 10^4 poise	4–30 x 10^4 poise
Comonomer Content	10–12 wt. %	2.8–4.0 wt. %
Crystalline M. Pt.	255–265°C	305–310°C
Onset of Thermal Degradation	380°C	440°C

"TEFLON"* PFA

"TEFLON"* FEP was commercialized in 1959. At that time there appeared to be no hope of synthesis of a thermoplastic perfluorinated polymer of any structure other than FEP. Within two years the situation had changed as the result of a series of unexpected discoveries. For several years, research on the synthesis of perfluorocarbon epoxides had been underway in Du Pont, culminating in the discovery of a route to hexafluoropropylene epoxide (HFPO)[5]. By 1960 sufficient HFPO had been made to permit exploration of its polymerization. This work shortly turned up an unusual polymerization system consisting of CsF initiator and polyethylene glycol ether solvent. Under the proper conditions this same system could be used to effect the condensation of HFPO with perfluorocarbon acid fluorides to produce perfluoroalkoxypropionyl fluorides [6]. These intermediates could be readily converted to perfluoroalkyl vinyl ethers by pyrolysis of their acid forms.

$$R_FCOF + CF_3\overset{\overset{O}{\frown}}{CF - CF_2} \longrightarrow R_FCF_2OCF(CF_3)\ COF$$

$$R_FCF_2OCF(CF_3)COF \longrightarrow R_FCF_2OCF = CF_2$$

Here were completely new tools for the study of perfluorinated polymers. It was evident that the carbon/oxygen bond would be equivalent to the carbon/fluorine bond in thermal and, perhaps, in chemical stability. Furthermore, the source of thermal instability in FEP was known to be the branch point, the result of steric crowding of the large CF_2 groups at these points. Oxygen at this location in the structure might well eliminate this instability, returning the polymer to performance comparable to that of TFE homopolymer.

Copolymerization of TFE with perfluoroalkyl vinyl ethers proved to be quite facile, either by dispersion techniques similar to that employed with FEP [7] or by a newly developed process employing a fluorocarbon solvent. However, from the earliest studies it was evident that some complications had to be overcome. The most significant of these was a tendency for the alkyl vinyl ethers to rearrange when exposed to free radicals. In the extreme case a chain reaction could be initiated which would result in incomplete rearrangement to the isomeric acid fluoride. During polymerization at temperatures low enough to prevent excessive reaction by this route, the process, nevertheless, competes effectively with free radical coupling as a termination mechanism.

*Du Pont's registered trademark for its fluorocarbon resin.

$R_FCF_2^{\bullet} + CF_2 = CFOCF_2R_F' \longrightarrow R_FCF_2CF_2\dot{C}FOCF_2R_F'$

$R_FCF_2CF_2COF \quad + R_FCF_2$

In contrast to the case of FEP, a relatively small amount of vinyl ether is required to develop adequate toughness. There also is no reactivity limitation to vinyl ether incorporation, but this is unimportant in view of the minor amount of comonomer required. Superficially it appears that the difference in HFP and perfluoroproprylvinyl ether (PPVE), for example, lies in the structure and size of the side chain. This effect is illustrated quite effectively by comparison of the toughness of "TEFLON"* FEP with that of "TEFLON"* PFA at comparable molecular weights (Figure 3). The difference illustrated occurs despite the remarkably lower PPVE content of PFA.

Thermal stability of PFA also is significantly improved, being in fact, almost as high as that of PTFE itself. However, this generalization is complicated by end group considerations. As noted before, "TEFLON"* PFA can be synthesized either by an aqueous dispersion process (as with FEP) or by a non-aqueous solution route. In the former case, it is best to employ ammonium persulfate initiator, a buffer and ammonium perfluorocarboxylate dispersing agent, together with ethane as chain transfer agent [8]. At temperatures below 80°C this produces a polymer with a minimum content of acid groups derived fron vinyl ether transfer and of ammonium carboxylate ends derived from the initiator. Instead, the chain transfer leads to both CF_2H and CH_3CH_2 ends. All of these contribute to a small reduction in thermal stability relative to the backbone structure. In the non-aqueous process which uses 1,1,2-trifluorotrichloroethane as solvent [9], the combination of perfluoropropionylperoxide initiator and methanol transfer gives a very different end group distribution. Again, temperature must be limited to avoid excessive vinyl ether transfer as well as transfer with the solvent. End groups under these conditions are primarily CF_3, CF_2H and CH_2OH, together with minor amounts of COF and CO_2CH_3.

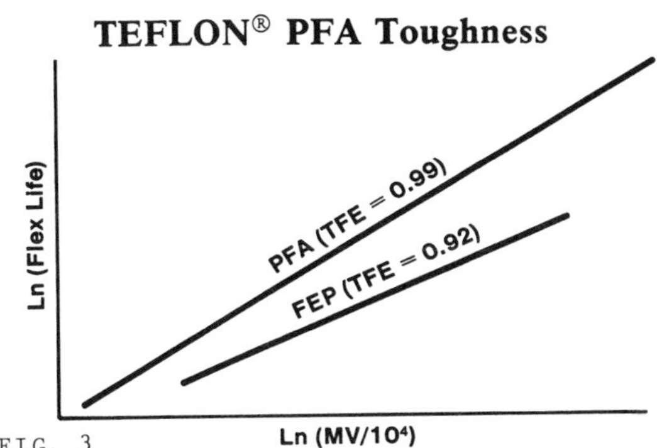

FIG. 3

*Du Pont's registered trademark for its fluorocarbon resin.

The choices of ethane chain transfer agent for aqueous polymerization and methanol for the non-aqueous system originally were dictated by the need to control molecular weight. Surprisingly these specific agents were nearly unique in their ability. Most probably they contribute to a narrowing of molecular weight distribution though this has not been demonstrated conclusively.

Thus, as the result of the lucky accident associated with the work on HFPO, the ultimate in thermoplastic perfluorocarbon polymers was commercialized in 1972, over ten years later. The long time between initial discovery and commercialization was only in part due to the enormous amount of new technology that had to be developed. Of equal importance was the lack of a clear cut market for this exotic polymer. Only when appreciation of its value in molding operations occurred did "TEFLON"* PFA begin to justify the effort and cost of its development.

"TEFZEL"

Well before study of perfluorinated copolymers began, copolymerization of TFE with hydrocarbon and partially fluorinated monomers had been thoroughly investigated. One such copolymer was that of TFE with ethylene. However, for many years this copolymer was considered of little interest because of apparent thermal stability and color deficiencies. Then, in the late 1960's, Dr. D. P. Carlson discovered that these deficiencies stemmed, not from inherent structural factors, but from the polymerization conditions employed in its synthesis. With this recognition it was possible for Dr. Carlson to radically change polymer performance [10]. When he carried out polymerization in trifluorotrichloroethane, using perfluoropropionylperoxide initiator and cyclohexane chain transfer agent, he found that the product, a copolymer with a strong tendency towards regular comonomer alternation, had a really attractive balance of physical, chemical and thermal properties.

However, there remained a critical deficiency. Early in development studies it was found that ETFE polymers, when subjected to thermal cycling at temperatures around 150°C, exhibited severe thermal stress cracking. As in the case of the perfluorocarbon polymers this toughness defect could be alleviated by increasing molecular weight, but this approach, even more than with FEP, led to a thermal stability limitation to melt processing. Carlson solved this problem by incorporating termonomers to reduce polymer crystallinity (Table III). As with perfluorinated materials, it was found that optimum effectiveness depended on side chain length. For example, comonomers such as hexafluoroacetone, perfluoropropylvinyl ether and 1,1,2-trihydrononafluorohexene give great improvement in toughness at relatively low levels of incorporation, whereas HFP leads to very little improvement even at higher levels.

TABLE III. "TEFZEL" Composition and Properties.

$CF_2{=}CF_2/CH_2{=}CH_2$	Ratio	51/49
$-CF_2CF_2CH_2CH_2-$	Content	88%
Termonomer	Content	2-6%
Crystalline M. Pt.		∠ 278°C

*Du Pont's registered trademark for its fluorocarbon resin.

ETFE polymers ("TEFZEL") modified with termonomers require great care in synthesis to insure optimum properties. Incorporation of ethylene near 50 mole % and control of degree of alternation near 90% must be maintained in order to prevent too high a concentration of hydrocarbon branches which result from radical rearrangement of ethylene diads and triads. This same type of rearrangement is the reason why such an unusual comonomer as hexafluoroacetone is effective in improving stress crack resistance. While the hydrocarbon branches also have a beneficial effect, they must be surpressed because they are sites for oxidative attack at processing conditions.

$$RCF_2CH_2CH_2CH_2CH_2^{\bullet} \longrightarrow RCF_2\dot{C}HCH_2CH_2CH_3$$

$$RCH_2CH_2C(CF_3)_2O \longrightarrow R\dot{C}HCH_2C(CF_3)_2OH$$

The remarkable properties of "TEFZEL" ETFE are best illustrated by comparison with the isomeric polydifluoroethylene. This difference can be thought of as the result of a difference in repeat unit structure. In fact, there is a remarkable similarity between ETFE and poly-1, 1, 4, 4-tetrafluorobutadiene in both chemical and physical properties, including nearly identical crystalline melting points.

RELATED POLYMERS

With commercialization of PFA and "TEFZEL" in 1972 a burst of enthusiasm for polymers combining the various structural characteristics ensued. Polymers developed included terpolymers of TFE/HFP/PPVE ("TEFLON"* EPE [11], "Hostaflon" TFA and "TEFLON"* FBE [12]). The former polymers have certain modest property advantages over blends of FEP with PFA. However, these advantages have not been sufficient to generate enthusiastic use in industry. The FBE copolymers are more interesting since, despite their hydrocarbon content, they have a balance of properties quite similar to those of PFA (Table IV). They are superior to PFA in two ways: improved radiation resistance and crosslinkability. Synthesis of these polymers is tricky due to a tendency to form mixtures of incompatible copolymers. This family has not yet been commercialized.

TABLE IV. Properties of "TEFLON"* FBE.

Property	PFA	FBE
TFE Mole Fraction	0.99	0.99
Crystalline M. Pt.	304°C	305°C
Melt Viscosity (372°C)	4.9×10^4	4.7×10^4
Tensile Strength, psi	3500	3700
Flex. Modulus, psi	80,000	90,000
Rate of Wt. Loss in Air at 380°C (hr^{-1})	0.002	0.08
Elongation Retention After 20K Rads	1%	40%

*Du Pont's registered trademark for its fluorocarbon resin.

This brings us to the present. Interest in new melt fabricable copolymers of TFE remains very high, not only in Du Pont, but in all the other companies engaged in fluoropolymer research and manufacture. The most important of these are Asahi Glass and Daikin in Japan, Allied Chemical in the U.S. and Hoechst and Montefluos in Europe. It is to be expected that workers in these organizations will continue to produce new polymers, though such materials will most likely see use in highly specialized applications rather than erode the major markets currently served by PTFE, FEP, PFA and "TEFZEL".

REFERENCES

1. M.I. Bro and R.W. Sandt, U.S. Patent 2,946,763, 7/26/60: Polymerization.
2. R.A. Morgan, E. I. du Pont de Nemours, Private Communication: Rheology.
3. R.C. Schreyer, U.S. Patent 3,085,083, 4/9/63: Stabilization.
4. S. Piekarski, E. I. du Pont de Nemours, Private Communication: Toughness.
5. H.S. Eleuterio and R.W. Meschke, U.S. Patent 3,358,003, 12/12/67: HFPO Synthesis.
6. C.G. Fritz, E.P. Moore, and S. Selman, U.S. Patent 3,114,778, 12/17/63: Vinyl Ether Synthesis.
7. R.A. Darby, E. I. du Pont de Nemours, Private Communication: Polymerization.
8. W.F. Gresham and A.F. Vogelpohl, U.S. Patent 3,635,926, 1/18/72: Chain Transfer.
9. D.P. Carlson, U.S. Patents 3,528,954 (9/15/70) and 3,642,742 (2/15/72): Non-Aqueous Polymerization.
10. D.P. Carlson, U.S. Patent 3,624,250 (11/30/71).
11. D.P. Carlson, U.S. Patents 3,528,954 (9/15/70) and 4,029,868 (6/14/77).
12. S.J. Fritschel, U.S. Patent 4,487,902 (12/11/84).

DEVELOPMENT OF KYNAR POLYVINYLIDENE FLUORIDE

JULIUS E. DOHANY
Pennwalt Corporation
900 First Avenue
King of Prussia, PA

ABSTRACT

KYNAR polyvinylidene fluoride was introduced in 1960. First commercial production began in 1965. For the first five years of its life, KYNAR resin was used almost exclusively in the electrical and electronic market. In 1965, the architectural finish, KYNAR 500®, was introduced followed by first plant expansion in 1969. In 1975, KYNAR 460 was introduced for high speed wire extrusion. In 1981, the Underwriter Laboratories listed KYNAR resins for use in plenum cables that can be used without conduit. In 1983, a new homopolymer series was introduced known as the series 700 and was followed by KYNAR FLEX® 2800 in 1984. To meet the demands of the growing markets, a new plant for KYNAR resins was built and placed in operation in 1985. Also that year, new coating resins, KYNAR SL® and KYNAR ADS®, were introduced. KYNAR polyvinylidene fluoride is a crystalline polymer that is produced by the emulsion polymerization technique. Typical properties are discussed.

INTRODUCTION

Polyvinylidene fluoride or PVDF has been known since 1948 when a patent was issued to Ford and Hanford (1) claiming the composition, but the polymer remained virtually uninvestigated until 1958 when researchers at Pennsalt Chemicals Company, as Pennwalt Corporation was then known, decided to synthesize the resin. Initial evaluation of this polymer showed that the polymer had attractive properties to justify full development.

The objective of the project was to develop a commercially feasible process for the manufacture of the monomer and polymer. This was accomplished in two years time and in early 1960, the PVDF was ready for market introduction under the trademark KYNAR®. Initial quantities of the KYNAR polymer, then called RC2525, were produced in a semiworks plant. In 1965, the first commercial plant for KYNAR resins began operations in Calvert City, Kentucky.

Because of KYNAR PVDF's excellent balance of properties, numerous applications for the resin soon developed. The markets for PVDF resins include:

-- Electric/Electronic Industry
-- Chemical Processing Industry
-- Architectural and Specialty Finishes
-- Specialty Products

Copyright 1986 by Elsevier Science Publishing Co., Inc.
High Performance Polymers: Their Origin and Development
R.B. Seymour and G.S. Kirshenbaum, Editors

GROWTH

For the first five years of its life, KYNAR PVDF was used primarily in the electrical and electronic markets. The product proved to have just the right combination of properties for insulation of computer wiring, and jacketing on cable constructions used in the military and aerospace applications.

One of its earliest applications, in fact, was the primary insulator on computer hookups used in back panels and installed by high speed, automatic wiring machines. While satisfactory electrical properties of the polymer were important, automatic wiring equipment also required a wire insulation material of exceptional physical toughness. KYNAR PVDF easily fit this requirement.

The versatile new fluoropolymer found use in other applications where tough and abrasion resistant properties were important. For example, corrosion engineers and cathodic protection service companies found that PVDF insulation solved many of the corrosion problems associated with underground anode-bed installations and down-well jacketing to protect the instrumentation cable in deep wells.

KYNAR fluoropolymer-based heat-shrinkable tubing is another example of an early commercial application. These non-burning, rigid sleeves offer resistance to most industrial fuels, solvents, and chemicals and are designed for applications requiring high mechanical strength up to temperatures of 175°C.

Another ingenious application of KYNAR fluoropolymer was the heat tracing tape that utilizes a carbon-filled KYNAR compound as the heating element which separates two parallel conductors. ETFE is applied as a high temperature protective jacket. The design is unique and the temperature is controlled automatically by the core compound and not by external thermostats. When energized, the carbon particles form a conductive path which generates heat. As the temperature of the trace tape approaches the specified design limits, the PVDF matrix expands which interrupts the flow of current by reducing the number of conductive paths. Thus, the self-limiting feature is always in control as the temperature goes up, and the output of the tape goes down.

The volume and variety of uses for KYNAR resins have also grown steadily in the chemical process and related industries that included applications in fluid handling systems such as pipe, valves, and pumps in vessel and column construction such as linings, trays, and packing; in filtration equipment, as woven mesh, such as the large cylinder covers, or face wire as they are called, used on bleach washer vacuum filter drums in the pulp and paper industry; as a multitude of diaphragms and small molded parts used in pressure and flow regulators, and in process industry instrumentation in general.

Since the middle 1970's, KYNAR resins have also been used for the manufacture of microporous and ultrafiltration membranes. KYNAR fluoropolymer was selected for this application because it is resistant to many chemicals both at room temperature and elevated temperature. Because PVDF is approved by the Federal Drug Administration (FDA) for articles or components of articles intended for repeated use in contact with food, KYNAR filters and fluid handling systems are being used by the chemical and the food industry and are also being used by the semiconductor industry as well as in biomedical applications. The latter two applications require particularly inert materials that will not contaminate the fluid with trace contaminants, and KYNAR PVDF does meet these extreme purity requirements.

In 1965, Pennwalt developed a modified grade of the PVDF resins that opened up still a third entirely different field of application -- the high quality factory applied metal finishes. Marketed under the trademark KYNAR 500®, this new resin grade proved to be a major step forward in providing architects and designers with long life exterior color coatings.

Coating manufacturers quickly saw the virtues of a tough and resistant premium metal coating system that could retain its original color better than any other finish in the face of normal weathering and atmospheric pollution.

These new KYNAR 500 based finishes were principally used first on industrial plants and power generation stations. Before long, the fluoropolymer coatings made the transition to prestigious sports arenas and major office towers.

Pennwalt supplied a growing number of paint company licensees with the KYNAR 500 resin that constitutes some 70 percent (by weight) of these architectural finishes.

The expansion of KYNAR 500 based finishes has been fueled by the development of new colors and textures. There are now metallic fluoropolymer coatings, stucco and cement like finishes, and even roofing tiles that simulate the look of terra cotta ceramics.

WHAT IS KYNAR PVDF?

PVDF belongs to the class of fluorinated ethylenes. With two fluorine atoms in the repeat unit, it contains 59.4% fluorine. The structure of the fluorinated ethylenes is shown in Figure 1. Of the four fluoroethylenes of this series, only polyvinyl fluoride (PVF), PVDF, and polytetrafluoroethylene (PTFE) are commercially significant and only two i.e. PVDF and PTFE, found extensive applications as engineering plastics.

FIGURE 1. Fluoroethylenes.

Poly(vinyl fluoride)	$-(CH_2 - CHF)_n-$	PVF
Poly(vinylidene fluoride)	$-(CH_2 - CF_2)_n-$	PVDF
Poly(trifluoroethylene)	$-(CHF - CF_2)_n-$	P3FE
Poly(tetrafluoroethylene)	$-(CF_2 - CF_2)_n-$	PTFE

The PVDF can be polymerized by free radical initiated polymerization in suspension or emulsion (2). Because it offers inherent property advantages to the resin, KYNAR polymers are made using the emulsion technique. The schematic outline of the manufacturing process is shown in Figure 2. The process is based on radical initiated polymerization in emulsion. The polymer recovery and work up follows well known techniques used by the plastics industry.

FIGURE 2. KYNAR Resin Process.

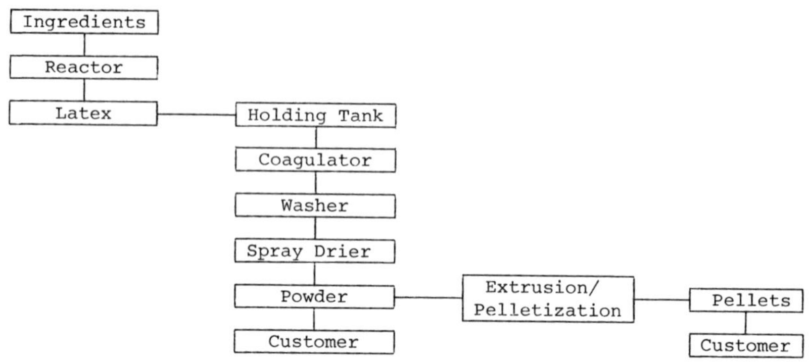

PVDF is a crystalline polymer that has offered interesting and fruitful subjects for analysis to many scientists in the entire domain on engineering plastics since about 1975. Both crystallization and morphology of the polymer are important features for the polymer properties and the richness of the morphological behavior is not surpassed by any other polymer (3). It has the characteristic stability of fluoropolymers when exposed to harsh thermal, chemical, and ultraviolet environments. The alternating CH_2 and CF_2 groups in the chain contributes to the unique polarity that influences its solubility and electric properties.

KYNAR PVDF homopolymers belong to two classes. A typical representative for one of the classes is KYNAR grade 460 and the other class that was introduced in 1981 is known as the series 700. In Table 1, the general property profile is shown with typical values for both KYNAR 460 and KYNAR 700 resin series.

TABLE 1. General Properties of KYNAR Grades.

Physical Properties	ASTM Method	Typical Values or Ranges	
		460	710 - 760
Melting Point, °C	D-3418	160	169
Density, g/cm³	D-792	1.76	1.78
Tensile Strength at Yield, MPa	D-638	28 - 41	34 - 48
Tensile Strength at Break, MPa	D-638	31 - 52	31 - 45
Elongation at Break, %	D-638	50 - 250	50 - 250
Tensile Modulus, MPa	D-882	1300	1500
IZOD Impact Strength	D-256		
Unnotched, kJ/m		0.8 - 2.1	1.0 - 4.2
Notched, kJ/m		0.11 - 0.2	0.15 - 0.40
Heat Distortion Temperature, °C			
Load 1.82, MPa	D-684	84 - 96	108 - 118

KYNAR 460 which was introduced in 1975 is, by design, particularly suitable for pressure extrusion of wire insulation at high extrusion rates. KYNAR 700 series grades introduced in 1981 are multipurpose grades and cover a broad range of melt flow from high to low, i.e. from grades 710 to 760. The strength and toughness typical for PVDF resins is reflected by the high tensile strength at yields of 28 to 48 MPa and the Izod impact strength ranging from 0.8 to 4.2 kJ/m. The melting point for the grade series 700 is higher than the melting point for the grade 460. The difference is due to a slight difference in crystallinity caused by the different polymerization conditions employed.

Compared to many thermoplastics, PVDF has excellent resistance to creep and fatigue. Yet in thin sections such as films, filament, and tubing, PVDF components are flexible and transparent. The dynamic mechanical spectrum of KYNAR PVDF in Figure 3 shows that its dynamic modulus of elasticity (E') is high and decays only gradually as the temperature is increasing.

FIGURE 3. Viscoelastic Spectrum of KYNAR PVDF (Rheovibron Viscoelastometer Model DDV II at 11 cps).

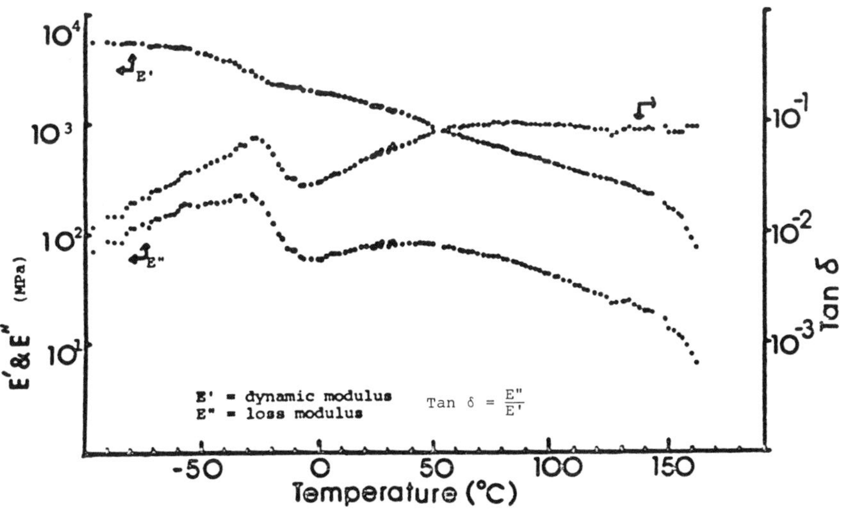

There are no abrupt transitions. The dynamic loss modulus or storage modulus (E"), which relates to the polymer's ability to absorb mechanical shock, and the mechanical loss factor tan delta (E"/E') show that glass transition of the polymer is at -30°C and that there is an alpha transition around 60°C. In general, the KYNAR homopolymer possesses useful mechanical properties from -80°C to 150°C.

As mentioned before, the alternating CH_2 and CF_2 groups influence PVDF's solubility and electrical properties. It is essentially insoluble in all nonpolar solvents but dissolves in strong polar solvents such as alkyl amides and is readily solvated at elevated temperatures by certain common solvents such as ketones and esters. This selective solubility was the key to the development of the now well-known KYNAR 500 finishes and corrosion coatings. The former are solvent based dispersions that are factory applied and baked on aluminum or galvanized steel, which form wall panels of industrial and monumental buildings. The corrosion coatings protect process equipment from corrosion.

The chemical resistance of KYNAR compared to several other plastics at 200°F (93°C) is graphically shown in Figure 4 as a "bull's eye" chart. KYNAR is essentially inert, i.e. "hits the bull's eye", in aliphatic, aromatic and chlorinated solvents, weak bases and salts, strong mineral acids, halogens, and strong oxidants. As noted before, the resin is solubilized by polar ester and ketone solvents. Strong inorganic or organic bases (caustic or primary amine, for example) attack the PVDF structure by dehydrofluorination, and application of KYNAR resins for such media is not advised without prior testing.

FIGURE 4. Chemical Resistance of KYNAR Resin vs. Other Well-Known Plastics at 93°C (200°F).

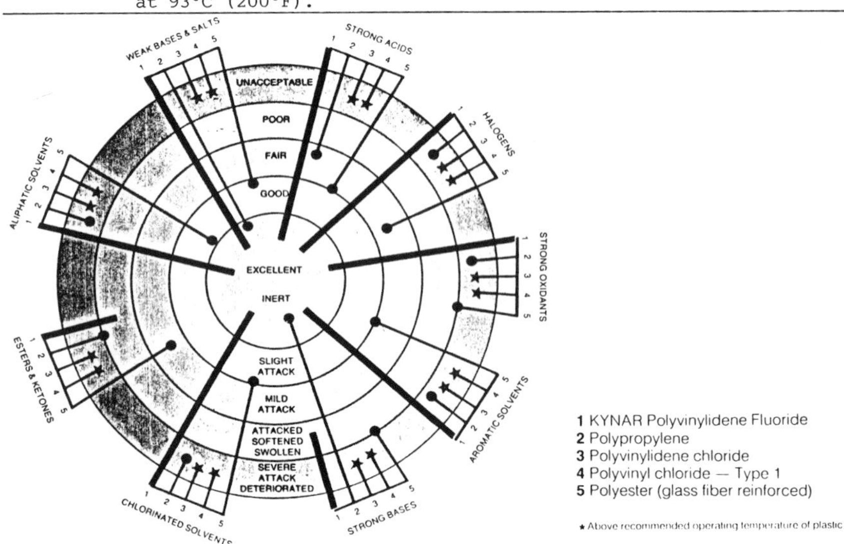

PVDF crosslinks when subjected to electron beam or high energy ionizing radiation from a gamma source without detrimental effects on physical properties. Addition of radiation sensitizers enhances crosslinking efficiency of PVDF. The different grades of KYNAR homopolymer are readily crosslinked and do not degrade when irradiated with moderate doses of high energy electron or gamma radiation. The efficiency of crosslinking is influenced by the grade, i.e., molecular weight variations are important. Figure 5 shows the response of various grades to high energy electron beam irradiation in terms of the amount of polymer that becomes insoluble in N,N-dimethylacetamide (DMAC), an excellent solvent for non-crosslinked PVDF resins. Examples of KYNAR fabricated products utilizing radiation technology are heat-shrinkable tubing and insulated wire capable of withstanding high temperatures.

FIGURE 5. Insoluble Fraction of Various Grades of KYNAR Homopolymer After Electron Beam Irradiation.

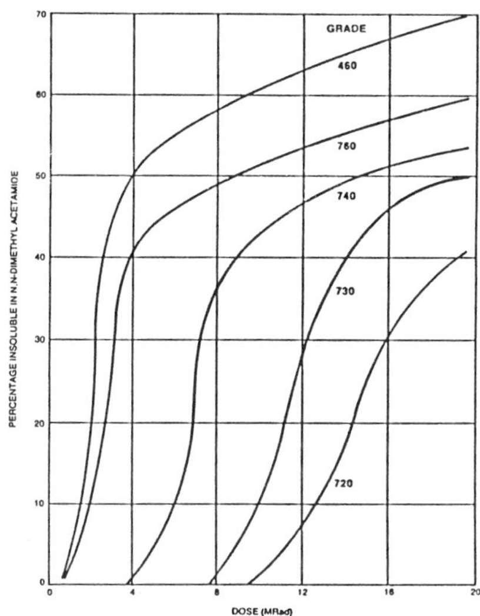

Because of its high polarity, PVDF has a high dielectric constant that is a function of frequency, temperature, thermal history, and orientation. The combination of high dielectric strength and excellent mechanical properties over a broad temperature range is the reason that KYNAR resin found so many uses in the wire and cable industry.

NEW APPLICATIONS -- NEW PRODUCTS

In 1981, the Underwriters Laboratory, based on the modified Steiner tunnel test per UL910, classified KYNAR resins for use as insulators and jackets for cables meeting the requirements of Article 760 of the National Electrical Code (NEC). This NEC code describes the electrical wiring that is installed in the return air handling space, i.e. plenum, between a suspended ceiling and the structural ceiling of commercial buildings.

KYNAR fluoropolymer, like the other fluoropolymers is more flame resistant and produces less smoke in the event of a fire than the PVC that has traditionally been used for this application. Plenum cable today is one of the most exciting and largest single markets that was ever developed for KYNAR fluoropolymers. Plenum cable constructions are used in several low voltage applications that replace the expensive to install metal conduit. Such low voltage cable includes fire alarm signal cable, sound masking cable, and voice and data transfer cable for telephone and computer networking.

In response to a market-driven need, Pennwalt developed a more flexible version of the fluoropolymer for certain applications. This fluoropolymer was introduced in 1984 under the trademark KYNAR FLEX 2800. KYNAR FLEX 2800 is a copolymer of vinylidene fluoride and hexafluoropropene (HFP).

This new copolymer compared to PVDF is impact modified and specifically designed for use as a jacket material in wire and cable construction requiring higher flexibility and improved resistance to impact. As shown in Table 2, KYNAR FLEX 2800 has a lower melting point and a flexural modulus that is about one-half of the modulus of the homopolymer. KYNAR FLEX 2800 has a significantly higher notched impact strength when compared to the homopolymer.

TABLE 2. Properties of KYNAR FLEX 2800.

Property	ASTM Method	Typical Values
Density	D-792	1.78 g/cm³
Melting Range	D-3418	141-146 °C
Tensile Strength at Yield	D-882	20-27 MPa
Tensile Strength at Break	D-882	21-48 MPa
Elongation at Break	D-882	400-650 %
Tensile Modulus	D-882	340-480 MPa
Flexural Modulus	D-882	480-620 MPa
IZOD Impact Strength (Notched)	D-256	530-1060 J/m

The semiconductor processing industry is one of the more recent applications for KYNAR resins. It is used for fluid handling systems for ultrapure fluids used in the production of microchips. Resin purity and inertness, and the total absence of any additives in the fabrication process, are absolute essentials for components handling the chemicals, etchants, and rinse waters used in the making of today's state of the art integrated circuits.

In 1985, Pennwalt reintroduced two fluoropolymers under the KYNAR trademark. These copolymers will broaden the KYNAR product line of high performance coatings. The two products were originally introduced in 1970 and 1973 respectively, but the market was not ready at that time for these products, and they were withdrawn in 1974. Now time has changed, and the market is ready for these products. The two new resins are called KYNAR SL® and KYNAR ADS®.

KYNAR SL is a copolymer of vinylidene fluoride and tetrafluoroethylene and is designed for low temperature bake coating applications. It complements the established KYNAR 500 based finishes that are baked on metallic substrates at relatively high temperatures. Typical properties of KYNAR SL are shown in Table 3. Like KYNAR FLEX, this copolymer is flexible but unlike the homopolymer and the former copolymer, it is much more soluble in common solvents.

TABLE 3. Properties of KYNAR SL.

Property	ASTM Method	Typical Value
Melting Point	D3418	122-126 °C
Density	D792	1.88 g/cm³
Refractive Index (n_D25)	D524	1.4026
Tensile Strength at Yield	D1708	14-20 MPa
Tensile Strength at Break	D1708	32-45 MPa
Elongation at Break	D1708	500-800 %
Tensile Modulus	D638	410-550 MPa
Solution Viscosity (15 wt% in MEK)		45±5 cps

KYNAR SL can be formulated either as a solution coating for drying at approximately 60°C, or as a dispersion coating that is baked at approximately 150°C (unlike KYNAR 500 based finishes that require bake temperatures of 232°C). Finishes of KYNAR SL have weathered more than fifteen years in Florida without showing appreciable signs of deterioration.

KYNAR ADS is a terpolymer of vinylidene fluoride, tetrafluoroethylene, and hexafluoropropene and is designed for solution coatings that dry at ambient temperature. Finishes formulated using this terpolymer are ideally suited for use as field applied topcoats over conventional maintenance coatings for added protection and long term color retention. Panels finished with KYNAR ADS coatings have completed ten years exposure on a Florida test fence with promising results. Some properties for this terpolymer are shown in Table 4.

TABLE 4. Properties of KYNAR ADS.

	ASTM Method	Values
Melting Point	D3418	86-90 °C
Specific Gravity	D792	1.846 g/cm³
Refractive Index (n_D25)	D524	1.384
Tensile Strength at Yield	D882	8 MPa
Tensile Strength at Break	D882	15 MPa
Elongation at Break	D882	990%
Tensile Modulus	D882	130 MPa
Solution Viscosity (20 wt% in MEK)		20±5 cps

The newest market for KYNAR resins is perhaps the most exciting and most challenging. This is the KYNAR® Piezo Film. By definition, a piezo electric material is one that can change polarization in response to mechanical stress. The CH_2- CF_2 repeat unit was found to exhibit the strongest piezo electric and pyro electric activity of all known polymers. This property was first reported in 1969 based on experiments using KYNAR PVDF. Since that time the piezo and pyro electric properties of PVDF have been the subject of many publications.

KYNAR Piezo film applications include a noise-cancelling telephone headset, infrared sensing, anti-intrusion devices, an ultrasonic gas flow respiration monitor, and a high-fidelity electric violin. Other potential uses now in development include computer switches, keyboards and printers, ultrasonic imaging systems for medical diagnosis, and robotic tactile sensors.

FUTURE

Major new market uses have been developed in each of the fluoropolymer's principal fields of use -- that is, in electrical/electronic, in the chemical process industries, and in the area of architectural coatings. To keep pace with the expanding demand for KYNAR resins in all its major markets, Pennwalt has just started up a new $65 million plant in Thorofare, New Jersey which is across the river from its headquarters in Philadelphia. It supplements our already increased production capability at Calvert City, Kentucky and doubles the company's overall KYNAR production capacity.

Among its distinctions, the new Thorofare plant has the latest in distributed control systems for automating much of the production process and operates the first hazardous waste incinerator to be approved by the state of New Jersey. The Thorofare plant represents the largest single capital expenditure ever undertaken by Pennwalt and this investment reflects the company's confidence in the future growth of this versatile high performance plastic during its silver anniversary year.

The first 25 years of KYNAR have been gratifyingly successful -- and at the current rate of new market and new product developments, the next 15 could be equally as exciting.

REFERENCES

1. T.A. Ford and W.E. Hanford (to E.I. DuPont de Nemours & Co., Inc.) U.S. Pat. 2,435,537 (1948)

2. J.E. Dohany and L.E. Robb in Kirk-Othmer: Encyclopedia of Chemical Technology, Volume II, 3rd Ed., 1980 by John Wiley & Sons, Inc.

3. A.J. Lovinger, Science, 220, 1115 (1983); in Developments in Crystalline Polymers, D.C. Bassett, Ed., Applied Science, London 1982, Vol. 1, 195 (Chapter 5).

Thermosets

HISTORY AND DEVELOPMENT OF EPOXY RESINS

John A. Gannon
CIBA-GEIGY Corporation
Ardsley, N.Y.

Epoxy resins were first offered commercially in 1946 and are now used in a wide variety of industries.

In recent years, ca. 45% of the total amount of epoxy resins sold were used in protective coatings while the remainder (55%) used were in structural applications such as laminates and composites, tooling, molding, casting, construction, bonding and adhesives.

The resins are characterized, after curing, by high chemical and corrosion resistance, good mechanical and thermal properties, outstanding adhesion to various substrates, good electrical properties and ability to be processed under a variety of conditions.

The resins also exhibit low shrinkage during cure.

Early History

The first commercial possibilities for epoxy resins were realized in Switzerland by Pierre Castan of De Trey Freres and in the Unites States by Sylvan Greenlee of DeVoe and Raynolds.

In 1936 Castan produced a bisphenol-A based epoxy resin via reaction with epichlorohydrin and subsequently prepared a thermoset composition after reaction of the resin with phthalic anhydride. The use of the hardened resin was foreseen in dental products but initial attempts to market the resin were unsuccessful. The patents were licensed to CIBA AG of Basel, Switzerland (now CIBA-GEIGY) and, in 1946, the first epoxy adhesive was shown at the Swiss Industries Fair (1). During 1939 Sylvan Greenlee produced a high molecular weight resin from bisphenol-A and epichlorohydrin which was subsequently esterified with unsaturated fatty acids to provide an air-drying coating. His first patent issued in 1948 (2).

Immediately after World War II, Greenlee and co-workers patented for DeVoe and Raynolds (later Celanese Chemical Company) a series of epoxy resin compositions encompassing resins and resin esters. Meanwhile, CIBA AG, further developed epoxy resins for casting, laminating and adhesive applications and CIBA Products Company was established in the United States.

In the late 1940's, Shell Chemical Company, the only supplier of epichlorohydrin at that time, and Union Carbide (then Bakelite) entered into the field of epoxy resins.

In the summer of 1955, the four basic epoxy resin manufacturers entered into a cross-licensing agreement. Subsequently, Dow Chemical Company and Reichold Chemicals entered the patent pool and introduced commercial lines of epoxy resins.

Development of Epichlorohydrin/Bisphenol-A Resins

The most important intermediate in the development of epoxy resins was the viscous liquid product derived from the epoxidation of bisphenol-A.

Copyright 1986 by Elsevier Science Publishing Co., Inc.
High Performance Polymers: Their Origin and Development
R.B. Seymour and G.S. Kirshenbaum, Editors

Epichlorohydrin assumes difunctionality in the synthesis of epoxy resins due to its readiness to dehydrohalogenate with alkaline reagents.

Schlack in 1933 (3) reported the first liquid diepoxide from bisphenol-A and an excess of epichlorohydrin:

$$\underset{\text{(excess)}}{\overset{\diagdown\!\!\!\overset{}{O}\!\!\!\diagup}{CH_2CHCH_2Cl}} + HO-C_6H_4-\overset{CH_3}{\underset{CH_3}{\overset{|}{\underset{|}{C}}}}-C_6H_4-OH \xrightarrow[\text{(catalyst)}]{NaOH}$$

$$ClCH_2\underset{OH}{\underset{|}{C}}HCH_2O-C_6H_4-\overset{CH_3}{\underset{CH_3}{\overset{|}{\underset{|}{C}}}}-C_6H_4-OCH_2\underset{OH}{\underset{|}{C}}HCH_2Cl \xrightarrow[\text{(stoichiometric quantity)}]{NaOH}$$

$$\underset{\diagdown O \diagup}{CH_2CH}CH_2O-C_6H_4-\overset{CH_3}{\underset{CH_3}{\overset{|}{\underset{|}{C}}}}-C_6H_4-OCH_2\underset{\diagdown O \diagup}{CHCH_2} + NaCl + H_2O$$

Minimization of the formation of high molecular weight species was accomplished by the use of an excess of epichlorohydrin; nevertheless, the commercial liquid epoxide resins produced were composed of a mixture of oligomers along with monomer with the latter ca. 80-85% by weight.

Greenlee as early as 1948 revealed the procedures for preparing high molecular weight solid epoxy resins by two techniques viz. (a) the direct reaction of bisphenol-A with epichlorohydrin and caustic using appropriate mole ratios to form the resinous products along with salt and water, referred to later as the "taffy" process and (b) the use of a liquid diepoxide resin as a starting material for reaction with bisphenol-A, thereby eliminating the need for removal of salt and water (the so-called advancement of fusion technique).

Production of very high molecular weight thermoplastic polymers from bisphenol-A and epichlorohydrin was reported in 1949 (4) using a one-step or two-step synthesis route i.e. starting with epichlorohydrin and bisphenol-A in 1:1 mole ratios or utilizing a purified diglycidyl ether of bisphenol-A resin and a stoichiometric quantity of bisphenol-A in a solution polymerization.

These materials, useful as films and fibers, were later designated as Phenoxy resins and were introduced in 1961 for coatings applications by Shell Chemical Company.

Subsequently, Union Carbide Corporation offered solutions, films and granular molding compounds.

Phenoxies, because of their molecular weight (Mm >30,000) bridge the gap between thermosetting and thermoplastic resins being used commercially in both forms.

The use of pre-catalyzed liquid epoxy resins for the preparation of solid resins and solid resin esters was disclosed in 1969 (5). Such compositions contain an alkyl-aryl phosphonium salt catalyst that specifically directs the epoxy-phenol reaction rather than the secondary alcohol-epoxy reaction which can occur near the end of the advancement procedure.

Commercial use of the pre-catalyzed liquid resins is for preparation of solid resin solutions or solid resins fatty esters for air-drying coatings.

The emergence of 100% solids, powder-coating applications has led to the development of low and medium molecular weight solid epoxy resins that exhibit good flow and fast reactivity with solid hardeners for decorative and functional applications.

Conventional solid epoxy resins have also exhibited improved degrees of purity stemming from requirements for lower hydrolyzable and total chlorine values, which are inherent in commercial epoxy resins.

A trend toward narrower specifications for epoxy resins has also occurred in recent years, resulting in specification epoxy values with a smaller range of values and a correspondingly narrower range of product viscosities.

Development of Multifunctional Resins

DOW Chemical Company provided the commercial availability of epoxidized phenol novolac resins in the late 1950's (6).

The novolacs are products of the condensation of an excess of phenol with formaldehyde under acidic conditions resulting in random p-p',o-p' and o-o' combinations.

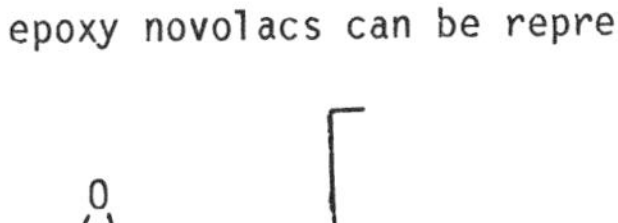

OH OH OH

CH_2 CH_2 n

where n=o to 4

Epoxidation of the multifunctional novolac is then accomplished by reaction of epichlorohydin and sodium hydroxide.

The epoxy novolacs can be represented by the general structure:

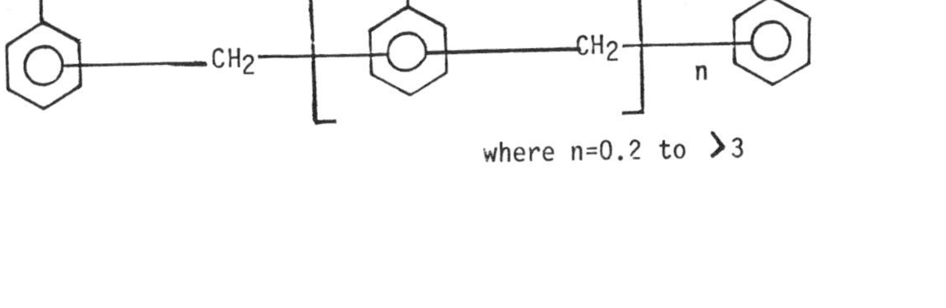

O O O

$O-CH_2-CH-CH_2$ $O-CH_2-CH-CH_2$ $O-CH_2-CH-CH_2$

CH_2 CH_2 n

where n=0.2 to >3

The commercial products are designated as D.E.N 431, 438 and 439 (DOW Chemical Company) as well as EPN 1139 and EPN 1138 (CIBA-GEIGY Corporation).

The commercial materials provide high functionality leading to high chemical resistance and heat distortion values after curing.

In a similar development, Koppers in the early 1960's provided novolacs based on o-cresol but the technology was later transferred to CIBA Products Company (now CIBA-GEIGY Corporation) who marketed a line of products (7).

The epoxy cresol novolacs may be presented as follows:

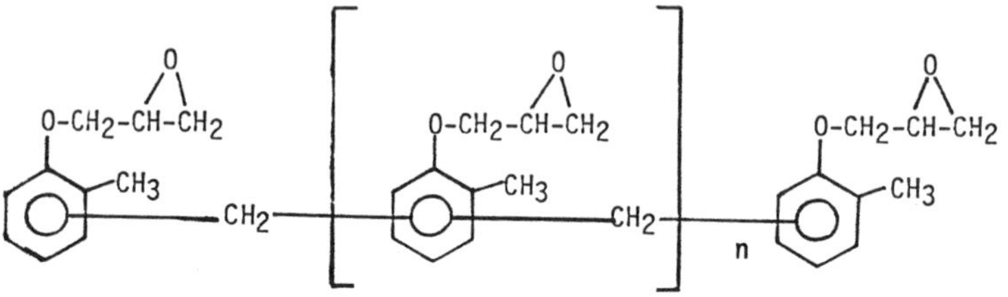

where n=0.7 to $>$ 3

The commercial products are designated as ECN resins 1235, 1273, 1280 and 1299.

Extensive development of both EPN (epoxy phenol novolac) and ECN (epoxy cresol novolac) resins has occurred to improve the quality of the resins by a reduction of ionic and reactive chlorohydrin impurities for application in electronics markets.

The lowest member of the series (n=o), the so-called Bisphenol-F resin, after glycylidation with epichlorohydrin, provided a low viscosity (4-6000 cP) liquid resin with slightly higher functionality and a more chemically resistant system after curing with polyamines.

A polyfunctional resin based on phenol and glyoxal forms the basis of a speciality epoxy resin: (8)

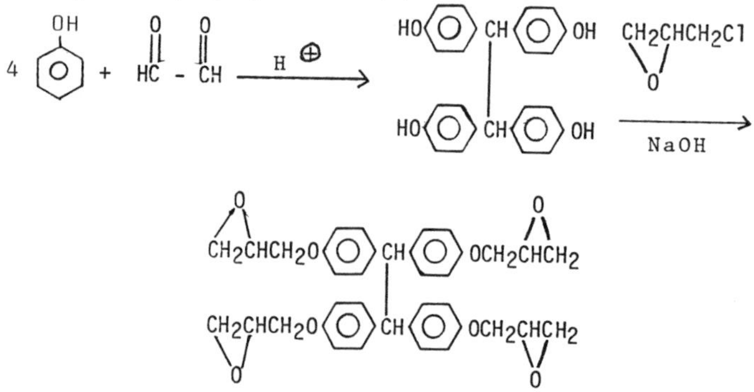

TETRAGLYCIDYLETHER OF TETRA-KIS (4-HYDROXY PHENYL) ETHANE (MP ~ 80°C)

The product was developed by Shell Chemical Company in the late 1950s and later commercialization of the product was adopted by both Shell and CIBA-GEIGY.

Union Carbide Corporation in the late 1950s (9) developed the triglycidyl derivative of p-aminophenol and the technology was later sold to CIBA Products Company (now CIBA-GEIGY).

The glycidylation of the aminophenol is carried out with a large excess of epichlorohydrin as the resin can selfpolymerize due the accelerating effect of the tertiary amine:

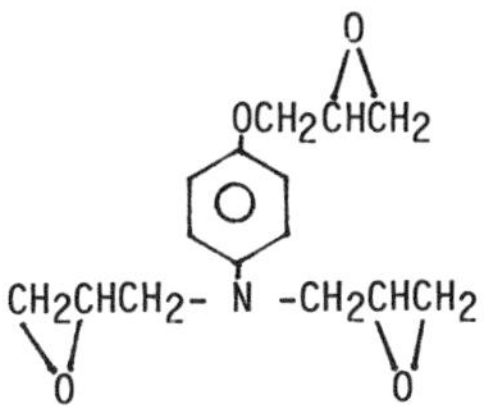

Triglycidyl p-Aminophenol
(viscosity ~ 3000-5000 cP @ 25°C)

The product is marketed as MY0500 and MY0510 resin.

A similar product, originating in the laboratories of Union Carbide also in the late 1950s, (10) viz. the tetraglycidyl ether of methylene dianiline, was manufactured and marketed by CIBA-GEIGY for many years:

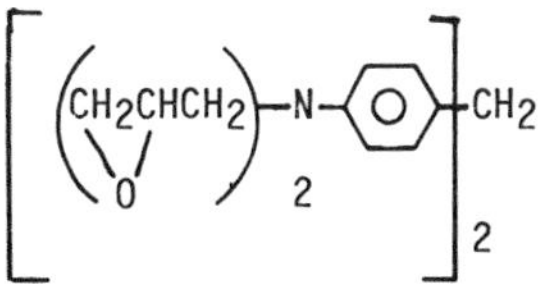

N,N,N',N'-Tetraglycidyl 4,4' diamino-diphenyl methane
(viscosity < 25,000 cP @ 50°C)

The resin has evolved into an important binder for graphite - reinforced composites for many military applications.

Early efforts by Henkel (11) and later, CIBA-GEIGY resulted in the commercialization of triglycidyl isocyanurate:

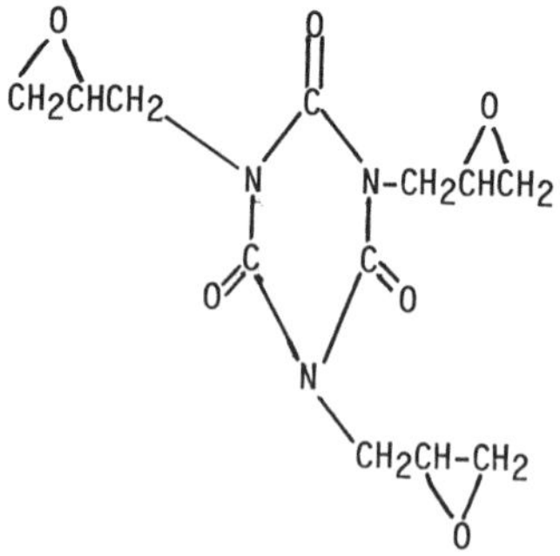

Triglycidyl Isocyanurate
(MP 85-110°C)

The product has established a position in the market due to the weatherability of the resin as a cross-linker for carboxyl-containing polyesters in powder coating applications.

Development of Cycloaliphatic Resins

Union Carbide in 1956 developed another route to epoxy resins viz. the peracetic acid epoxidation of olefins: (12)

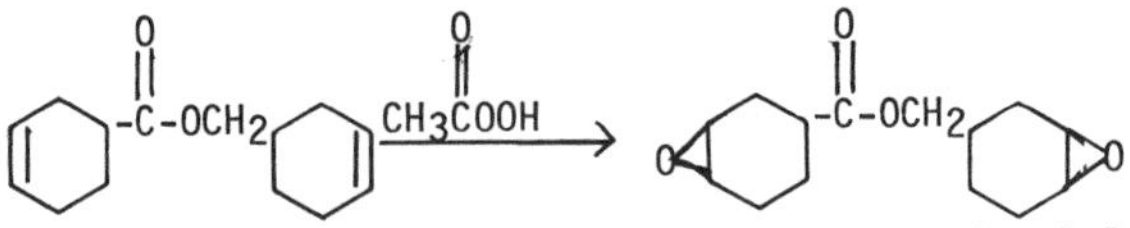

CH_3COOH

3-Cyclohexylmethyl-3 Cyclohexane Carboxylate

3,4 epoxy cyclohexylmethyl-3, 4-Epoxy Cyclohexane Carboxylate

In the early 1960s (13) in Europe cycloaliphatic products were developed and offered by CIBA commercially and were introduced in the United States through CIBA Products Company in about 1963.

At the present time, only a few products remain commercially available viz. the aforementioned product plus:

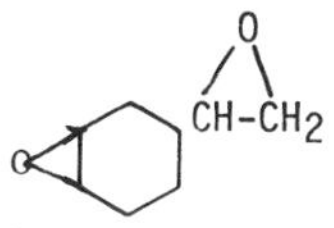

O
CH-CH2

Vinyl Cyclohexane Dioxide

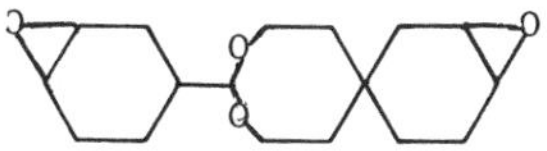

2(3,4 epoxy-cyclohexyl)
5-5 Spiro 3,4 Epoxy
Cyclohexane-meta
Dioxane

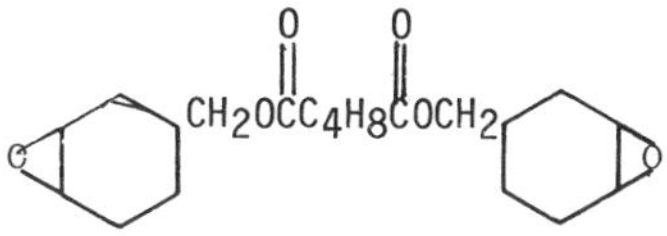

$CH_2OC(=O)C_4H_8C(=O)OCH_2$

bis(3,4 Epoxy-Cyclohexyl) Adipate

The products are liquids with generally low viscosity ($<$ 1000 cP @ 25°C).

In about 1960, FMC Corporation began marketing epoxidized polybutadiene via peracetic acid but the material was discontinued in 1965.

Glycidyl Esters

Glycidyl esters of cycloaliphatic structures were initially investigated by Shell and Bayer in Europe. (14)

A typical commercial product is based on hexahydrophthalic acid:

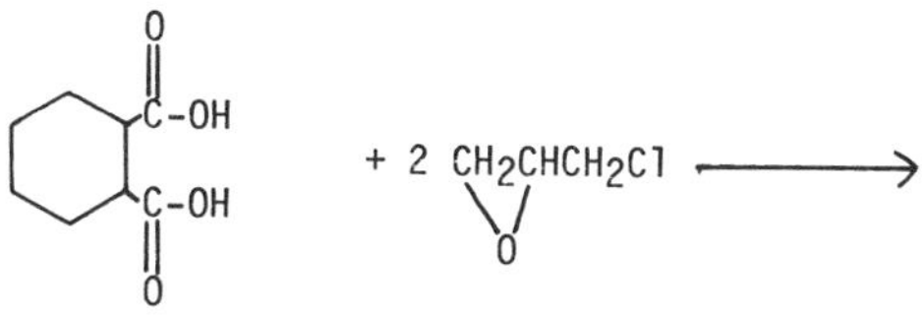

$$\text{C}_6\text{H}_{10}(\text{COOH})_2 + 2\ \text{CH}_2\text{CHCH}_2\text{Cl} \longrightarrow$$

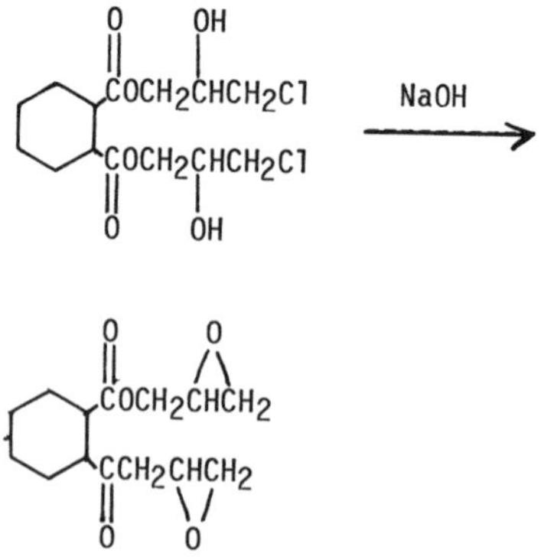

$$\text{C}_6\text{H}_{10}(\text{COOCH}_2\text{CH(OH)CH}_2\text{Cl})_2 \xrightarrow{\text{NaOH}}$$

$$\text{C}_6\text{H}_{10}(\text{COOCH}_2\text{CH}(\text{O})\text{CH}_2)_2$$

Diglycidyl Ester of Hexahydrophthalic Acid

Viscosities of these products are low, e.g. ca. 500 cP at 25°C and the reactivities resemble the standard bisphenol-A epichlorohydrin resins.

Diglycidyl esters are currently marketed by Mobay and CIBA-GEIGY.

Recent Developments in Epoxy Resins

A number of recent developments in epoxy resins have been the result of modifications of epoxy resins.

An important example is the preparation of a waterdispersible product by graft polymerization of, for example, styrene-methacrylic acid copolymer onto the aliphatic backbone of a high molecular weight epoxy resin (15).

The acrylic modified epoxy copolymer can be neutralized via the pendant carboxyl groups to provide a water-dispersible system with good hydrolytic stability. The grafting is presumed to occur onto the methylene hydrogens of the glycidyl group in the epoxy resin:

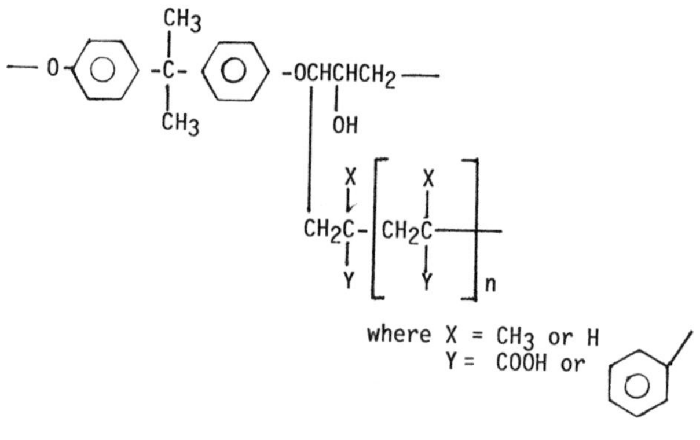

$$-\text{O}-\text{C}_6\text{H}_4-\text{C}(\text{CH}_3)_2-\text{C}_6\text{H}_4-\text{OCH}(-\text{CH}_2\text{C}(\text{X})(\text{Y})-[\text{CH}_2\text{C}(\text{X})(\text{Y})]_n-)\text{CH}(\text{OH})\text{CH}_2-$$

where $X = CH_3$ or H

$Y =$ COOH or C_6H_5

The product is a mixture of ungrafted epoxy resin as well as acrylic copolymer with pendant carboxyl groups and cure is effected via amino-formaldehyde resins to produce coatings for beer and beverage cans.

Another example of modified epoxy resin is the area of cathodic electrodeposition whereby a low molecular weight epoxy resin is reacted via the epoxy terminal groups with a primary or secondary amine to form an adduct which is solubilized with an acid (16):

$$\underset{\diagdown O \diagup}{CH_2CHCH_2O}-\langle\bigcirc\rangle-\underset{CH_3}{\overset{CH_3}{C}}-\langle\bigcirc\rangle-O\left[CH_2\underset{OH}{CH}CH_2O-\langle\bigcirc\rangle-\underset{CH_3}{\overset{CH_3}{C}}-\langle\bigcirc\rangle-O\right]_n CH_2\underset{\diagdown O \diagup}{CHCH_2}$$

$$+ R_1R_2NH \longrightarrow$$

$$\underset{R_2}{\overset{R_1}{N}}-CH_2\underset{OH}{CH}CH_2O-\langle\bigcirc\rangle-\underset{CH_3}{\overset{CH_3}{C}}-\langle\bigcirc\rangle-O\left[CH_2\underset{OH}{CH}CH_2O-\langle\bigcirc\rangle-\underset{CH_3}{\overset{CH_3}{C}}-\langle\bigcirc\rangle-O\right]_n CH_2\underset{OH}{CH}CH_2\underset{R_2}{\overset{R_1}{N}}$$

$$\xrightarrow{CH_3COOH} \sim\sim\sim\underset{R_2}{\overset{R_1}{N}}H^{\oplus} \quad OOCCH_3^{\ominus}$$

Cure of the solubilized epoxy resin is then effected by blocked isocyanates at elevated temperatures.

1. Swiss Patent 211,116 (1940) and U.S. Patent 2,324,483 (1943)
2. U.S. Patent 2,456,408 (1948)
3. U.S. Patent 2,131,120 (1938)
4. U.S. Patent 2,602,075 (1952)
5. G. Somerville ORG. COAT. PLAST. CHEM. Preprints 29 (1):55 (1969)
6. D.D. Applegath, R.F. Helmreich, G.A. Sweeney, SPE Journal 15 38 Jan. 1959
7. R.E. Burge and J.R. Halstrom, SPE Journal 20 75-79 (1964)
8. U.S. Patent 2,806,016 (1957); U.S. 2,857,362 (1958)
9. U.S. Patent 2,951,825 (1960)
10. U.S. Patent 2,897,179 (1959)
11. U.S. Patent 2,951,825 (1960)
12. U.S. Patent 2,716,123 (1953), U.S. Patent 2,745,847 (1953) U.S. Patent 2,750,395 (1954)
13. French Patent 1,233,231 (1958) Belgian Patent 596,268 (1959) Belgian Patent 396,269 (1959)
14. U.S. Patent 2,895,947 (1954) German Patent Application 1,211,177 (1964)
15. U.S. Patent 4,212,781 (1980)
16. U.S. Patent 3,984,299 (1976)

CYANATE ESTERS - HIGH PERFORMANCE RESINS

Richard B. Graver
Celanese Specialty Resins, 9800E Bluegrass Parkway, Louisville, KY

ABSTRACT

Cyanate esters of poly phenolic compounds were first prepared by E.Grigat in the early 1960's. The crystalline, highly pure monomers were made by reacting phenols with cyanogen halide in the presence of a base. Cyanate esters, like isocyanates, undergo a number of interesting reactions which lead to thermoset polymers. One of the most interesting polymerization reactions of cyanate esters is the formation of triazine rings by cyclotrimerization under relatively mild conditions (150-200°C). Polymers formed by cyclotrimerization have many commercially attractive properties. These properties and commercial application potential of these thermoset polymers are discussed.

INTRODUCTION

In the early 1960's Grigat (1) and Martin (2) and Jenson (3) reported the synthesis of stable aryl cyanates. Grigat (1) prepared these new materials by reacting phenolic compounds with a cyanogen halide.

$$\mathrm{ArOH + ClCN \xrightarrow[-BASE\bullet HCl]{BASE} ArOCN}$$

Martin (2) and Jenson (3) used thermolysis of thiatriazoles.

$$\mathrm{C_6H_5{-}CS{-}Cl \xrightarrow{NaN_3} C_6H_5{-}O{-}C{-}S} \longrightarrow$$

$$\xrightarrow[-S\ ;\ N_2]{\Delta} \mathrm{C_6H_5OCN}$$

Grigat reported that the carbon atom of the -OCN group is strongly electrophilic. Thus cyanates, like isocyanates (-NCO) should be reactive with nucleophilic reagents under mild conditions. Cyanates do react with active hydrogen containing materials such as polyols, amines and carboxylic acids under fairly mild conditions. The adducts with active hydrogen co-reactants are chemically different than those obtained from isocyanates. These are termed imidocarbonates, isoureas, bis(imidocarbonate) imides, etc. Polymeric products of this type made from cyanates have relatively poor hydrolytic, thermal and chemical stability as compared to analogous

Copyright 1986 by Elsevier Science Publishing Co., Inc.
High Performance Polymers: Their Origin and Development
R.B. Seymour and G.S. Kirshenbaum, Editors

polymers made from isocyanates. In general, it should be noted that cyanates are less reactive than isocyanates but the reaction products made by nucleophilic addition tend to be less stable.

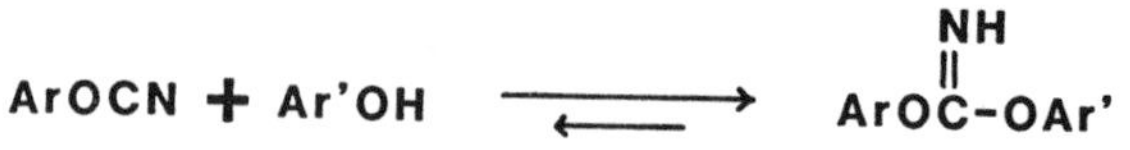

$$\mathrm{ArOCN + Ar'OH \rightleftharpoons ArOC({=}NH){-}OAr'}$$

IMINOCARBONATE

$$\mathrm{ArOCN + Ar'NH_2 \rightleftharpoons ArO{-}C({=}NH){-}NHAr'}$$

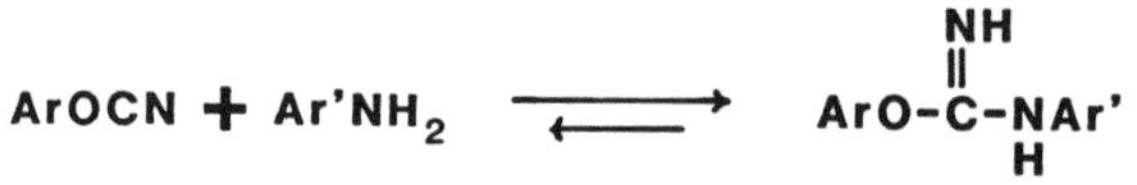

IMINOCARBAMATE

(ISO UREA)

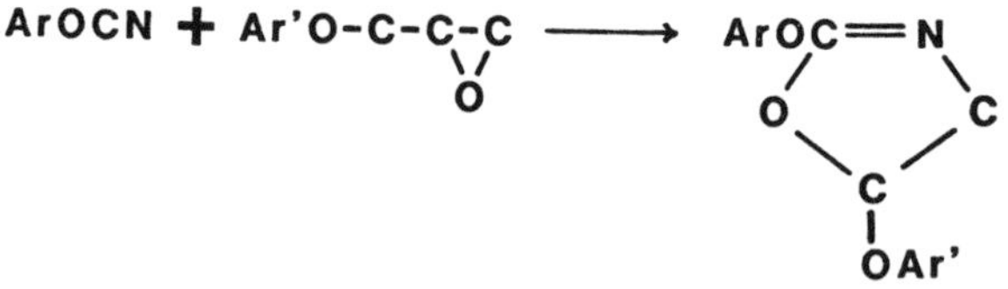

$$\mathrm{ArOCN + Ar'O{-}C{-}C{-}C\ (epoxide) \longrightarrow oxazoline\ ring\ (ArOC{=}N{-}C{-}C(OAr'){-}O)}$$

OXAZOLINE

Commercially valuable polymers are those formed by cyclotrimerization of di (or poly) functional aryl cyanate esters to s-triazines. This reaction occurs when aryl cyanate esters are heated to 150-200°C. The reaction can be controlled to yield resins (prepolymers) which are reproducible over a wide molecular weight range. These amorphous resins may be liquids, solids or solutions.

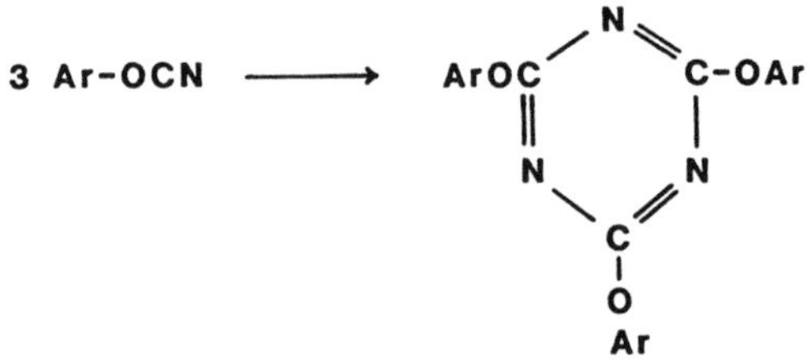

$$\mathrm{3\ Ar{-}OCN \longrightarrow (ArOCN)_3\ (2,4,6\text{-}tris(aryloxy)\text{-}s\text{-}triazine)}$$

TRIAZINE

One of the most interesting cyanate esters is derived from Bisphenol A. Bisphenol A is widely used as a starting material for epoxy resins, polycarbonate resins, polysulphones, polyetherimides and polyarylates. Bisphenol A dicyanate can be commercially produced by reacting bisphenol A with cyanogen chloride in the presence of a tertiary amine such as triethyl amine.

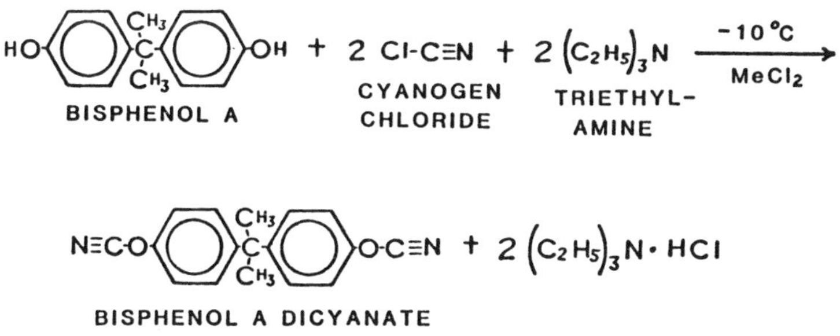

$$HO-C_6H_4-C(CH_3)_2-C_6H_4-OH + 2\ Cl\text{-}C{\equiv}N + 2\,(C_2H_5)_3N \xrightarrow[MeCl_2]{-10\,^{\circ}C}$$

BISPHENOL A — CYANOGEN CHLORIDE — TRIETHYLAMINE

$$N{\equiv}C\text{-}O-C_6H_4-C(CH_3)_2-C_6H_4-O\text{-}C{\equiv}N + 2\,(C_2H_5)_3N\cdot HCl$$

BISPHENOL A DICYANATE

Cyclotrimerization

Aryl dicyanate esters, NCO-Ar-OCN, appear to have several unique and interesting features:

- o Crystalline solids
- o High purity
- o Low melting point
- o Low toxicity
- o Widely soluble
- o Widely compatible

POLYFUNCTIONAL CYANATES

(Grigat - 1967)

Cyanate	M.P. °C.
1,3 DICYANATO BENZENE	80
1,4 DICYANATO BENZENE	115-116
1,3,5 TRICYANATO BENZENE	102
4,4' DICYANATO BIPHENYL	131
2,2-DICYANATO-1,1' BINAPHTHYL	149
2,2- BIS 4-CYANATOPHENYL PROPANE	82
4,4'- DICYANATO DIPHENYL SULPHONE	169-170
2,2'- DICYANATO-3,3,3',5'-TETRAMETHYL DIBENZYL ETHER	107-108
$(NCO-C_6H_4-O-CH_2-Si(CH_3)_2)_2O$	Liquid (unstable)

The incorporation of s-triazine ring structures into thermoset resins has been practiced for a number of years by incorporation of melamine formaldehyde condensates into the formulations. Curing occurs through a further condensation reaction with the release of volatile material. Aryl cyanate esters cyclotrimerize to s-triazine structures upon heating above 150°C without the release of volatiles.

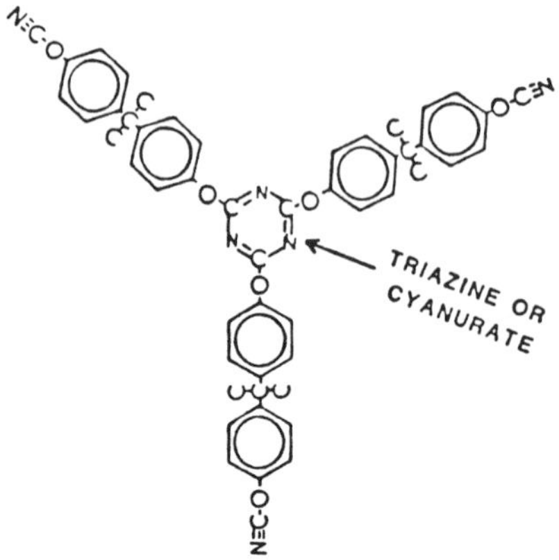

TRIAZINE OR CYANURATE

The resulting polymers of intermediate molecular weight can be recovered as liquids, solids or solutions which can be further reacted to a fully thermoset matrix linked by s-triazine ring structures. This network can be more accurately classified as a polycyanurate comprising the esters of bisphenol A with cyanuric acid.

POLYCYANURATE (BADCy) NETWORK

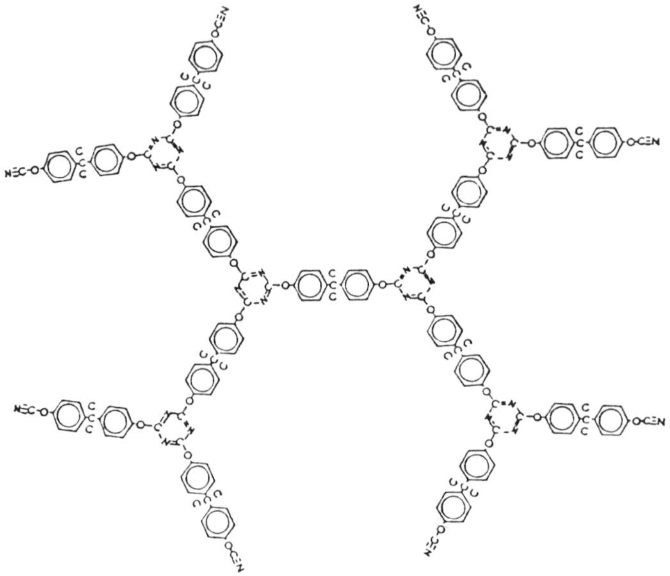

The trimerization process appears to proceed in a stepwise manner and n values of 3, 5, 7, 9, etc. can be readily identified by GPC or LC separation methods.

PREPOLYMER MOLECULAR FRACTION ANALYSIS BY GEL PERMEATION CHROMATOGRAPHY

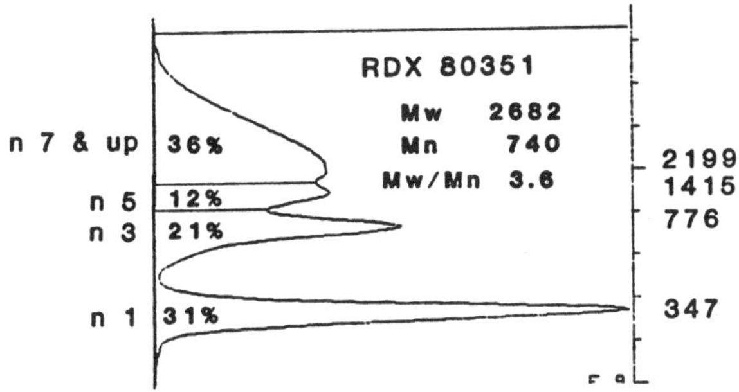

The partially trimerized, soluble resins can be blended with other resins, catalysts, additives, etc. to formulate commercially acceptable coatings, adhesives, laminates and structural composites.

<u>Applications</u>

Aryl cyanate esters such as bisphenol A dicyanate can be processed much like epoxy resins for many applications. This may be implied by comparing the monomer structure of bisphenol A diglycidyl ether to bisphenol A dicyanate.

Bisphenol A dicyanate

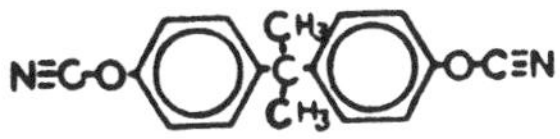

$$N{\equiv}C{-}O{-}C_6H_4{-}C(CH_3)_2{-}C_6H_4{-}O{-}C{\equiv}N$$

Bisphenol A Diglycidyl Ether

$$\overset{O}{CH_2{-}CH}{-}CH_2{-}O{-}C_6H_4{-}C(CH_3)_2{-}C_6H_4{-}O{-}CH_2{-}\overset{}{CH{-}CH_2}(O)$$

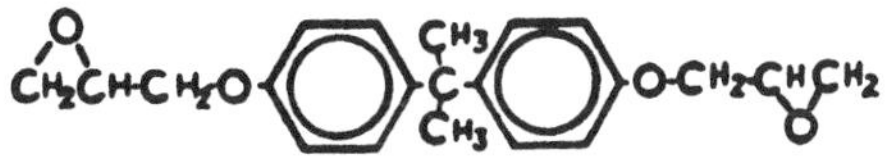

Similarities to epoxy resin include no volatile polymerization by-products, stable B-stage resins possible, low shrinkage, high adhesion and void free structures. High purity makes cyanate esters very attractive for use in electronic applications. The performance of bisphenol A dicyanate resins in printed wiring board laminates was described by Weirauch et al(4). Mobay Chemical (and Bayer) introduced products of this type for PWB applications in the United States in 1976-78. These "Triazine A" resins were not commercially successful and were withdrawn from the market about 1979.

The wide compatibility of aryl cyanate esters witn both thermosetting and thermoplastic resins has been reported in patents and other publications. Approximately sixty US patents have issued which describe blends and combinations with other resins and polymers.

An interesting blend with thermoplastic resins to form semi IPN's is described by Wertz and Prevorsek (6). They mixed monomer with engineering thermoplastic resins and then cured the blend following injection molding or prepregging to obtain a tough, chemical resistant structure. End use applications mentioned included electronics, printed wiring boards and aircraft canopy covers.

50/50 THERMOPLASTIC/BISPHENOL A DICYANATE SEMI IPN'S

THERMOPLASTIC	TENSILE STRENGTH psi	TENSILE MODULUS psi	BREAK ELONG. %
Copolyester/Carbonate	12,250	311,000	17.6
Polycarbonate	12,300	299,000	17.3
Polysulfone	10,600	297,000	12.7
Dibutylphthalate	420	19,200	70.0
PET	11,100	355,000	12.5
Polyether sulfone	10,400	340,000	9.6

• Wertz & Prevorsek -1984

Mitsubishi Gas Chemical (7) has developed a family of products which are stated to be a coreactive blend of triazine resin (terminology used to describe bisphenol A dicyanate and its prepolymers) and a bismaleimide resin. Properties of these resins are described in patents (8) and product literature (7). End uses suggested include electronic encapsulation, printed wiring board and structural composites.

Kubens (9) found cyanate ester resins were compatible and reactive with epoxy resins. We have confirmed this in our laboratories. Such combinations can be formulated to yield cured materials containing both triazine and oxazoline ring structures. If brominated epoxy resins are employed in such blends, flame retardant materials can be obtained containing as low as 12% bromine versus 20% bromine required to flame retard epoxy systems. Such products show promise for use in printed wiring board laminates.

TYPICAL LAMINATE PROPERTIES

	EPOXY	TRIAZINE	BT RESIN
Tg, °C.	110	255	310
Flexural, Kg/mm^2	60	60	60
Peel Strength, Kg cm	1.7-1.9	1.7-1.9	1.5-1.7
(150°C.)	0.7-0.8	1.6-1.8	1.4-1.6
Dielectric Constant	4.8	4.2	4.1
Dissipation Factor	0.02	0.003	0.002
Volume Resisitivity, ohm-cm	1×10^{14}	1×10^{15}	1×10^{15}
(150°C.)	1×10^{12}	5×10^{14}	5×10^{14}

Mitsubishi Gas Chem. -BT RESIN (3rd Edition)

Delano (10) studied the use of cyanate esters in carbon fiber prepregs for advanced structural composites. He found that state of the art mechanical properties versus epoxies could be obtained in composites even though the cyanate were not fully crosslinked. Using new catalyst technology (unpublished) we have determined that fully crosslinked systems can be obtained at 170-200°C. These materials have improved thermal, mechanical, electrical, toughness and moisture resistance properties over epoxy resins.

NOVEL CYANATE RESINS

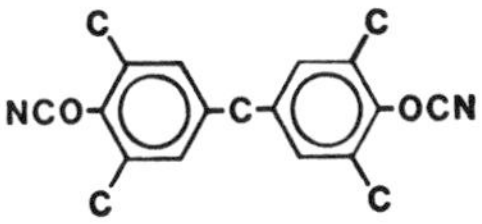

4,4'- DCTMDM

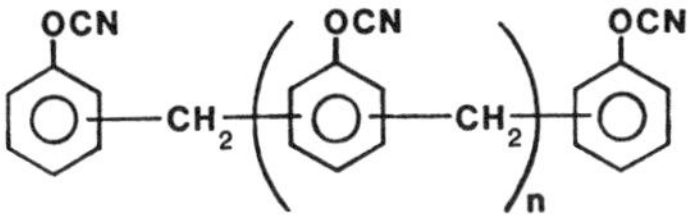

NOVOLAC CYANATES

Brand and Harrison (11) reported that carbon fiber composites based on triazine resins could be toughened by modification with monocyanate esters and/or with butadiene-nitrile rubbers. These modifications gave up to 50% greater fracture toughness and improved moisture resistance even though they reported the systems were not fully crosslinked. Our studies indicate that further improvements in toughness can be obtained by selection of the proper catalyst package. Cercena (5) proposed an alternative to blending of thermoplastic resins with triazine resins. She end-capped a low molecular weight polycarbonate with an aryl cyanate. When cyclotrimerized and cross-linked a tough, light colored and solvent resistant polymer was obtained.

Future Trends

Three major trends appear to be developing for cyanate esters and triazine resins:

- o Blends with thermoplastic resins
- o Blends with thermoset resins
- o Development of novel cyanate ester monomers

Properties of some new experimental cyanate esters are listed.

Although current commercial usage of these resins is not large, the proliferation of patent literature and published data indicates there is considerable potential for these materials.

STRUCTURE / PROPERTY RELATIONSHIPS IN DICYANATES

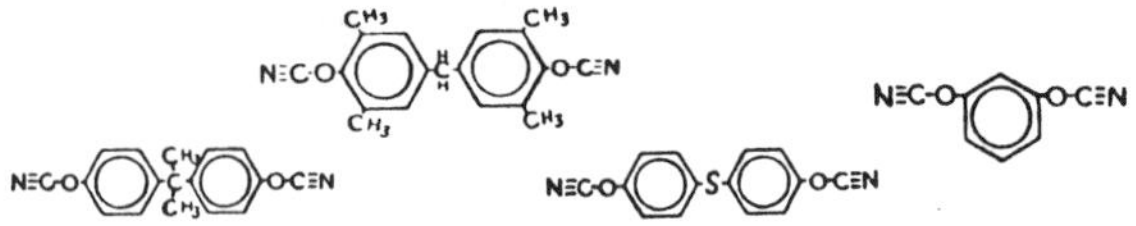

Monomer	BADCy	METHYLCy	THIOCy	RESORCy
MELT POINT (°C)	79	106	94	78
°C AT 0.1 TORR VP	182	208	200	114
TOXICITY	LOW	LOW	LOW	IRRITANT
Homopolymer				
HDT (°C)				
DRY	255	236	257	251
H_2O SAT.	185	230	181	170
% H_2O @ SAT.	1.7	1.1	1.5	2.3
TENSILE				
STRENGTH (PSI)	12,000	10,000	11,400	17,300
STRAIN (%)	3.6	2.5	3.6	2.9
MODULUS (10^6PSI)	0.48	0.47	0.46	0.70
% $MeCl_2$ ABSORPTION	5.8	9.4	0.6	< 0.1
FLAMMABILITY, UL 94	BURNS	BURNS	V-0	INTUMESES

REFERENCES

1. E. Grigat and R. Putter, Angew Chem. Internat. Edit 6, 206 (1967).
2. D. Martin, Angew Chem. Internat. Edit 3, 311 (1964).
3. K. A. Jenson and A. Holm, Acta Chem. Scand. 18, 826 (1964).
4. K. K. Weirauch, P.G. Gemeinhardt, a. L. Baron, Soc. Plast. Eng., Tech Paper 22, 317-21 (1976).
5. J. L. Cercena, S. J. Huang, Polymer Preprint (ACS Div. Polm. Chem.), 25 (1), 114.
6. D. H. Wertz, D. C. Prevorsek, Plast. Eng., 40 (4), 31-3 (1984);US Pat.
7. Mitsubishi Gas Chemical "BT Resin Product Bulletin 3rd Edition".
8. M. Guku, K. Suzuki, K. Nakamichi, US Patent 4,110,364 (1978).
9. R. Kubens, H. Schulthier, R.W. Stammhein, E. Grigat, H-D. Schminke, US Patent 3,562,214 (1971).
10. C. B. Delano, A. H. McLeod, C. J. Kiskiras, NASA Contract Final Report FR-80-42/AS (Contract NAS3-22025).
11. R. A. Brand and E. S. Harrison, NASA Contract Report 3615 (Contract NAS1-15859).

POLYIMIDES

John J. King and Byung H. Lee
CIBA-GEIGY Corporation
Ardsley, New York

I. INTRODUCTION

Polyimides have been known to be one of the best materials for use in high temperature applications. To date the volume of polyimides used is small but growing very rapidly due to the increased performance characteristics demanded in various areas. These polyimides exhibit outstanding resistance to irradiation, mechanical deformation, and solvent attack, while maintaining an excellent balance of mechanical and electrical properties. However, the rigid structure and low concentration of hydrogen in the polyimide chains, which impart properties such as thermooxidative stability to the material, are also responsible for their processing difficulties. With the development of new classes of polyimides and considerable research to overcome some of the inherent difficulties of these polymers, they are finding applications in a number of end-uses including adhesives, laminates, sealants, coatings, foams, electronic films and fibers. This paper discusses the history and development of polyimides with recent development in the field of addition-type polyimides.

II. Synthesis of Polyimides

The first synthesis of polyimides was reported by Bogert and Renshaw[1], who observed that, when 4-aminophthalic anhydride or dimethyl 4-aminophthalate is heated, water or alcohol is evolved and a polyimide is formed [1]:

H_2N O O O $-H_2O$ O N O n [1]

H_2N $COOCH_3$ $COOCH_3$ $-CH_3OH$

Polyimides with aliphatic units in their main chain are generally obtained by thermal polycondensation by heating salts of aromatic tetracarboxylic acids and aliphatic diamines. Edwards and Robinson[2] were the first to describe the preparation of an aliphatic polyimide by the fusion of a diamine with a diester of a tetracarboxylic acid. The reaction for the polyimide formation may be represented as follows [2]:

Copyright 1986 by Elsevier Science Publishing Co., Inc.
High Performance Polymers: Their Origin and Development
R.B. Seymour and G.S. Kirshenbaum, Editors

$$\text{HOOC, COOH, }CH_3OOC\text{, }COOCH_3\text{ (benzene)} + H_2N-(CH_2)_m-NH_2$$

$$\downarrow$$

$$\text{HOOC, }CH_3OOC\text{, }COOCH_3\text{ (benzene)}-\overset{\ominus\oplus}{COONH_3}-(CH_2)_m-NH_2 \qquad [2]$$

$$\downarrow$$

$$\left[-N \langle (C{=}O)_2 C_6H_2 (C{=}O)_2 \rangle N-(CH_2)_m - \right]_n$$

The melt polycondensation method for the preparation of polyimide has limited applicability. The melting points of the polyimides obtained must be below the reaction temperature in order to keep the reaction mixture in a fused state during the polycondensation process. Only in this case is it possible to achieve a reasonable molecular weight. This method can therefore be used successfully only for aliphatic diamines containing a certain number of methylene units. Aromatic polyimides, on the other hand, are generally infusible, so that when aromatic diamines are used, the reaction mixture solidifies too early to allow the formation of a high molecular weight product.

Polyimides with aromatic units in the main chain are generally prepared by a two-step polycondensation method. The first reports on the two-step polycondensation method for the preparation of polyimides appeared in 1959 in several patents assigned to DuPont Company, and later in various papers.[3-17]

The first step in the preparation of polyimides is the formation of polyamic acid by the acylation reaction of a diamine with a dianhydride of a tetracarboxylic acid in a suitable polar solvent as shown in equation [3]:

$$O\langle (C{=}O)_2 \rangle Ar \langle (C{=}O)_2 \rangle O + H_2N-Ar'-NH_2$$

$$\downarrow \qquad [3]$$

$$\left[-NH-OC(HOOC)Ar(COOH)CONH-Ar'- \right]_n$$

The second step of the reaction - the dehydrocyclization of the polyamic acid (imidization) - proceeds as shown in equation [4]:

$$\left[\begin{matrix} HOOC & & CONH-Ar' \\ & Ar & \\ -NH-OC & & COOH \end{matrix} \right]_n \xrightarrow{-2n\ H_2O} \left[-N \overset{O \quad O}{\underset{O \quad O}{\langle \; Ar \; \rangle}} N-Ar'- \right]_n \qquad [4]$$

and is carried out thermally or chemically.

The mechanism of the formation of polyamic acids and the effect of various factors on the course of this first step in the preparation of polyimides, (e.g., solvents[6], the order of addition of the reagents[6], the ratio of the two starting materials[10], the concentration of the reagents in the solvent[18], the presence of moisture[10], the manner of the reaction [heterogeneous or homogeneous][17b] and the degree of purity of the monomers[17b]) have been investigated in great detail.

The conversion of polyamic acids into polyimides - imidization or dehydrocyclization - consists of the intramolecular formation of water from the polyamic acid to form a cyclic polyimide. It can be carried out in two ways, thermally and chemically.

The thermal imidization method[5,6,12b] generally consists of heating the dried polyamic acid with a continuous or stepwise increase in temperature. The imidization process can be followed by means of the changes taking place in the infrared or ultraviolet spectra of the polyamic acid and also by the measurement of water evolved during the imidization process[19].

The chemical imidization method[6,8] consists of a treatment of a polyamic acid film or powder with dehydrating agents. Acetic anhydride or other lower aliphatic acids, such as propionic, valeric acid or others, can be used for this purpose. Ketenes and dimethyl ketenes can also be used as dehydrating agents. Tertiary amines are used as catalysts for the chemical imidization. The chemical imidization may also be carried out in a solution of the polyamic acid in aprotic solvent[20] by adding acetic anhydride and tertiary amine.

Various efforts[21,22,23] have been made to overcome the difficulties associated with the above synthesis. One approach is to modify the polymer structure to increase the solubility of the polyamic acid, for example, introducing some methylene groups via the use of an aliphatic diamine together with the aromatic diamine or the use of bipyridyl diamines. The other approach has been to use a different reaction to form the imide linkage. It is now known that polyimides can be formed from tetraacids and their dianhydride by condensation with diisocyanates with subsequent loss of CO_2 and H_2O [5].[24-27]

$$\text{(dianhydride: O=C–Ar–C=O cyclic anhydrides)} + OCN\text{-}Ar'\text{-}NCO \xrightarrow{CO_2} \left[-N(CO)_2Ar(CO)_2N\text{-}Ar'- \right]_n \qquad [5]$$

III. Developments

As discussed above, numerous polyimides have been synthesized in many different methods and characterized. The direct production of high molecular weight aromatic polyimides in a one-step polymerization could not be accomplished because the polyimides are usually insoluble and infusible. The polymer chains precipitate from the reaction media (whether soluble or melt) before high molecular weights are obtained. Commercial polyimides are, therefore, classified by the several different processes they were prepared to afford reasonable processibility in the final product.

A. Condensation-Type Polyimides

Processing of the polyimide can only be accomplished after the first step at which point it is still soluble or fusible. Due to the intractability of fully cyclized polyimides, they are sometimes supplied as precursor solutions of the polyamic acid. The final polyimide is produced by applying heat to remove the solvent and initiate the polycondensation reaction forming the fully imidized product. Two commercially available condensatiion-type polyimide precursor solutions are the Skybond resins, manufactured by the Monsanto Chemical Company, and Pyralin, a product of the E.I. DuPont de Nemours and Company. A third resin which falls into this category, NR-150, also by DuPont, has been discontinued commercially.

1. Skybond Resins

Skybond resins are based on 3,3',4,4'-benzophenonetetracarboxylic dianhydride (BTDA) and various aromatic diamines. The reaction sequence used to produce the polyamic acid precursor for Skybond 703, one type of skybond resin, is shown in equation[3].[28] The polyimide is then formed by dehydrocyclization of the precursor resin.

2. Pyralin

Pyralin polyamic acid precursor solutions are similar to the skybond resins, but they are based on pyromellitic dianhydrides. The reaction sequence to form the Pyralin polyamic acid precursor and then the fully imidized product is the same as that for the skybond resins discussed previously.

B. Thermosetting Polyimides

Aromatic polyimides have been tested extensively as matrix resins for fiber composites because of their attractive properties. Usually the polyamic acid precursor resins in polar solvents like N-methylpyrolidone, dimethylformamide or dimethylacetamide are impregnated in fibrous materials. Unfortunately these solvents show an extremely good affinity to the resin and are therefore very difficult to remove from the prepreg and the demolded composite. Residual solvent acts as a plasticizer and when present in a laminate prevents the development of good high temperature physical properties. In addition, the condensation type cyclization reaction of the soluble intermediate prevents the formation of void-free laminates. Because of this failure of aromatic polyimides or matrix resins for advanced composites, a wide variety of thermosetting-type polyimides was developed with the aim of improving the processing properties. In general the objective of all thermosetting-type polyimides is the preparation of an imide ring-containing low molecular weight backbone that carries terminal polymerizable end groups, and therefore the prepoly- mer provides real processability and can be fully cured through the end-groups in the later stage.[29-31]

1. Norbornene End-capped Polyimides

H.R. Lubowitz of TRW[32,33] discovered polyimide systems that were end-capped with norbornene anhydride groups. The synthetic route used was the classical amic acid precursor method whose molecular weight could easily be adjusted by varying the molar ratio between the reactants. The prepolymer (P13N, CIBA-GEIGY) showed a real melt transition at around 270°C and could therefore be processed in laminates.

The norbornene concept was further developed by NASA Lewis Research Center and culminated in the so-called PMR-concept, a novel class of addition-type polyimides formed by in situ polymerization of monomeric reactants.[34,35] The reaction sequence[36-37] involves aromatic diamine, 5-norbornene-2,3-dicarboxylic acid monoalkyl ester and dialkyl ester of an aromatic tetracarboxylic acid which produces methanol, water, and the following intermediates [6].

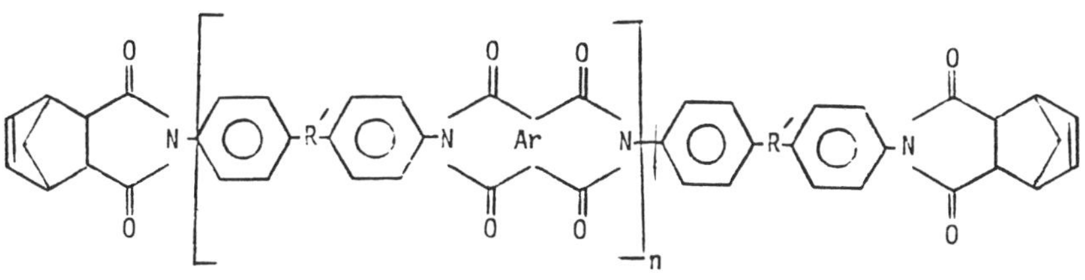

[6]

The cure of the PMRs is generally accepted to proceed via the reverse Diels-Alder reaction of the norbornenyl end-groups to generate maleimide end groups and cyclopentadiene in the temperature range of 270-300°C, followed by the immediate copolymerization of cyclopentadiene with maleimide end groups to form a cross-linked structure without the evolution of volatiles.

2. Acetylene-Terminated Polyimides

The cyclotrimerization of ethynylbenzene leads to 1,3,5-triphenylbenzene, and consequently polycyclotrimerization of bisethynylbenzene yields a high temperature stable polyphenylene resin.[38] This chemical concept has been extended to develop the first ethynyl-terminated polyimide in the literature.[39-41] The following figure [7] outlines the synthetic route characterized by simply replacing some of the aromatic diamine with an ethynyl-substituted aromatic monoamine to provide a ethynyl-terminated polyimide resin.

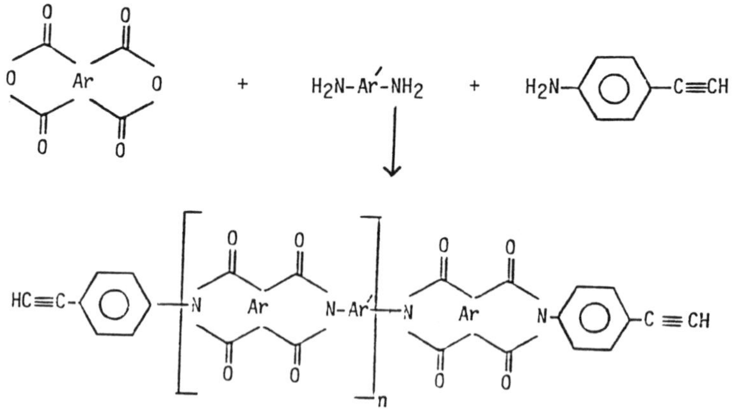

O O O Ar O O O + $H_2N-Ar'-NH_2$ + $H_2N-C_6H_4-C{\equiv}CH$

↓ [7]

$HC{\equiv}C-C_6H_4-$[N (O O) Ar (O O) N-Ar'$]_n$-N (O O) Ar (O O) N-$C_6H_4-C{\equiv}CH$

One resin based on this chemistry became commercially available under the trade mark of Thermid 600 (National Starch and Chemical Corp.), which provides a material with a high degree of thermal stability upon heating presumably by the self polymerization of terminal acetylene groups.[42]

C. Addition-Type Polyimides (Maleimide-based Polyimides)

One of the promising high temperature thermosets is based on bismaleimides. The usual synthesis of the monomers[43,44] involves the reaction of maleic anhydride with aromatic diamines forming a bismaleamic acid as an intermediate which undergoes cyclodehydration in the presence of acetic anhydride and sodium acetate or tertiary amine to form the bismaleimide[8].

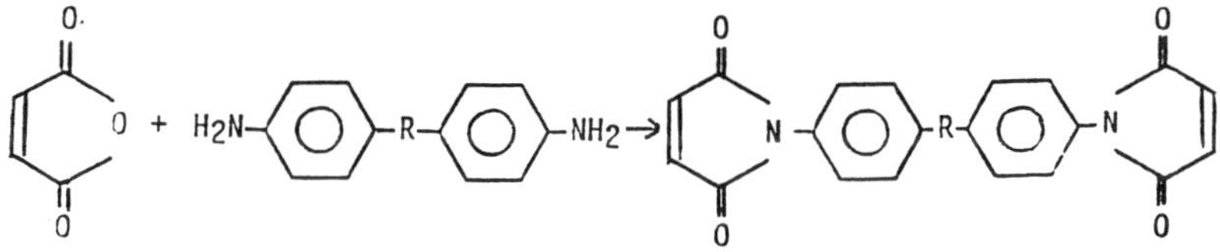

[8]

The double bond of the maleimide is very reactive. Grundschober[45] first reported their homo and copolymerization which can be achieved by simply heating the monomer to temperature between 150-400°C. It can also undergo a chain extension reaction with primary and secondary amino compounds, known as Michael addition, which is used to synthesize a well-known commercial resin, Kerimid 601 of Rhone Poulenc[46], by reacting 4,4'- bismaleimidodiphenylmethane with methylene dianiline to form a polyaminobismaleimide[9].

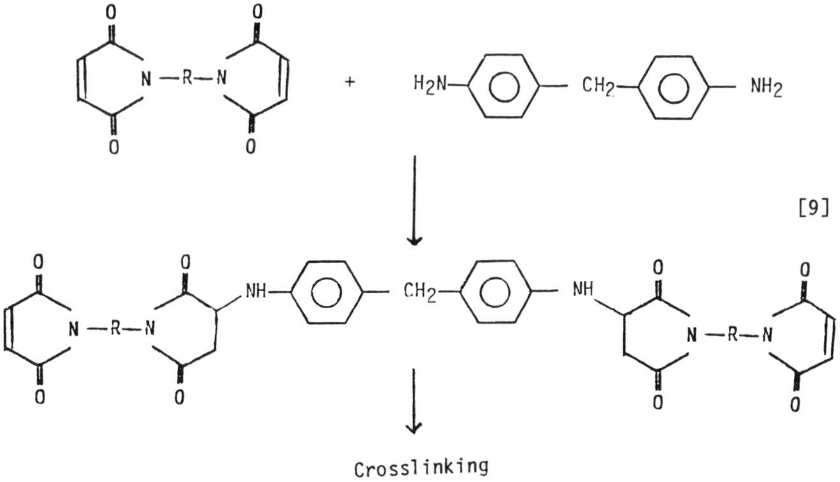

[9]

Crosslinking

However, the major disadvantage of all the bismaleimide resins is their brittleness. They are high modulus, low strength materials with a very low elongation at break resulted mainly by their high crosslink density. In order to overcome those problems, extensive work was done to control the molecular weight of the prepolymer by reacting the diamine with m-aminobenzoic acid chloride[47] or by reacting the bismaleimide with m-aminobenzoic acid hydrazide.[48] But all these prepolymers have to be processed from high boiling polar solvents.

Recently, CIBA-GEIGY introduced unique crosslinking chemistry into bismaleimide system by utilizing so-called "ene" reaction to react the maleimide double bond[10][44,49-51]

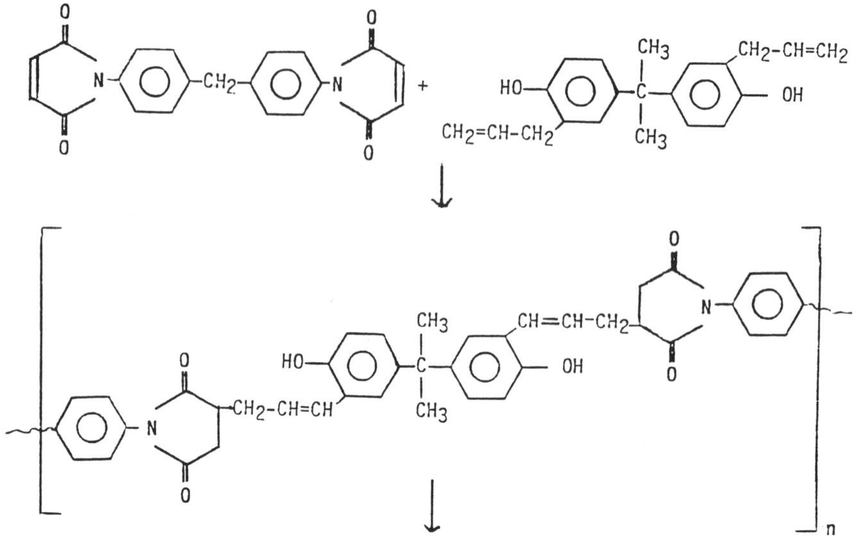

Polymerization

[10]

This system, Matrimid™ 5292, has been recognized for its excellent handling and processing characteristics in addition to its ability to provide excellent toughness and thermal/mechanical properties to the cured products.

D. Thermoplastic Polyimides

Three fully imidized thermoplastic aromatic polyimides are available today. Two, Upjohn's polyimide 2080 and CIBA-GEIGY's XU 218, are commercially available. An experimental thermoplastic polyimide, Larc-TPI, is not commercially available, but a vast amount of data has been developed on it.

1. Polyimide 2080

The polyimide is produced by the polycondensation reaction of a aromatic diisocyanate with benzophenonetetracarboxylic dianhydride (BTDA) in suitable solvent[5].

2. XU 218

XU 218 is produced by the reaction of an aromatic cycloaliphatic diamine, 5[6]-amino-1(4-aminophenyl) -1,3,3-trimethylindane (DAPI) with BTDA as shown below[11]. The intermediate polyamic acid is chemically imidized by acetic anhydride/tertiary amine.[52]

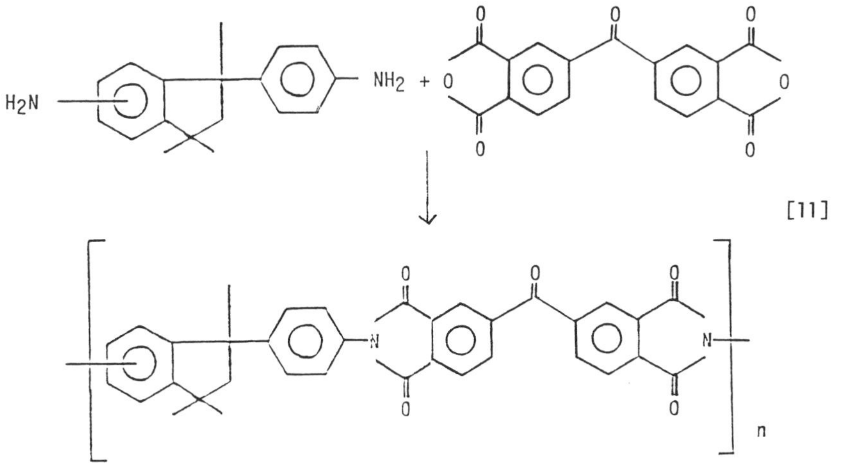

H2N NH2 + O O O O O O O O

[11]

O O O O O N N n

3. Larc-TPI (NASA)

This material is produced by the reaction of BTDA with 3,3'-diaminobenzophenone in suitable solvent. The resin can be used as is or precipitated in water or methanol.[53] Its reaction sequence is shown below[12].

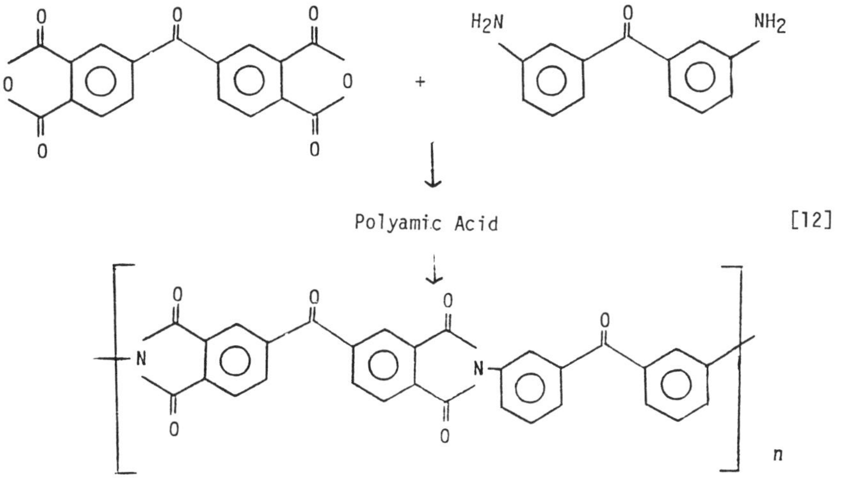

O O O O O O O + H2N O NH2

Polyamic Acid

[12]

O O O O O N N n

IV. PROCESSING

Although polyimides are attractive engineering plastics possessing excellent thermal, physical and mechanical properties, processing problems and cost have limited usage. This processing problem was somewhat overcome by introducing new classes of polyimides with improved processability. This section discusses processing difficulties, improvements in processing and processing methods.

A. Processing Problems and Improvements

1. Condensation-Type Polyimides

Skybond and Pyralin, which were early condensation polyimides, have a number of processing problems associated with them. In this class of polyimides, the polyamic acid precursor undergoes imidization during processing. As a by-product of ring closure reaction, water is produced. The presence of low molecular weight volatiles such as water can lead to void formation.[31]

A second problem associated with this class of polyimides is poor flow. That is, the temperatures required for cyclization of the polyamic acid precursors and fusion are quite close. When these steps occur during final processing any precipitation strongly affects the flow characteristics of the polyimide.[54] Poor flow properties can adversely affect consolidation of the polyimide material.

2. Thermosetting/Addition-Type Polyimides

Some of the problems associated with condensation-type polyimides were alleviated with the introduction of thermosetting-type polyimides. These materials are low molecular weight, preimidized oligomers which are terminated with reactive end groups. Processing is simplified, since these materials can polymerize with a minimum of volatile by-products. However, not all processing problems are eliminated. In the case of the Thermid, the acetylene end-group reaction occurs quite quickly, requiring rapid heating rates to produce a melt before full polymerization. Recent advances in addition-type polyimides technology indicate that these problems are, however, being overcome. In addition, some thermal stability is sacrificed with thermosetting polyimides.[55]

3. Thermoplastic Polyimides

In fully imidized thermoplastic polyimides, the material is fully imidized before final processing. By its very nature, a fully imidized thermoplastic, i.e., CIBA-GEIGY's XU 218 and Upjohn's Polyimide 2080, remain soluble and fusible. As such, fully imidized thermoplastic polyimides can be molded into final parts without undergoing any further chemical reaction. Some thermal stability is, however, sacrified.

B. Processing Methods

Autoclave molding usually involves heat (up to 260°C), pressure (35-50 psi) and vacuum. This method is particularly attractive for processing condensation-type polyimides and some processable polyimides.[31]

Compression molding involves placing a material, usually in powder form, into the cavity of a heated mold (250°C) and applying heat (up to 350°C) and pressure (3000-5000 psi).[56] This processing method can be used for the addition-type and thermoplastic polyimides.

Direct forming uses powder metallurgy techniques to cold-form a part followed by sintering technique at 360-380°C.[57]

Lamination involves impregnating materials, i.e., glass, quartz and graphite fabric and boron, graphite and aramid fibers with a solution of the polyimide, assembling the impregnated material, and compressing using pressure.[56]

Injection molding is done by placing the polymer in powder or pellet form into the heated cavity of an injection molding machine, allowed to soften, followed by pushing forward into a cooled cavity by a ram or reciprocating screw where it remains until it is below its Tg or Tm to ensure that the part retains the desired form.[58]

Fibers are produced by continuous solution spinning, washing, drying, and drawing using commercial wet and dry spinning equipment. This method is particularly attractive for the soluble thermoplastic polyimides.[53]

Films of polyimides can be cast by pouring the material in liquid form onto a roller which process it to the desired thickness.[56]

V. PROPERTIES

Key properties of aromatic polyimides are their outstanding thermooxidative stability and high resistance to radiation and to deformation under load at elevated temperatures. This section discusses the properties of polyimides in general.

The zero strength temperatures of polyimides are much higher than the 550°C value given for aluminum, generally running around 800-900°C. Polyimide films remain thermally stable under vacuum or an inert atmosphere to over 500°C. Beyond this temperature there is a sharp increase in weight loss, leveling out at approximately 35%.[5,12b] No further weight loss is apparent up to 1000°C. The ultimate thermal stability of any individual polyimide is highly dependent upon its chemical structure.

Mechanical properties of polyimide films based on pyromellitic dianhydride and bis(4-aminophenyl) ether have been correlated. Optimum polyimide film properties were obtained from a polyamic acid solution which had a molecular weight average of 140,000-200,000 and a number average of 40,000-50,000. Maximum tensile strengths were approached at a molecular number of 40,000. Maximum elongation, tear strength, and impact strength were exhibited by materials having a molecular number above 50,000. Elongation and tear strengths were still increasing with the highest number obtained at molecular number 60,000.[59] Generally, these results may be applied to all polyimide-type systems.

Aromatic polyimides have good elongations, 7-10%, indicating toughness and good machinability.

Dielectric properties are excellent, ranging 3.3-3.5 at 60 Hz. Polyimides also have a low coefficient of thermal expansion, $28\text{-}34\times10^{-6}$in./ in.F^{o}. Use of fillers, e.g., glass, graphite and asbestos, reduces dimensional changes to even lower levels.

All polyimides tend to be very poor in ultraviolet stability with substantial deterioration of properties occurring probably at 4000 hr exposure.

Aromatic polyimides are resistant to most organic solvents and dilute acids, however, long term exposure to caustic solution is not desirable. They are also soluble in highly polar solvents.

Overall, the general properties of polyimides are outstanding. However, it is best to keep in mind that polyimides, depending upon their chemical structures, have widely divergent properties, but overall, their ability to maintain structural integrity and good mechanical and electrical properties at elevated temperatures are the main reasons why polyimides are utilized today.

VI. Applications

Polyimides are primarily used in aerospace, automotive and electronics industries where a need exists for materials with long term, high temperature capabilities.

Autoclave molded parts are extensively used in structural applications despite the high void contents of these materials due to their excellent retention of properties at elevated temperatures.[31]

Molded polyimides are widely used as bearing materials because of their high temperature and compressive capabilities, chemical resistance, and natural lubricity. These bearings are found in jet engines, appliances, and office equipment. Polyimides are also used in rotary-vane compressors in piston rings and in non-lubricated seals.[60,61] They also replaced PTFE rotary seal rings in heavy-duty transmissions for automotive and off-the-road equipment.[57]

Because of their excellent radiation resistance, aromatic polyimides are used for valve seats and seals, thermal and electrical insulators in nuclear applications.

Polyimide films are used as insulation for electric motors, aircraft parts, missile wire cable, magnetic wire, and flat flexible cable.[60]

Polyimides have been used as high temperature structural adhesives for both primary and secondary bonding applications.[62]

Fibers are used to make high temperature, flame resistant fabrics. They are also used in electrical insulation and as reinforcement materials.[60]

Polyimides are used in the construction of multilayer printed circuit boards because of their low coefficient of thermal expansion, high bond strength/peel strength of copper foil at elevated temperature, and no resin smear.[63]

REFERENCES

1. M.T. Bogert and R.R. Renshaw, J. Am. Chem. Soc., 30, 1140 (1908).
2. (a) W.M. Edwards and I.M. Robinson, U.S. Patent 2,710,853 (1955)
   (b) ibid., 2,867,609 (1959)
   (c) ibid., 2,880,230 (1959)
3. W. M. Edwards and I.M. Robinson, U.S. Patent 2,900,369 (1959)
4. Belg. Patent 589,179 (1960)
5. J.I. Jones, F.W. Ochynski and F.A. Rackley, Chem. Ind., 1686 (1962)
6. G.M. Bower and L.W. Frost, J. Polym. Sci. A,1, 3135 (1963)
7. R.J. Angelo, U.S. Patent 3,073,785 (1963)
8. Brit. Patent 903,271 (1962); 942,025 (1963)
9. L.E. Amborski, ACS, Div. Polym. Chem., Preprints, 4(1), 175 (1963)
10. L.W. Frost and I. Kesse, J. Appl. Polym. Sci., 8,1039 (1964)
11. French Patent 1,365,545 (1964)
12. (a) C.E. Sroog, A.L. Endrey, S.V. Abramo, C.Z. Berr, W.M. Edwards and K.L. Oliver, ACS, Div. Polym. Chem.,Preprints, 5(1), 132 (1964)
    (b) ibid., J. Polm. Sci. A, 3, 1373 (1965)
13. W.M. Edwards, U.S. Patent 3,179,614 (1965); 3,179,634 (1965)
14. W.M. Edwards and A. L. Endrey, U.S. Patent 3,179,631 (1965)
15. A.L. Endrey, U.S. Patent 3,179,630 (1965); 3,179,633 (1965)
16. H.H. Levine, P.M. Hergenrother and W.J. Wrasidlo, "High Temperature Adhesives", Contract No. W-63-0420-C, (1964)
17. (a) C.E. Sroog, J. Polym. Sci. C, 16, 1191 (1967)
    (b) M. L. Wallach, ACS, Div. Polym. Chem., Preprints, 6(1), 53 (1965)
    (c) ibid., 8(1), 656 (1967)
    (d) ibid., 8(2), 1170 (1967)
    (e) W.L. Wallach, J. Polym. Sci. A-2, 5, 653 (1967)
    (f) R. Ikeda, J. Poly. Sci. B, 4, 353 (1966)
18. S.A. Zakoshchikov, K.N. Vlasova, G.M. Zolotareva, N.M. Krasnova and G.A. Ruzhentseva, Plasticheskie Massy, 14 (1966); C.A. 64, 15994f, (1966)
19. W. Wrasidlo, P.M. Hergenrother and H.H. Levine, ACS, Div. Polym. Chem., Preprints, 5(1), 141 (1964)
20. W.R. Hendrix, U.S. Patent 3,249,561 (1966)
21. (a) K. Kurita and R.L. Williams, J. Polym. Sci. Polym. chem. Ed., 11, 3125, 3151 (1973)
    (b) ibid., 12, 1809 (1974)
22. L.F. Charbonneau, J. Polym. Sci. Polym. Chem. Ed., 16, 197 (1978)
23. W.M. Alvino and L.E. Edelman, J. Appl. Polym. Sci., 22, 1983 (1978)
24. R.A. Meyers, J. Polym. Sci. A-1, 7, 2757 (1969)
25. P.S. Carleton, W.J. Farrissey and J.S. Rose, J. Appl. Polym. Sci., 16, 2983 (1972)
26. L.M. Alberino, W.J. Farrissey and J.S. Rose, US Patent 3,708,458 (1973)
27. W.M. Alvino and L.E. Edelman, J. Appl. Polym. Sci., 19, 2961 (1975)
28. H. Stenzenberger, Kautschuk and Gummi-Kunststoffe, Jahrgang, Nr.8, 477 (1976)
29. E.A. Burns, R.J. Jones, R.W. Vaughn and W.P. Kendrick, NASA Contract NA 3-12412, (Jan. 1970)

30. T.T. Serafini, P. Delvigs and W.B. Alston, 27th National SAMPE Symposium, 320, (1982)
31. T.T. Serafini, Handbook of Composites, George Lubin (Ed.), Van Nostrand Reinhold, New York, 89 (1982)
32. H.R. Lubowitz, ACS, Org. Coat. Plast. Chem., 31(1), 561 (1971)
33. H.R. Lubowitz, U.S. Patent 3,528,950 (1970)
34. T.T. Serafini, P. Delvigs and G.R. Lightsey, J. Appl. Polym. Sci., 16, 905 (1972)
35. T.T. Serafini, P. Delvigs, J. Appl. Polym. Sci., Appl. Polym. Symp., 22, 89 (1973)
36. L.E. Lorensen, SAMPE Quarterly, P.1 (Jan. 1975)
37. R.D. Vannucci and W.B. Alston, 31st Annual Technical Conference, Reinforced Plastics/Composite Institute, SPI Inc., Section 20A, P.1 (1976)
38. H. Jabloner, L.C. Cessna and R.H. Mayer, ACS, Org. Coat. Plast. Chem., 34(1), 198 (1974)
39. A.L. Landis, et. al., ACS, Polymer preprints, 15(2), 537 (1974)
40. N.A. Bilow, U.S. Patent 3,845,018 (1974)
41. A.L. Landis, N. Bilow, L.B. Keller, R.H. Boschan and A.A. Castillo, ACS, Polymer preprints, 19(2), 23 (1978)
42. N. Bilow, A.L. Landis, R.H. Boscham and J.G. Fashold, SAMPE Journal, 18(1), 8 (1982)
43. (a) N.E. Searle, U.S. Patent 2,444,536 (1948)
(b) H.N. Cole and W.F. Gruber, U.S. Patent 3,127,414 (1964)
44. B. Lee, M.A. Chaudhari and T. Galvin, 17th National SAMPE Technical Conference, 172 (1985)
45. F. Grundschober and J. Sambeth, U.S. Patent 3,380,964 (1964)
46. M. Bergain, et.al., British Patent 1,190,718 (1968)
47. H. Stenzenberger, British Patent 1,501,606 (1976); U.S. Patent 4,239,883 (1980); ibid., 4,269,966 (1981)
48. H. Stenzenberger, U.S. Patent 4211861 (1980)
49. J. King, M. Chaudhari and S. Zahir, 29th National SAMPE Symposium, 392 (1984)
50. T.J. Galvin, M. Chaudhari and J. King, Annual AICHE, (1984)
51. M. Chaudhari, T. Galvin and J. King, 30th National SAMPE Symposium, 735 (1985)
52. J.H. Bateman, W. Geresy, Jr. and D.S. Neiditch, ACS, Org. Coat. Plastic Chem., 35(2), 77 (1975)
53. A.K. St.Clair and T.L. St. Clair, 28th National SAMPE Symposium, 165 (1981)
54. R.A. Dine-Hart and W.W. Wright, RPG Conference, New and Improved Resin Systems (Sept. 18-19, 1973)
55. F.P. Darmory, Plastics Design and Processing, 18 (Sept., 1974)
56. F. Billmeyer, Jr., Textbook of Polymer Science, 2nd Ed., Wiley (1971)
57. J.N. Anderson, SAE Technical Paper Series 810968 (1981)
58. W.V. Titow and B.J. Lanham, Reinforced Thermoplastics, Applied Science Publishers, London (1975)
59. M.L. Wallach, J. Polym. Sci. A2, 6, 953 (1968)
60. S. Oldham, SAMPE Quarterly, 1 (Jan. 1979)
61. J.D. Hensen, Plastics Engineering, 20 (Oct., 1977)
62. P.S. Blatz, Adhesive Age, 39 (Sept., 1978)
63. L. Hayes and R.E. Mayfield, Circuits Manufacturing, 52 (Aug., 1978)

UV/EB CURING TECHNOLOGY: A SHORT HISTORY

B. CHRISTMAS, **R. KEMMERER**, F. KOSNIK
Celanese Specialty Resins, 9800 East Bluegrass Parkway, Jeffersontown, Kentucky

ABSTRACT

Since the 1960's the initiation of free radical polymerization by ultraviolet light (UV) or electronic beam (EB) has had significant industrial impact. Initially, this curing technique was limited to certain furniture coating applications and the availability of materials and equipment was rather limited. Since those early days, a significant amount of research and development time has been expended in order to more fully exploit this technology. The result has been that a wide variety of materials and equipment are now available. Furthermore, the applications which employ this technology now extend to adhesives, coatings, and inks for several new specialty fields. This paper gives an overview of this technology with emphasis on recent developments. A short summary of the chemistry involved is given and formulational techniques for various end uses are outlined.

INTRODUCTION

Radiation curing is defined as the polymerization (crosslinking) of a coating, ink, or adhesive directly on a substrate (such as paper, metal, or plastic) via interaction with incident radiation. Although a number of radiation forms have been used, current technology is dominated by ultraviolet (UV) and electron beam (EB) curing. Polymerization usually takes place by a free radical mechanism; cationic polymerization is also practiced although limited to UV initiation.

UV/EB coatings (formulations) are traditionally 100% reactive solids so that the need for solvents is eliminated. This feature as well as speed (cure literally in a second or so) and minimal heat buildup (applicable to a wide variety of substrates) make UV/EB an attractive alternative to bake systems.

Compared to that of most paint and coatings technologies, the history of UV/EB curing is indeed short. Developed commercially in the late 60's and early 70's, this technology has seen steady growth in terms of the number of commerical users and the markets served. The growth rate for radiation curable materials is over 20% per year [1]. The types of applications being pursued today (curing 3-D parts, magnetic tape binders, fiber optics, conductive coatings, structural adhesives and binders, etc.) bear little resemblance to the relatively simple wood fillers and topcoats developed in the 60's.

The growth of UV/EB curing has been spurred most certainly by the oft cited environmental and energy advantages but, more importantly, by rapid advances in the development of raw materials and equipment. These developments have resulted in performance advantages over other technologies which is, in the final analysis, the key to growth.

Copyright 1986 by Elsevier Science Publishing Co., Inc.
High Performance Polymers: Their Origin and Development
R.B. Seymour and G.S. Kirshenbaum, Editors

This history will review some of the early applications and describe the evolution of raw materials and equipment which have enabled expansion into new application areas. Raw material developments are discussed with respect to the problems they addressed such as oxygen inhibition, adhesion, pigmentation, and cure speed. There are still limitations to the technology which have prevented penetration into certain application areas. These are discussed as they relate to future growth.

EARLY APPLICATIONS

Although there were earlier R&D efforts [2], UV/EB curing became a viable, commercial technology in the late 60's and early 70's. Development activities focused in two distinct arenas. Commercial UV applications were developed in Europe whereas EB was pioneered in the U.S.

UV-curable unsaturated polyester-styrene wood sealers and topcoats were introduced in Europe (Germany) in the 60's [3]. This application is still practiced on a large scale today. These early efforts were beset by problems which limited the acceptance of UV curing and thus its growth. Since high styrene concentrations were required to get a workable viscosity, there was little formulating latitude. The coatings were brittle and oxygen sensitivity was a problem. Volatility (odor) was also a drawback.

Commercial UV curing in the U.S. began in 1970. The first application was, again, particle board filler. Twenty lines were in operation the first year and this is still probably the largest single volume application in the U.S. Several other applications were also developed around the same period. In 1972, metal decorating by UV was being seriously pursued by several can manufacturers [4]. Patents appeared in 1972 on the use of acrylated urethanes to replace moisture cure urethanes for floor tiles (no wax) [5]. Overprint varnishes for paper also began to see intense development. After considerable development by Sun and Inmont, the first UV ink production began in June 1970. Lithographic inks were introduced in Europe in 1972 [5]. Substrates included carton board, paper, and metal.

While UV was getting its start in Europe (late 60's), Ford developed an EB curing process in the U.S. [6]. In 1971 a commercial line was set up to paint interior automotive parts (dashboards and trim). At Boise Cascade, a wood finishing line utilizing the Ford scanning unit, was also operational. There were originially many safety hazards associated with the equipment, and the required nitrogen blanketing was very expensive on these large scales. Styrene and methyl methacrylate were the workhorse monomers in combination with unsaturated polyesters or acrylic polymers. Volatility of monomers was undoubtedly a problem also. Both applications were abandoned in the mid-70's due to poor economics. In 1972, Svedex in Holland developed an EB system, based on U.S. technology, to coat and cure paint applied to hardboard flush doors [7]. This system remains in operation today.

At the same time that UV/EB developments were emerging, a world energy crisis materialized which sent finishers searching for energy efficient alternatives to thermal processes. UV/EB was a natural (it's commonly cited that UV utilizes about 10% of the energy required for thermal cure) and this crisis precipitated considerable development activity. Raw material and equipment availability increased significantly, and the types of resins and monomers also expanded. Efforts began in earnest to tackle limitations of the technology. A journal dedicated solely to the technology first issued in 1973 (J. of Radiation Curing). A conference highlighting UV/EB was also spawned during the period, AFP/SME's Radcure, in 1974. Predictions for enormous growth were heard repeatedly. For many reasons that didn't happen, but there is no doubt that the development activity during this period spurred technological growth.

RAW MATERIALS

Monomers

The limited availability of raw materials in the 60's is apparent from the early reviews of radiation curing [8]. As previously mentioned, the first UV/EB applications utilized primarily polyesters diluted with styrene. The first change in this technology came with the use of multi-functional acrylates (MFA's) as total or partial substitution for styrene. These low volatility reactive diluents overcame many of the styrene problems. These monomers, produced via esterification of polyols, included hexanediol diacrylate (HDODA), trimethylol propane triacrylate (TMPTA), tetraethylene glycol diacrylate (TTEGDA), and pentaerythritol triacrylate (PETA).

Although the availability of MFA's added new dimensions to formulating, they also added a new problem, skin irritation. A new set of handling precautions needed to be established but problems persisted, especially in certain ink applications such as lithography. Volumes remained relatively small until other applications, such as electronics, took off.

In May 1974, Celanese announced plans to build a large scale plant to manufacture these monomers. It was in production later that year thereby significantly changing the availability of MFA's.

Thus, in addition to the continuing search for monomers with unique performance, there was a major thrust in the late 70's and early 80's to produce monomers with less skin irritation potential. These efforts led to a second generation of monomers which are esters of alkoxylated alcohols. This process adds molecular weight apparently rendering the monomers less penetrating. There are, of course, performance trade-offs. A large variety of ethoxylated and propoxylated monomers are now available. Decreased skin irritation properties are reported.

Whereas the number of monomers commercially available in the early years was very limited, there are now products available for virtually every formulating task.

Acrylate Oligomers

Many laboratories and commercial users produced their own oligomers for some of the first applications. While polyesters were the preferred oligomers in Europe, all acrylic systems (monomers and oligomers) were being developed in the U.S. These employed a variety of acrylated (or methacrylated) resins.

Acrylated epoxies comprise one of the largest classes of UV/EB oligomers. These resins were described in patents as early as 1968 [9]. Originally developed for thermal cure applications, the diacrylate of the diglycidyl ether of bisphenol A, remains the largest volume oligomer of this class used today. Acrylated epoxies were much faster curing than polyesters and performed extremely well in air [10]. Early suppliers (in 1974) were Dow and Shell. Celanese offered commercial products in 1976. When acrylated epoxies are combined with MFA's, a wide range of performance properties can be developed.

If there is one drawback to epoxy acrylates, it is that they are characterized by very high viscosities due to H-bonding of secondary hydroxyls. Monomer dilution is essential, and attaining application viscosities can be a problem.

Polyester acrylates produced by direct esterification or trans-esterification of residual hydroxyls are lower viscosity and thus offer a handling advantage. Due to the strong polyester use in Europe, these products were primarily developed by European suppliers. Properties of these resins can vary significantly depending on the raw material selection. Cure speed is generally slower than epoxy acrylates possibly due to their inability to form carbinyl radicals via hydrogen abstraction.

Urethane acrylates were first utilized in floor tile coatings around 1972. As with polyesters, a large number of compositional variants can be synthesized. There is considerable captive production of these resins as well as a number of commercial offerings. This resin class finds particular utility in plastics applications.

Thiol/ene and Cationic Chemistries

In addition to acrylated systems, two other technologies bear mentioning. A mercaptan - olefin (thiol/ene) system was developed by W. R. Grace [11] in the early 70's. This system is primarily used to make photopolymer printing plates. It is estimated that over half the news-papers in the U.S. are printed using plates produced by this process. High cost has limited growth of thiol/ene in other applications.

Cationic polymerization of epoxies is well known and during the late 70's advanced rapidly as a viable technology for UV curing [12][13]. This progress was due to development of new classes of photoinitiators.

Aryldiazonium salts as UV initiators were described as early as 1973 [14]. Cationic polymerization with these initiators is not oxygen sensitive. This first class of salts had poor pot life and had problems due to N_2 liberation. The next generations, diaryl iodonium and diaryl sulfonium salts, overcame most of these problems. They were developed about the same time by 3M and General Electric [15-18]. Cationic systems utilizing cycloaliphatic epoxies exhibit the interesting feature that once polymerization begins, it can continue in the dark. Compared to free radical methods, use of cationic UV cure remains relatively small, but has found certain niches, such as metal varnishes.

Photoinitiators

Free radical photoinitiator development has had considerable impact on the growth of UV. Several advancements have opened up new applications. One of the first classes developed was benzoin ethers which form radicals via a Norrish I type mechanism. These had stability problems and were very oxygen sensitive. Oxygen sensitivity was addressed through several developments.

The use of benzophenone with an alkyl amine to overcome oxygen inhibition was an important discovery [19]. A hydrogen alpha to nitrogen is abstracted in this process. Benzil alkyl ketals comprise another efficient class of photinitiators and are also less oxygen sensitive. Stability is much better than with the benzoin ethers.

Pigmentation of UV coatings has always presented curing problems. Thioxanthones were shown to have utility in these applications. This photoinitiator class absorbs in longer wave length regions of the UV spectrum where pigment absorptions are less intense.

Several UV applications require good weatherability, i.e., non-yellowing. Photoinitiators based on alpha-substituted acetophenone show good performance in these applications. Hydroxy cyclohexyl phenyl ketone is a popular commercial product of this type.

Like monomers and oligomers, there is now a large choice of photo-initiators available which are designed to perform a variety of tasks [20].

EQUIPMENT

UV

The first UV processes employed low pressure mercury lamps for curing. These emitted 1-2 watt/inch of energy. Higher energy mercury lamps became available in 1970 and generally had outputs of 200 watt/inch. This development was key to the commercialization of UV in the U.S. Although widely used, medium pressure mercury (MPM) lamps were fairly inefficient (13% of the emitted energy was UV) and they had limited spectral output mercury lines. These units ran hot (up to 700° C) and life expectancy of the early models was short.

In 1974 electrodeless lamps powered by microwaves were developed. Output was 36% UV and bulb life was considerably longer. Spectral output was more flexible due to different bulb compositions. Power was up to 300 watt/inch. Although pulsed xenon lamps offer another alternative, MPM lamps and electrodeless units account for most of the UV curing done today [21].

EB

The original Ford electron beam units were of the scanning beam design. This type dominated the early EB years. The growth of EB was helped significantly by the introduction in 1971 [22] of electron accelerators which employ linear cathodes which do not need to be scanned. This so-called electron curtain process was commercially developed by Energy Sciences International and Radiaton Polymer Company, and is the state-of-the-art today. Besides the need for nitrogen blanketing, reliability (maintenance), and equipment cost remain major considerations when contemplating EB curing.

FUTURE

We've witnessed tremendous technical growth over the last 10-15 years and the future for UV/EB technology looks bright. Several applications were mentioned during the course of this review and, of course, many others are being practiced.

Some recent developments include:

. EB transfer metallization
. Coat and cure 3-D parts - furniture, automotive
. Coatings for fiber optics
. Coatings and adhesives for optical disks
. Release coatings/pressure sensitive adhesives
. Abrasion/scratch resistant plastic coatings
. Binders for magnetic tape
. Coatings for metallized plastics
. Laminating adhesives for packaging
. Caul sheets
. PWB - conformals, conductive inks, etc.

It is probably fair to say that UV/EB has been tried in virtually every conceivable coating and adhesive application. Research remains very active in all those areas as commercialization is pursued or process/product improvements sought. Continued development of new materials and equipment to support these efforts will ensure UV/EB growth.

REFERENCES

1. K.R. Lawson, Products Finishing, Nov. issue, 75 (1983).

2. W. Deminger and M. Patheiger, J. Oil Chem. Assoc., 52, 930 (1969).

3. A. Vrancken, Proceedings: Radiation Curing V, Sept., 13 (1980).

4. R.B. Mesrobian, SME Technical Paper #FC 75-301 (1975).

5. J.W. Prane and C. Bluestein, SME Technical Paper #FC 76-527 (1976).

6. W.J. Burlant, U.S. Patent 3,437,514 (To Ford Motor Co.), April 8, 1969.

7. K. O'Hara, SME Technical Paper #FC 75-321 (1975).

8. T.J. Miranda and T.F. Huemmer, J. Paint Technology, Vol. 41, No. 529, 118 (1969).

9. F. Fekete and W.J. Plant, U.S. Patent 3,373,075 (To H.H. Robertson Co.), March 12, 1968.

10. R. Darbenko, C. Friedlander, G. Gruber, P. Prucnal, and M. Wismer, Progress in Organic Coatings, 11, 71 (1983).

11. C.R. Morgan and A.D. Ketley, ACS Div. of Org. Coating & Plastics Preprints, 165th/Mtg, 33 281 (1973).

12. J.V. Crivello, UV Curing: Science and Technology (Ed. S. Peter Pappas), pp. 24-77, Technology Marketing Corporation (1978).

13. C.L. Olson, J. Radiation Curing, 3 (3), 2 (1976).

14. J.H. Feinberg, U.S. Patent 3,711,390, Jan. 16, 1973.

15. J.V. Crivello and J.H.W. Lam, 4th International Symposium on Cationic Polym., Akron, Ohio, June 24, 1976.

16. J.V. Crivello and J.H.W. Lam, Macromolecules, 10 (6) 1307 (1977).

17. J.V. Crivello, U.S. Patent 3,981,897 (to General Electric), Sept. 21, 1976.

18. G.H. Smith, Belgium Patent 828,841 (to 3M), Nov. 7, 1975.

19. C.L. Osborne and D.J. Trecker, U.S. Patent 3,759,807 (to Union Carbide), Sept. 18, 1973.

20. L.R. Gatechair and D. Wostratzky, paper published in: Radiation Curing VI Conference Proceedings, pp. 1-1 - 1-24 (1982).

21. W.R. Schaeffer, AFP/SME Technical Paper #FC 85-768 presented at Finishing '85 (1985).

22. L.A. Wasselle, "Electron Beam Curing of Polymers," Stanford Research Institute Report, July, 1977.

Fibers

CARBON FIBERS, FROM LIGHT BULBS TO OUTER SPACE

ROGER BACON AND CHARLES T. MOSES
Union Carbide Corporation, Parma Technical Center
P. O. Box 6116, Cleveland, OH

ABSTRACT

Carbon fibers, first used as light bulb filaments in 1879, trace their modern history from about 1944. In 1959, the first important textile grade carbon fiber was manufactured commercially from rayon cloth and found use as a reinforcement in phenolic-matrix composite exit cones for rocket engines. Key technological developments in rayon-, polyacrylonitrile-, and pitch-based manufacturing processes during the period 1964-1975 resulted in very high-strength and high-modulus fibers and led to ever-widening applications in missile components, aircraft structures, and sports equipment. Since 1980, ultrahigh-strength fibers (from polyacrylonitrile) and ultrahigh-modulus fibers (from pitch) have been developed and are expected to be used extensively in aircraft and spacecraft structures.

INTRODUCTION

The basic element of carbon fibers is a long, thin filament composed mainly of carbon. The internal structure consists of undulating graphitic layers stacked in parallel arrangement; hence, "graphite fiber" is often used in place of "carbon fiber" but the latter term is preferred.

Commercial forms of carbon fibers include continuous filament yarns or tows, woven fabrics, discontinuous filament mats and felts, and (rarely) single filaments.

Carbon fibers are manufactured by pyrolysis and thermal treatment of organic precursor fibers, viz., rayon, polyacrylonitrile (PAN), or pitch. In pilot plant quantities, carbon fibers have been grown by chemical vapor deposition from an organic vapor such as methane or benzene; in the laboratory, they have also been grown by physical vapor deposition.

The important properties of carbon fibers include mechanical stiffness (Young's modulus), tensile and compressive strength, thermal stability, thermal conductivity (along the fiber axis), thermal resistivity (of bulk fibers), low coefficient of thermal expansion, and electrical conductivity.

Early carbon fibers - The first known use was as an electrically conducting filament in the first Edison incandescent light bulb in 1879 [1, 2]. Although tungsten largely displaced carbon in lamp filaments beginning around 1910, carbon filaments were still used in U.S. Navy ships as late as 1960, because they withstood ship vibrations better than tungsten. During the latter half of the 1950's, carbon fibers were "reinvented" in more versatile forms and began finding uses as fire-proof industrial fabrics, electrical resistance heaters, high temperature

Copyright 1986 by Elsevier Science Publishing Co., Inc.
High Performance Polymers: Their Origin and Development
R.B. Seymour and G.S. Kirshenbaum, Editors

packing and gasketing, and as reinforcing fibers in phenolic resins for rocket nozzles and missile parts [3].

High-performance carbon fibers - In the late 1950's, one of the present authors demonstrated that graphite whiskers, grown by physical vapor deposition in a 3900°K high-pressure carbon arc, were capable of exhibiting a tensile strength as high as three million psi and a Young's modulus in excess of 100 million psi [4]. These whiskers were "laboratory curiosities"; however, less than ten years after their initial discovery in 1955, two processes for manufacturing high-strength and high-modulus carbon fibers from organic precursor fibers (rayon and PAN) were invented almost simultaneously. Five years still later, a high modulus carbon fiber made from pitch was invented. These "high-performance" carbon fibers were used as reinforcements in structural composites, finding applications in sporting goods, aircraft structures, missiles, and space satellites. Matrix materials for these composites included resins (epoxy, polyester, thermoplastics), metals (aluminum, magnesium, copper), and carbon.

Because of their military usefulness, nearly every major development in carbon fiber technology was strongly supported by government agencies. (Boron fibers, invented in the late 1950's, and the first practical "high-performance" filamentary reinforcement, was already used in military aircraft structures). In the U.S., rayon-, PAN- and pitch-based carbon fiber processes were developed under sponsorship of the Air Force Materials Laboratory. In England, the Royal Aircraft Establishment and the Atomic Energy Research Establishment carried out much of the early PAN-based carbon fiber research and development as did the Government Industrial Research Institute in Japan. The development of applications for carbon fibers for aerospace use has been almost exclusively financed by government military and space agencies in the U.S., U.K., Japan, France, and other countries.

HISTORICAL DEVELOPMENT OF CARBON FIBER TECHNOLOGY

Nearly 8 million pounds of carbon fibers are now being produced each year. Commercial production is based on the pyrolysis of organic precursor fibers: either rayon, polyacrylonitrile (PAN), or pitch.

Each of the three major precursor processes has its own special place in history and will be discussed in turn. Rayon-based fibers were first in commercial production (in 1959) and led the way to earliest applications, which were primarily military. PAN-based fibers have proven to be superior to rayon-based fibers in several respects, notably in tensile strength, and have largely dominated the explosive growth of the industry since 1970. Pitch-based fibers, however, are uniquely capable of achieving extremely high axial Young's modulus and thermal conductivity and, therefore, have an assured place in critical military and space applications.

After tracing the historical developments of each of the three major precursor processes, a section on "other precursors" will briefly discuss various attempts to develop carbon fibers based on alternate organic precursors.

Finally, a section on carbon fibers grown by chemical vapor deposition (CVD) processes will be included. Although such fibers have never reached a commercial stage, they have demonstrated unique properties and current inventors are predicting future commercialization.

Carbon Fibers from Rayon and Other Cellulosic Precursors

In the beginning - The first commercially produced carbon filament made from a cellulosic precursor was the incandescent lamp filament used in the first Edison light bulb, beginning in 1879 [1, 2]. Cotton threads and, later, bamboo fibers were formed into filaments of desired size and shape and then baked to a substantially all-carbon replica of the original filament. Numerous patents describe improvements that were made in light bulb filaments for the next 30 years, after which tungsten wire drawing technology advanced to displace carbon filaments from the growing electric light bulb industry.

Carbonization of rayon (and PAN) yarns and fabrics was briefly investigated by Union Carbide Corporation during World War II as a possible substitute material for control grids in vacuum tube power amplifiers [5]. However, the first commercial venture into multifilament carbon fibers was about 1957 by Barnebey-Cheney, a licensee of W. F. Abbott's Carbon Wool Corporation, which briefly manufactured carbon fiber tows, mats, and batting materials from cotton and rayon precursors [6]. These materials were developed for use as high temperature thermal insulation, as particle filters for hot, corrosive gases or liquids and as activated carbon fibers.

"Graphite" fabrics - In 1958, Union Carbide submitted carbon fabrics to the U.S. Air Force for trials to replace glass fabrics in reinforced phenolic resin composites for rocket nozzle exit cones and re-entry heat shields [7, 8]. Later, carbon cloth strips were tried as heating elements for wall panels and for "instant" hot water heaters. These fabrics represented a more useful form for many purposes than the heavy tows, mats, and batting materials previously available. Another advancement in the art was the use of a high temperature "graphitization" step (heat treatment to ~ 2500°C or more) which rendered the carbon fibers more stable against heat and chemical attack than fibers which had only been "carbonized" (to ~ 1000°C).

These rayon-based carbon fabrics found immediate success in the exit cone application as well as in heat shields. In subsequent developments, the phenolic resin matrix was itself "carbonized" and "graphitized" to further enhance heat resistance capability and such "carbon-carbon" composites, containing rayon-based carbon fabrics, are found today on the nose and the wing leading edge of the space shuttle.

Other early entrants in the rayon-based carbon fabric business were HITCO and 3M [8], as well as Great Lakes Carbon and Carborundum [9].

Carbon yarn for composites - In 1963, Union Carbide began commercial production of continuously processed carbon yarn, permitting for the first time the development of carbon fiber composites made by filament winding or by lay-up of prepreg tapes. This was the initial entry of carbon fibers into the "advanced composites" industry which had previously been dominated by glass and boron fibers. These continuously processed yarns required very rapid pyrolysis of the rayon precursor compared with previous schedules used for batch processing. Pre-impregnation of the rayon by phosphoric acid derivatives, known fire-retardants for cotton fabrics, provided the necessary "catalyst" for pyrolysis [9, p. 16]. These continuous yarns quickly found uses as packing and gasketing materials.

Until 1965, all commercial carbon fibers were of relatively low strength and low modulus. The potential of the graphite structure to exhibit extremely high tensile strength as well as high modulus (in the layer plane direction) had been demonstrated in a study of vapor-grown whiskers [4]. Also, significantly higher strengths and moduli than those of rayon-based fibers had been demonstrated in PAN-based fibers in a 1961 Japanese study, largely unknown to Western workers at the time [10].

High-modulus yarn for advanced composites - In 1964, the first true high modulus fibers were prepared by the Royal Aircraft Establishment in England from PAN (in April) [11] and by Union Carbide Corporation from rayon (in August) [12]. The latter fibers were prepared by "stress-graphitizing" or "hot-stretching" rayon-based carbon yarn during graphitization at temperatures exceeding 2800°C [9, p. 20]. A series of increasingly high modulus fibers was soon produced commercially, beginning with "Thornel" 25 in late 1965 and continuing until 1978, when the high cost of the process forced its shutdown. Another producer during this period was HITCO [9, p. 21], a licensee of Union Carbide. Rayon-based high-modulus carbon fiber process development was supported by the U.S. Air Force (AFML) from 1965 through 1970 [13].

These high-modulus rayon-based fibers were used extensively in the U.S. as well as in France and Germany in the development of epoxy-matrix composites for space structures. They were also used to develop a new generation of high-strength and stiffness carbon-carbon composites for rocket nozzle throats and missile nose-tips. In recent years, the less costly high-modulus PAN- and pitch-based carbon fibers have been substituted for the rayon-based materials.

Carbon Fibers from Polyacrylonitrile (PAN)

Soon after the development of polyacrylonitrile fibers by DuPont Company in the 1940's, the remarkable thermal stability of this material was recognized. At the time, cellulosic fibers and PAN (homopolymer) fibers were the only fibrous substances which did not melt below their charring temperature and were, therefore, capable of retaining their fiber identity as carbon materials. Winter, of Union Carbide's National Carbon Co. Research Laboratories, investigated DuPont's early PAN fibers as precursors for carbon fibers near the end of World War II; some of these fibers had been thermally stabilized (apparently early versions of "black Orlon") by W. D. Coxe of DuPont [5].

The first public report that acrylic fibers could be rendered fireproof by heat-treating them in air or inert atmosphere at ~ 200°C was published by Houtz in 1950 [14]. The material was popularly called "black Orlon". Several companies in the U.S. patented fireproofed fabrics based on this process, including DuPont, Johns-Manville, and Carborundum [15]. Various textile forms of this material are commercially available today [16].

First process study of PAN-carbon fibers - The first extensive study of carbonization and graphitization of PAN fibers was made in 1961 by Shindo [10] at the Government Industrial Research Institute, Osaka, Japan. Shindo recognized the importance of an oxidative heat treatment step prior to carbonization in reducing the processing time and improving the carbon yield from PAN. He demonstrated good tensile strengths and moduli over 20 million psi, about three times those available from rayon precursor carbon fibers of that time. Shindo's work was quickly taken up by other Japanese workers; two carbon companies, Tokai Electrode Co. and Nippon Carbon Co. licensed the process and began commercial production on a pilot scale about 1964.

The key to high modulus and strength: stretching - These early fibers, however, were neither high-strength nor high-modulus by today's standards. One problem was that most acrylic fibers of that time were really not suitable for converting to carbon fibers: they either contained the wrong comonomer, possessed the wrong internal structure, or contained too many impurities. In England, a suitable fiber (except for an impurity problem which has since been corrected) did exist and was used as the basis for experiments by Watt and co-workers at the Royal Aircraft Establishment [11]. They discovered that the application of tension during the initial oxidation step at ~ 220°C was important to maintaining or enhancing molecular alignment and thus developing an oriented, high-modulus and high-strength carbon fiber after heat treatment to ~ 2500°C [17]. This high-modulus fiber was subsequently designated "Type I". The RAE workers also discovered that an optimum final heat treatment temperature of only 1000-1500°C would yield a lower modulus but much higher strength carbon fiber, designated "Type II" [18]. The "high strength fiber" has become the dominant fiber and has been used widely in the fiber reinforcement of epoxy or polyester resin composites in sporting goods (golf clubs, tennis rackets, fishing rods, skis, etc.) and for aircraft secondary structures. The high modulus Type I fiber has been produced in relatively small quantities but, in recent years, it has replaced high-modulus rayon-based carbon fibers, to which it is quite similar in properties, for "modulus-critical" applications, such as space antennas, and for carbon-carbon missile and rocket engine parts.

The British carbon fiber technology, to which workers at Rolls Royce [19, 20] and at the Atomic Energy Research Establishment [21] contributed significantly, was licensed to three manufacturers. Morgan Crucible Co., a carbon products and refractory ceramic company, was first, in 1967, to produce carbon fibers commercially ("Modmor I") but has since dropped out of the business. Courtaulds, the producers of the acrylic precursor fiber, extended their fiber technology to include carbonizing and is still a major producer of carbon fibers. Rolls-Royce, who developed the jet engine fan blades which unfortunately failed due to bird-strike impact and led to bankruptcy of the Rolls-Royce engine manufacturing business, has, under re-structuring, continued its development of carbon fiber applications in jet engines, but no longer produces its own fiber.

Higher strength from better precursors - The British licensees licensed, in turn, manufacturers in the U.S.A. and Japan. However, once the key elements of the British processes were known, the most important factor in the further advancement of PAN-carbon fibers was the development of superior precursors. It was in this endeavor that the Japanese PAN fiber industry excelled: Toray Industries developed a precursor far superior to any previously available and offered it to Union Carbide in the U.S. under a technical exchange agreement (1970) under which Toray obtained Union Carbide's carbonizing technology. The resulting high-strength, moderate-modulus (Type II) carbon fibers were produced briefly by Union Carbide in the U.S. in 1971-72 and simultaneously by Toray in Japan, where it became the dominant carbon fiber sold throughout the world for the next dozen years. Other Japanese fiber companies soon followed Toray in producing excellent quality precursors. Sumitomo offered their precursor to Hercules, who had been producing carbon fibers under license from Courtaulds, and that partnership has led to a series of high strength fibers which are now widely used in the U.S. Another successful U.S.-Japanese partnership has been that between Celanese (later BASF) and Toho Beslon; several other such partnerships have formed.

Strength still climbing - The tensile strength of Type II, high-strength carbon fibers has steadily risen from 2.7 GPa in the early 1970's to 3.5 or 4.0 GPa in the early 1980's. Three years ago, a new, ultrahigh-strength carbon fiber, possessing somewhat higher modulus than the Type II fiber, was introduced by Hercules/Sumitomo [22]; similar fibers are now available from several manufacturers [16]. It is usually called "intermediate modulus carbon fiber." Tensile strengths over 5.5 GPa (800 ksi) are now achieved. The processes which have made this remarkable development possible are still largely proprietary but, since all such fibers have in common a smaller-than-normal filament diameter (4-6 μm vs. 7 μm), it appears that additional stretching in some part of the process is a key element [23].

The motivation behind the development of increased strength carbon fibers has been the assertion by airframe manufacturers, notably Boeing, that carbon fibers possessing very high strain-to-failure (2%) would permit their use in primary structures of future aircraft. Heretofore, carbon fibers have been used in wing skins and control surfaces where strength requirements were not critical. Another application for these new fibers will be light-weight rocket motor cases where high burst pressures are required. It appears that this new generation of extremely strong carbon fibers has an important future in the aerospace industry.

Carbon Fibers from Pitch

Petroleum- and coal-derived pitches are basic raw materials for the manufacture of carbon and graphite. It is not surprising, then, that pitch was considered as a promising material for melt spinning into fibers for later conversion into carbon fibers. Pitch contains more than 90% carbon, much more than rayon or PAN. It is a complex mixture of aromatic hydrocarbon molecules of wide molecular weight distribution and, therefore, behaves like a glass in its ability to be drawn into a fiber.

Low-modulus carbon fibers - The first reported success in producing a carbon fiber from pitch was that of S. Otani of Gunma University, Japan [24], about 1964. Since the solid pitch fiber would only melt again if heated to carbonizing temperatures, Otani thermoset his fibers by heating in ozone and/or air. This process was further developed to a commercial stage by Kureha Chemical Company, who used petroleum pitch as a starting material [25, 26]. These inexpensive fibers are of low modulus and modest strength and are used for industrial applications such as furnace insulation, fuel cell electrode substrates, bearing and friction materials, packing and gasketing, and fiber fillers for plastics to provide electrical conductivity.

Coal-derived pitch, prepared by a solvent-refining process, was used as a carbon fiber precusor for a short time by Britain's National Coal Board [27]. The process depended upon extensive filtration to remove mineral matter which would otherwise harm the carbon fiber's tensile strength; this fact may explain why the process never reached commercialization. Other coal-based processes have been recently reported in Japan, slated for commercialization in 1986 or 1987.

The pitch-based carbon fibers described above are derived from an isotropic liquid pitch which remains isotropic during spinning, thermosetting, and carbonizing. An isotropic carbon fiber is by nature low in Young's modulus (less than 70 GPa) and rather low in tensile strength (less than 1.0 GPa). The fiber is capable of being converted to a high modulus "graphite" fiber by "stress-graphitization", a process which had been used to make high modulus carbon fibers from rayon. However, this conversion, reported by workers at the University of British Columbia [28], was never commercialized, due to its high cost.

Key to high modulus: liquid crystal - The solution to producing high modulus carbon fibers from pitch was found by Singer of Union Carbide in 1970 [29, 30]. The key idea was to advance the pitch raw material to a mesophase, or liquid crystal pitch, which was known to be capable of orientation through flow and shear [31]. This mesophase pitch could be melt spun and drawn into fine filaments; the molecular constituents were highly oriented and retained that orientation throughout all subsequent processing.

This process, which was developed with partial funding from U.S. Air Force (AFML) and, later, U.S. Navy (NSSC), was commercialized by Union Carbide in 1974, when a random mat fiber was produced, and in 1975, when a continuous filament yarn ("Thornel P-55") was produced. Ultrahigh-modulus fibers, with moduli of 690 to 830 GPa, were subsequently developed and brought to commercial production in 1980-82. Mesophase pitch fibers convert to carbon fibers with 80 to 90% yields by weight and very easily develop high modulus merely by heat treatment, requiring no stretching. These fibers are much more "graphitic" than PAN- or rayon-based fibers and exhibit very high values of thermal conductivity, electrical conductivity, oxidation resistance, and dimensional stability (low thermal expansion), all of which make them particularly suitable for many aerospace applications, including rocket nozzle throats, missile nose-tips, and satellite structures. They are also used extensively in aircraft brakes.

Other companies which have developed carbon fibers from mesophase pitch include Exxon Enterprises (who sold their carbon fiber technology to DuPont) and many Japanese companies. The Exxon work was based on extensive research at Rennselaer Polytechnic Institute by Riggs and Diefendorf [32]. No commercial product, other than Union Carbide's, has appeared on the market except in sample quantities. However, based on technology developed by the Government Industrial Research Institute, Kyushu, Japan, fourteen Japanese companies are trying to develop products of similar nature [33].

Carbon Fibers from Other Precursors

For completeness, after discussing the histories of carbon fibers derived from cellulose, PAN, and pitch, the category of "other precursors" should be covered. The tremendous activity in carbon fiber research and development is reflected in the large number of precursors which have been converted into carbon fibers. Besides the "big three", the list [34] includes phenolic polymers, phenol formaldehyde resin, furan resins [35], polyacenaphthalene, polyacrylether, polyamide, polyphenylene, polyacetylene, polyimide, polybenzimidazole, polybenzimidazonium salt, polytriazoles, modified polyethylene/polypropylene, polyvinyl chloride, polyamide, polymethyl vinyl ketone, polyvinyl alcohol, and polyvinyl acetate. Nearly all of these precursor routes to carbon fibers were reported within the ten-year period between 1966 and 1976 but none of them seems to represent "breakthrough" technology, since none has been successfully commercialized.

Lignin, a waste product of the pulp and paper industry, appeared for a time to be a possible starting material for a low-cost, low-modulus carbon fiber [36]. This substance was dry spun and reportedly converted with high yield to carbon fiber without any thermosetting step. During its brief commercial existence (it was produced by Nippon Kayaku Co. of Japan), it was said to be lowest priced of all carbon fibers. It may have been too weak to compete successfully.

Carbon Fibers by Chemical Vapor Deposition (CVD)

Carbon fibers or "whiskers" have been discovered on numerous occasions growing, sometimes in great profusion, in high temperature furnaces containing carbonaceous atmospheres. Although pilot plants have been built and operated to produce carbon fibers by chemical vapor deposition, no sustained commercial success has yet occurred. Possibly, the earliest published account of these CVD fibers appeared in Comptes Rendus in 1890 [37]. A more recent study (1944) was that of Gibson et al. who grew fibers from methane at 1000°C which were highly oriented, as shown by X-ray diffraction [38]. The growth mechanism sometimes involved tiny catalyst particles (from the furnace's refractory lining) which seeded the growth of tubular leader filaments; subsequently, the filament thickened through the deposition of additional layers of pyrolytic carbon [39]. Sometimes, very strong, high modulus fibers were produced [40]. An attempt to commercialize this type of process was made during the mid 1950's by the Pittsburgh Coke and Chemical Company [41]; their pilot plant operated only for a short period of time.

More recently, M. Endo, T. Koyama, and co-workers at Shinshu University, Japan, have produced CVD carbon fibers by pyrolysis of benzene [42, 43]. After a very high temperature heat treatment, these fibers have exhibited high crystalline perfection equivalent to physical vapor deposited whiskers; their axial electrical conductivity approaches that of single crystal graphite [44] and their axial thermal conductivity is substantially higher than that of copper [45]. These fibers, as well as similar fibers developed recently at the General Motors Research Laboratory, are moving into pilot plant production at the present time.

PROPERTIES OF CARBON FIBERS

Table I shows representative physical properties of various classes of carbon fibers and compares them with other fibers or wires.

Low weight is an important property for any aerospace material. Carbon fibers are second only to organic fibers, such as Kevlar, in being the lowest density structural fibers. They range from 1.3 to 2.2 g/cc.

Stiffness, often combined with high strength, is a unique attribute of all carbon fibers except the so-called "low-modulus" varieties. Even the lowest-modulus of the "high-performance" carbon fibers possesses a Young's modulus (230 GPa) greater than that of steel (210 GPa); on a weight specific basis, its modulus is nearly five times that of steel. Ultrahigh-modulus fibers possess Young's moduli as high as 900 GPa, close to the theoretical limit of 1020 GPa for the graphite single crystal.

TABLE I. Representative properties of carbon fibers and comparisons with other materials.

Fiber Type	Density (g/cc)	Young's Mod. (GPa)	Tensile Strength (GPa)	Elect. Resist. (μohm-m)	Thermal Conduct. (watts/m-K)
Low-modulus carbon fibers					
(ex-rayon, ex-pitch)	1.3-1.7	40-60	0.6-1.0	30-100	7-28
High-performance carbon fibers					
High-strength (ex-PAN)	1.7-1.8	230-250	2.8-4.0	12-30	7-10
Ultrahigh-strength (ex-PAN)	1.7-1.8	260-290	4.1-5.7	14-20	7-9
High-modulus (ex-PAN, ex-mesophase pitch)	1.8-2.0	350-550	1.7-3.5	5-10	60-200
Ultrahigh-modulus (ex-mesophase, CVD)	2.0-2.2	600-900	2.1-2.4	1-4	400-2500
Kevlar	1.45	130	3.6	$>10^{15}$	0.04-0.08
Glass	2.5	70-90	2.0-4.0	$>10^{15}$	0.5-1.0
Steel	7.8	210	0.6-2.5	0.1	35-65
Copper	8.9	120	220	0.02	400

Tensile strengths of commercial carbon fibers range as high as 5.7 GPa (830 kpsi), unmatched by any other commercial fiber.

Dimensional stability is an important attribute of space materials. High-performance carbon fibers possess a small, negative coefficient of thermal expansion at all temperatures below 375°C. That property, combined with their high modulus, permits a composites engineer to design a near-zero thermal expansion material by combining carbon fibers with one of a variety of matrices.

Electrical and thermal conductivities of carbon fibers cover wide ranges. For the most graphitic, ultrahigh-modulus fibers, the electrical conductivity on a weight-specific basis is within a factor of three of equalling steel. The absolute thermal conductivity (room temperature or below) exceeds that of copper [46].

SUMMARY

Table II shows the key technical and commercial events in the historical development of the carbon fiber industry.

TABLE II. Key Events in the History of Carbon Fibers

Date	Precursor	Fiber Development	Significance
1879	Cotton	Light bulb filament, U.S.A. [1, 2]	First commercial carbon filament
1890	CVD	Filamentary carbon by CVD, France [37]	Earliest known research on CVD carbon fibers
1944	CVD	Oriented vapor-deposited carbon filaments, U.K. [38]	First observation of oriented carbon fiber structure
1944	PAN	Carbon fiber research, "black Orlon", U.S.A. [5]	Probably first carbon fibers from PAN
1958	PVD	Graphite whiskers, U.S.A. [4]	Demonstrated ultrahigh-strength and modulus
1959	Rayon	Graphite fabrics, U.S.A. U.S.A. [7]	First sustained cf product. Sparked use in aerospace
1961	PAN	Carbon fiber research, Japan [10]	First major research on PAN-carbon fiber
1964	Pitch	Carbon fiber research, Japan [24]	First major research on pitch-carbon fiber
1964	PAN	High-modulus PAN-carbon fiber, U.K. [11]	Sparked high-performance cf industry in U.K and Japan
1964	Rayon	High-modulus rayon-carbon fiber, U.S.A. [9]	Sparked high performance cf industry in U.S.A.
1965	Rayon	"Thornel 25" U.S.A. [12]	First commercial high-modulus continuous carbon yarn
1966	PAN	High-strength carbon fiber, U.K. [18]	Sparked rapid growth of carbon fiber applications
1967	PAN	"Modmor I", U.K.	First commercial high-modulus PAN-based carbon fiber
1970	PAN	Improved PAN precursor, Japan	Sparked Japanese dominance of carbon fiber industry
1970	Pitch	High modulus cf from mesophase, U.S.A. [30]	First mesophase pitch-based carbon fiber
1975	Pitch	"Thornel P-55", U.S.A.	First commercial high-modulus pitch-based carbon yarn
1980	Pitch	"Thornel P-100", U.S.A.	Ultrahigh-modulus carbon yarn, for space applications
1983	PAN	Ultrahigh-strength cf, U.S./Japan [22]	Will spark heavy usage in aircraft and spacecraft

Edison's light bulb filament is the first known example of filamentary carbon, made in 1879. Except for early observations of filamentary carbon growth by CVD, it took another 65 years before significant further developments in carbon fiber technology occurred.

From 1944 to 1964, low modulus carbon fibers were discovered or invented and, ultimately, used in such applications as bearings, packing and gasketing, thermal insulation, and electrical conductors. Their first major successful application, however, was as fiber reinforcements in phenolic resin and carbon matrix composites for military vehicles; thus, their potential future in the composites industry was quickly recognized by military agencies.

In the late fifties and early sixties, graphite whiskers and CVD fibers demonstrated the intrinsic capabilities of highly oriented graphite fibers (strength, modulus, electrical conductivity); however, the production methods for these fibers were uneconomical.

From 1964 to 1975, spurred by the growing awareness of the potential properties of carbon fibers, high-modulus and high-strength carbon fibers from rayon, PAN, and pitch were invented, developed, and commercialized, opening up an explosive growth in the "high performance" composites industry.

From 1975 to the present, a surprisingly rapid improvement in PAN-based carbon fiber strengths developed, motivated by the needs of aircraft and rocket motor case manufacturers. At the same time, ultrahigh-modulus pitch-based fibers were developed, responding to the needs of space structure designers for rigid, light weight, and dimensionally stable materials.

The full history of carbon fibers has yet to be written: the industry is barely out of its infancy.

FURTHER READING

Readers wishing to learn more about carbon fibers are referred to the following technical reviews:

"Carbon Fibres from Mesophase Pitch" by B. Rand, in Handbook of Composites, Vol. 1 - Strong Fibres, W. Watt and B. V. Perov, eds. (Elsevier, New York 1985) pp. 495-575.

"Carbon Fibers" by J.-B. Donnet and R. C. Bansal, Marcel Dekker, New York, 1984 (291 pages).

"Graphite Fibers and Composites," by D. M. Riggs, R. J. Shuford, and R. W. Lewis in Handbook of Composites, Geo. Lubin, ed. (Van Nostrand Reinhold, New York 1982) pp. 196-271.

"Technology of Carbon and Graphite Fiber Composites," by John Delmonte (Van Nostrand Reinhold, New York 1981) 452 pages.

"Structure and Physical Properties of Carbon Fibers" by W. N. Reynolds, in Chemistry and Physics of Carbon <u>11</u>, P. L. Walker and P. A. Thrower, eds. (Marcel Dekker, New York 1973) pp. 1-67.

"Carbon Fibers from Rayon Precursors," by R. Bacon, in Chemistry and Physics of Carbon 9, P. L. Walker and P. A. Thrower, eds. (Marcel Dekker, New York 1973) pp. 1-102.

Information about carbon fiber products and manufacturers, world-wide, may be found in "Carbon and High Performance Fibres, Directory-3," compiled by D. R. Lovell, (Pammac Directories, High Wycombe, England 1985).

A descriptive compilation of hundreds of carbon fiber U.S. patents through 1979 may be found in "Carbon and Graphite Fibers, Manufacture and Applications", Marshall Sittig, ed. (Noyes Data, Park Ridge, New Jersey, 1980).

REFERENCES

1. Bright, A. A. Jr. The Electric Lamp Industry (Macmillan, New York 1949) pp. 64-67.
2. Vanderbilt, B. M. Thomas Edison, Chemist, (American Chemical Society, Washington, D. C. 1971) pp. 52-60.
3. Cranch, G. E., Proceedings of the Fifth Conference on Carbon, Vol. 1, (Macmillan, New York 1962) pp. 589-94.
4. Bacon, R. J. Appl. Phys. 31, 283 (1960).
5. Winter, L. L. Union Carbide Corp., National Carbon Research Laboratories, Biweekly Reports, 1944-45 (unpublished work).
6. Product Engineering, Sept. 23, 1957, p. 5; Chemical Week, April 26, 1958, p. 56.
7. Anon., J. Electrochem Soc. 106, 147C (June, 1959).
8. Schmidt, D. L. and W. C. Jones Chemical Engineering Progress 58, October 1962, pp. 42-50.
9. Bacon, R. Carbon Fibers from Rayon Precursors, in Chemistry and Physics of Carbon 9, P. L. Walker and P. A. Thrower, eds. (Marcel Dekker, New York 1973) pp. 1-102.
10. Shindo, A. Studies on Graphite Fiber, Rept. No. 317, Gov't. Ind. Res. Inst., Osaka (1961); also J. Ceram. Assoc. Japan 69, C195 (1961).
11. Watt, W. et al. The Engineer (London) 221, 815 (May 27, 1966).
12. Bacon, R. et al. Society of the Plastics Industry 21st Technical and Management Conference, Section 8E (February, 1966).
13. Union Carbide Corporation, Techn. Rept. AFML-TR-66-334 Parts I-V, 1966-1970.
14. Houtz, R. C. Textile Res. J. 20, 786 (1950).
15. Delmonte, J. Technology of Carbon and Graphite Composites (Van Nostrand Reinhold, New York 1981) pp. 11-14.
16. Lovell, D. R. Carbon and High Performance Fibers Directory, 3rd Edition, D. Pamington, ed. (Pammac Directories, High Wycombe, U.K. 1985) p. 327.
17. Watt, W. and W. Johnson in High Temperature Resistant Fibers from Organic Polymers, J. Preston, ed., Appl. Polymer Symp. 9 (1969) pp. 229-43.
18. Moreton, R. et al. Nature 213, 690 (1967).
19. Standage, A. E. and R. Prescott, Nature 211, 169 (1966).
20. Johnson, J. W. Appl. Polymer Symp. 9, 229 (1969).
21. Logsdail, D. H. Appl. Polymer Symp. 9, 245 (1969).
22. Hercules Corp., Orange County SAMPE Seminar, Feb. 10, 1983.
23. Pepper, R. et al. (Fiber Materials, Inc.) U.S. Patent 4,526,770 (1984).
24. Otani, S. Carbon 3, 31 (1965).

25. Anon., The Engineer, 13 Nov. 1969, p. 37.
26. Araki, T. and S. Gomi, Appl. Polymer Symp. 9, 331 (1969).
27. Anon., The Engineer, 2 July 1970, p. 30.
28. Hawthorne, H. M. et al. Nature 227, 946 (1970).
29. Singer, L. S. (Union Carbide Corp.) U.S. Patent 4,005,183 (1977).
30. Singer, L. S. Carbon 16, 408 (1977).
31. White, J. L. et al. Carbon 5, 517 (1967).
32. Riggs, D. M. and R. J. Diefendorf, "Elongational Characteristics of Mesophase-Containing Pitches," paper presented at the 31st Pacific Coast Regional Meeting of the American Ceramic Society, October 1978.
33. Anon., Japan Economic Journal, October 16, 1984, p. 16.
34. J.-B. Donnet and R. C. Bansal, Carbon Fibers (Marcel Dekker, New York 1984) pp. 7-8.
35. Yamada, S. and M. Yamamoto, Appl. Polymer Symp. 9, 339 (1969).
36. Fukuoka, Y. Japan Chem. Quarterly 5, 63 (1969).
37. P. and L. Schutzenberger, C. R. Acad. Sci. (Paris) 111 774-778 (1890).
38. Gibson, J. et al. Nature 154, 544 (1944).
39. Hillert, M. and N. Lange, Z. Krist. 111, 24 (1958).
40. Bourdeau, R. G. and F. E. Papalegis (Union Carbide Corp.) U.S. Patent 3,378,345 (1968).
41. Anon., Chemical Engineering, October 1957, pp. 172-4.
42. Koyama, T. Carbon 10, 757 (1972).
43. Endo, M. et al., Jpn. J. Appl. Phys. 15, 2073 (1976).
44. Chieu, T. C. et al. Phys. Rev. B 26, 5867 (1982).
45. Piraux, L. et al. Solid State Comm. 50, 697 (1984).
46. Nysten, B. et al. J. Phys. D 18, 1307 (1985).

HISTORY AND DEVELOPMENT OF POLYBENZIMIDAZOLES

E. J. POWERS AND G. A. SERAD, PBI Business Unit, Celanese Specialty Operations, P.O. Box 32414, Charlotte, N. C.

ABSTRACT

Aromatic polybenzimidazoles were synthesized by H. Vogel and C. S. Marvel in 1961 with anticipation, later justified, that the polymers would have exceptional thermal and oxidative stability. Subsequently, NASA and the Air Force Materials Laboratory (AFML) sponsored considerable work with polybenzimidazoles for aerospace and defense applications as a non-flammable and thermally stable textile fiber and as high temperature matrix resins, adhesives and foams. The route to fiber used solutions of high molecular weight polymer. Structural applications used low temperature melting pre-polymers that were cured (polymerized) in place. Applications of polybenzimidazoles were not implemented in the 60's and 70's since the polymers' tetraamine precursors were not commercially available.

In 1983, Celanese Corporation commercialized PBI fiber, spun from solutions of poly[2,2'-(m-phenylene)-5,5'-bibenzimidazole], for a wide range of textile applications. And, with a unique, new polymer commercially available for the first time, Celanese also undertook the development and evaluation of other forms of the polymer. Process and application development of PBI films, fibrids, papers, microporous resin, sizing, coatings and molding resins have been started. Applications for PBI utilize its unique chemistry, a polymeric secondary amine, as well as its thermal and chemical stability.

INTRODUCTION

Polybenzimidazoles are a class of thermally stable polymers, typically condensed from aromatic bis-o-diamines and dicarboxylates. The namesake polymer repeat unit, which contains a 1,3-dinitrogen heterocycle, is benzimidazole. As a monomer, benzimidazole, has a melting point of 170°C and a boiling point of >360°C. Benzimidazole's hydrocarbon analog, indene, has a melting point of -2°C and a boiling point of 183°C. With only a molecular weight difference of 2, the differences in thermal behavior between benzimidazole and indene foretells some of the unusual thermal characteristics to be found in benzimidazole polymers.

Poly[2,2'-(m-phenylene)-5,5'-bibenzimidazole] is the polybenzimidazole on which most attention has been focused over the past 25 years. It is the only commercially available polybenzimidazole and it is the specific structure that is typically intended when the acronym PBI is used. PBI was selected as the preferred polybenzimidazole because of its combination of thermal stability and processability.

PBI is a thermoplastic polymer with a Tg of 435°C. It does not burn, contribute fuel to flames or produce much smoke. It forms a tough char in high yield (to 80%). It is soluble in a few solvents and resists most other solvents. And, if it is cross-linked it resists even solvents that dissolved it originally. PBI's TGA decomposition temperature in air is

Copyright 1986 by Elsevier Science Publishing Co., Inc.
High Performance Polymers: Their Origin and Development
R.B. Seymour and G.S. Kirshenbaum, Editors

580°C. And, although PBI's resistance to long term exposure in heated air does not match that of some polyimides, new data, which is reported at the end of this paper, indicates that phosphated PBI is very thermo-oxidatively stable. PBI is hydrophilic, which lends to its excellent comfort as a textile and which affords unique chromatographic separation characteristics. For other applications, PBI's hydrophilicity can be moderated by annealing.

PBI development was started 25 years ago for specific aerospace and defense applications. With the unique polymeric material now commercially available, new forms of the polymer and radically different applications are possible.

POLYBENZIMIDAZOLE EARLY DEVELOPMENT, CHEMISTRY AND PROPERTIES

Polybenzimidazoles first appeared in US Patent 2,895,948 in 1959. In 1961, Vogel and Marvel opened the field of high temperature polymers [1] with their studies of the thermal stability of aromatic polybenzimidazoles [3]. Subsequently, AFML and NASA funded programs for basic studies and for structural and textile applications of polybenzimidazoles to meet new aircraft and aerospace material needs. Several reviews were published [4-7]. Structural applications of polybenzimidazoles are still in developmental stages which is discussed below. One polybenzimidazole polymer (PBI) which was developed for fiber applications is now commercial. Details of the polymer synthesis and fiber process are described in the next section.

<u>Polybenzimidazoles are synthesized</u> from aromatic tetraamines (bis-o-diamines) and dicarboxylates (acids, esters or amides). Generalized monomers are shown in FIGURE 1 and polybenzimidazoles from those monomers are shown in FIGURE 2. In either figure, R(1) and R(2) can be aromatic or aliphatic and may contain other groups as ether, ketone, sulfone, etc. The structures shown do not exhaust all possible benzimidazole polymers. Naphthyl groups could be substituted for phenyl groups in all of the structures. And, amine and carboxylate monomers could be mixed to prepare a host of possible copolymers. At least fifteen amines and sixty acids have been condensed to form polybenzimidazoles over the last 25 years [1].

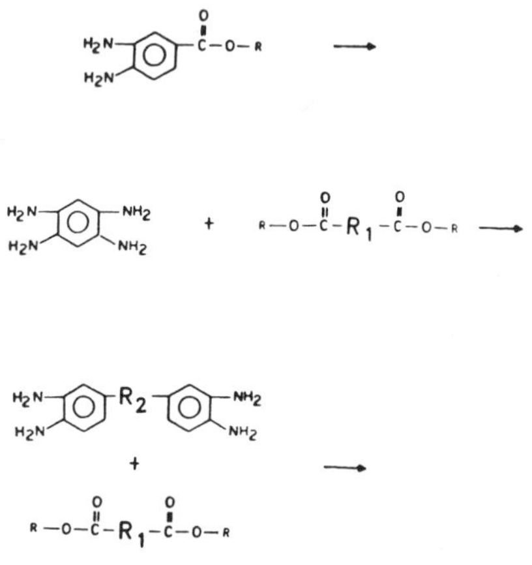

FIG. 1. Monomer types used to synthesize polybenzimidazoles.

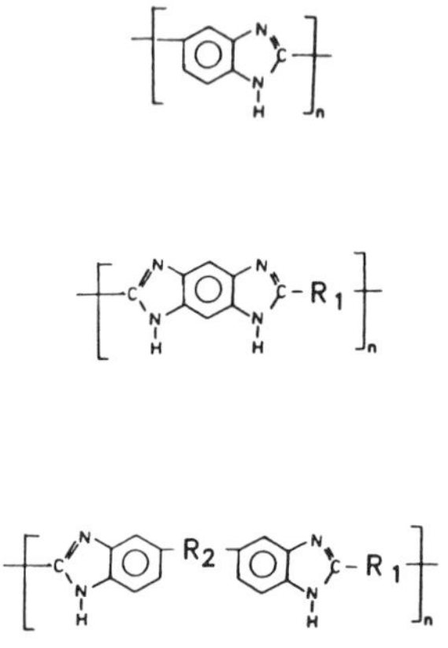

FIG. 2. Polybenzimidazole structures.

Polybenzimidazole polymers can be formed in the melt or in solution. Mechanistic studies have proposed Schiff's base or aminoamide intermediates. Recent texts on high temperature polymers discuss benzimidazole reaction mechanisms with references to original works [1,2].

The melting temperatures and thermal stabilities of several polybenzimidazoles are shown in TABLE I. For all aromatic polybenzimidazoles, melting points are over 600°C. Generally, the presence of other structural elements will reduce melting points and thermal stabilities. For example, TABLE I includes polybenzimidazoles condensed from either a tetraamine ether or a diacid ether. Strained polymer structures also reduce polybenzimidazole stability as evidenced by the low melting point of a polymer formed with a biphenyl-2,2'-dicarboxylate. Not surprisingly, the aliphatic, adipic acid forms a much less stable polybenzimidazole than the aromatic examples above it in the table.

TABLE I. Structures and stabilities of polybenzimidazoles [6].

TETRAAMINE	ACID	MP(°C)	WGT. LOSS % IN NITROGEN *	WGT. LOSS % IN AIR *
3,4-DIAMINOBENZOIC		>600	0.4	
BIPHENYL	TEREPHTHALIC	>600	0	
BENZENE	TEREPHTHALIC	>600	1.0	
BIPHENYL	ISOPHTHALIC	>600	0.4	5.2
BENZENE	ISOPHTHALIC	>600	0.3	
DIPHENYLETHER	ISOPHTHALIC	>400		
BIPHENYL	PHTHALIC	>500	0.4	7.0
BIPHENYL	4,4'-OXYDIBENZOIC	>400		
BIPHENYL	BIPHENYL-4,4'-DIACID	>600	0.8	
BIPHENYL	BIPHENYL-2,2'-DIACID	430	8.0	
BIPHENYL	ADIPIC	450d		
BENZENE	ADIPIC	490d		

*Weight loss after one hour at 500°C, after heating 1 hr at 400 and 450°C.

A few, available, comparisons of polybenzimidazole weight losses in nitrogen to weight losses in air (TABLE I) indicate that polybenzimidazoles are less oxidatively stable than they are resistant to purely thermal degradation. This indication is confirmed in studies of composite and adhesive properties, which are discussed below. Some fundamental studies of PBI's oxidative mechanism have been conducted [8-10].

The most frequently studied polybenzimidazole composition is poly-[2,2'-(m-phenylene)-5,5'-bibenzimidazole, which is shown in FIGURE 3. It corresponds to the structures at the bottom of FIGURES 1 and 2 where R(2) is a bond, forming bibenzimidazole and R(1) is m-phenylene. This polybenzimidazole, which is synthesized from diphenylisophthalate and tetraaminobiphenyl, offered a combination of thermal stability and processability. This is the same polymer from which Celanese produces fiber.

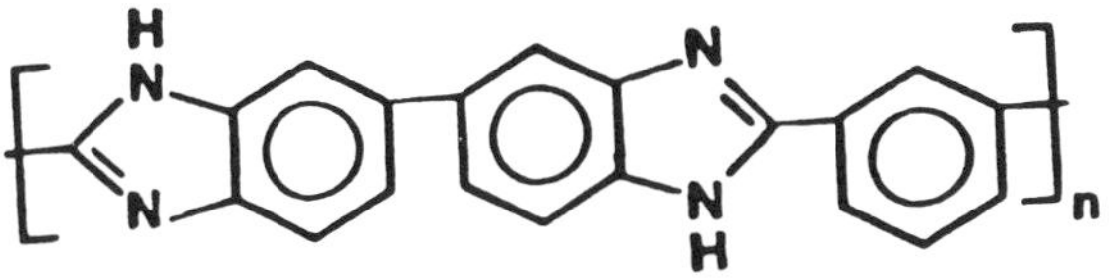

FIG. 3. Structure of poly[2,2'-(m-phenylene)-5,5'-bibenzimidazole].

The development of PBI fiber by Celanese, initially under AFML contracts, was straight-forward although many development efforts ran concurrently: a reproducible process for a high molecular weight linear thermoplastic polymer, solutioning methods, solution stabilization techniques, fiber spinning and processing, textile processing, fabric construction preparation and testing, textile marketing. Many of these topics are covered in the next section.

Structural applications of PBI followed a different process approach. In 1968, a patent, issued to Levine, assigned to the Whittaker Corporation, taught that PBI polymers having inherent viscosities between 0.03 and 0.10 (measured at 0.5% in dimethyl sulfoxide) were fusible and were capable of becoming infusible upon further heating [11]. Essentially, PBI prepolymers were melted to flow and wet substrates. Subsequent heating of laminates, adhesives etc., polymerized the PBI, in place. TABLE II shows the melting behavior of PBI prepolymers, taken from Levine's patent. In his 1969 review article [4], Levine discusses the properties of PBI with references to products which were available from the, then, Narmco Division of the Whittaker Corporation: Imidite 1850 for laminates, Imidite 850 for adhesives, and Imidite Foam Compounds. Similar products are available, today, from the Aerotherm Division of Acurex Corporation. Their literature lists Polybenzimidazole Prepolymers 2801 and 2803, PBI 1850 Laminating Material, a PBI 850 Adhesive, and PBI SA and PC which are syntactic foam materials using silica and phenolic micro balloons, respectively [12].

TABLE II. Inherent viscosity versus melting point [11] and processing options for poly-2,2'(m-phenylene)-5,5'-bibenzimidazole.

INHERENT VISCOSITY*	MELTING POINT (°C)	APPLICATION CHARACTERISTICS
0.030	175	<------
0.032	120	¦
0.038	95	MELTS - FLOWS,
0.042	85	POLYMERIZES IN PLACE
0.050	105	
0.062	200	COMPOSITES, ADHESIVES
0.075	240	& FOAMS
0.100	320	<------
0.150	460	
0.200	>500	INJECTION MOLD ?
0.800		
	ESSENTIALLY NON-MELTING	HIGH MW POLYMER FOR FIBER MANUFACTURING
	COMPRESSION MOLDABLE, FORMABLE FROM SOLUTION:	
	SHEETS, SHAPES, FILM, FIBRID, PAPER,	SIZING, COATINGS, MICROPOROUS RESIN

* Inherent viscosities were measured at 0.5% in dimethylsulfoxide.

Structural applications of PBI have yet to find large scale applications. However, PBI for aerospace structures is being pursued in the 80s as the only known, available polymeric matrix capable of maintaining load bearing properties for an adequate length of time at 1200°F [13-15].

Factors which have retarded the use of PBI for structural applications include difficult process conditions, the need for technique to control voids from offgases, and hygiene requirements in handling the prepolymer.

Earlier concerns for raw material availability should no longer apply, since the tetraamine precursor, TAB, is available from Celanese.

A recently published process [13] for curing a PBI and carbon fiber composite, showing cycle times, temperatures and pressures is outlined in TABLE III. Risks of voids arise as the prepolymer cures and the byproducts of the PBI condensation reaction are released. When the monomers are the typical diphenylisophthalate and tetraaminobiphenyl, water and phenol are released. However, it should be noted that low void parts have been made and PBI properties appear to be less sensitive to voids than other resin matrices [15].

TABLE III. Typical cure sequence for poly[2,2'-(m-phenylene)-5,5'-bibenzimidazole] prepolymer in a carbon fiber prepreg [13].

TIME (HOURS)	ACTIVITY	TEMPERATURE (°F)	PRESSURE (PSI)
	HEAT UP		200
1			
	HOLD	350	
2			
	HEAT UP		500
3		450	
			700
4		650	
	HOLD	800	
5			
			1500
6			
7	COOL DOWN		

The hygiene requirements in handling PBI prepolymers arise because, at low molecular weight, the presence of detectable quantities of the PBI monomer, tetraaminobiphenyl (TAB) are probable. TAB is structurally similar to aromatic amines which are carcinogenic and, accordingly, no-contact methods of handling TAB have to be employed unless and until safe exposure levels for TAB can be established.

<u>Neat resin properties of PBI made from prepolymer</u> are shown in TABLE IV. Note that PBI's Tg and Heat Distortion Temperatures are over 800°F coupled with good impact resistance. This is a desirable property combination which explains some of the interest in PBI structures for aerospace and military applications. TABLE V shows the unusual property retentions of PBI composites versus temperature. Additional studies have shown that PBI composites retain 26% of their short beam shear strength at 1200°F whereas polyimides would have essentially no shear strength under similar conditions [15]. However, for long term high temperature applications in air, PBI is less stable than some polyimides. For example, a PBI/glass laminate had over twice the flexural strength of a polyimide/glass laminate when tested at 300°C. But, after 1000 hours at 300°C, the polyimide laminate's strength, at 300°C, was twice that of PBI [2, pg.297].

<u>PBI adhesives</u> from prepolymers also provide performance over wide temperature ranges. An Acurex Technical Bulletin reports an ultimate shear stress of 3000 psi from -400°F to -100°F which gradually drops to 1200 psi at 1000°F [12]. However, PBI adhesives loose properties on longer term aging in air faster than some polyimides [2, pg. 303].

TABLE IV. Some physical properties of molded poly[2,2'-(m-phenylene)-5,5'-bibenzimidazole] [4, 33].

PROPERTY	VALUE
Tg	815°F
TENSILE STRENGTH	17 - 25 KSI
TENSILE MODULUS	0.7 - 1.3 MSI
COMPRESSIVE STRENGTH	
YIELD	54 KSI
ULTIMATE	66 KSI
IZOD IMPACT - NOTCH	
AT 73°F	1.5 LB/IN
AT -320°F	1.5 LB/IN
HEAT DEFLECTION ASTM D 648-56	802 - 817°F

TABLE V. Properties of carbon fiber laminates of poly[2,2'-(m-phenylene)-5,5'-bibenzimidazole versus temperature [13].

TEMPERATURE (°F)	TENSILE STRENGTH (KSI)	FLEXURAL STRENGTH (KSI)	TENSILE MODULUS (MSI)	FLEXURAL MODULUS (MSI)
RT	185	220	22	24
200	160	220	20	22
400	150	205	18	20
600	140	180	16	18
800	110	140	14	15
1000	70	90	11	13
1200	30	40	8	9

The low temperature adhesive properties, as well as the impact properties at -320°F for molded PBI in TABLE IV are noteworthy. Gillham's Torsional Braid Analysis observed a low temperature dampening maximum which indicates that relaxation mechanisms are available to dissipate energy [17].

Foams from PBI prepolymer indicate that it is an ideal insulating material. It does not melt or contribute fuel to flames. It converts to carbon in high yield forming a tough char while maintaining its form [18]. It generates almost no smoke and has excellent thermal characteristics [19].

Syntactic foams with a PBI matrix (made from prepolymer) have excellent compressive strength and ablation characteristics [2]. As mentioned earlier, such materials are available from the Aerotherm Division of Acurex Corporation [12].

PBI FIBER DEVELOPMENT AND TEXTILE APPLICATIONS

PBI fiber is dry spun from a solution of high molecular weight poly-[2,2'-(m-phenylene)-5,5'-bibenzimidazole]. Here, prepolymer is an intermediate which is not isolated. A schematic of Celanese's PBI fiber process is shown in FIGURE 4.

PBI synthesis chemistry is shown in FIGURE 5. Tetraaminobiphenyl and diphenylisophthalate are melt and solid state polymerized in two stages under an inert atmosphere. As polymerization occurs, the melting point of the polymer drops at first and then increases as shown in TABLE II. In the

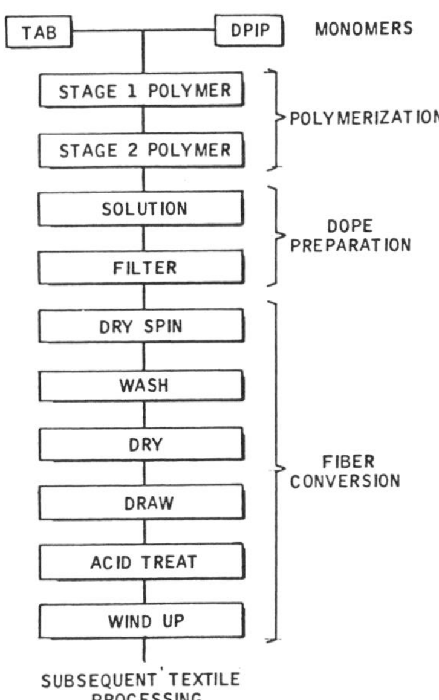

FIG. 4. Schematic of Celanese's PBI fiber process.

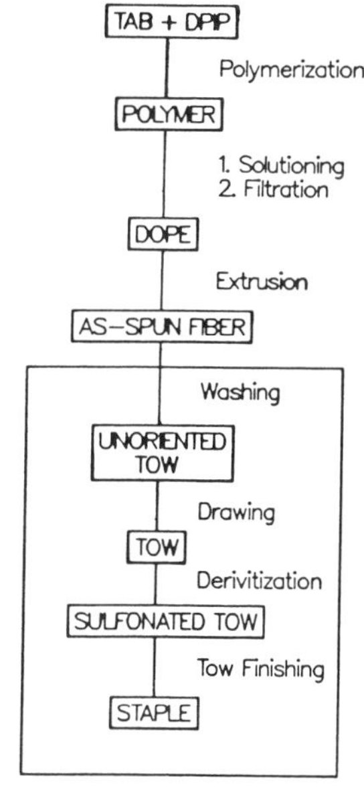

FIG. 7. Accessable materials from Celanese's fiber process.

first stage of polymerization, melted low molecular weight material foams as the phenol and water byproducts evolve. After it cools, the foam is crushed. In the second stage, the PBI polymer powder is heated to advance the polymer's molecular weight. At the end of stage two, Celanese PBI polymer must show a negative response in a test for free TAB, before it is processed further. Other publications can be reviewed for other process detail [18,20,21].

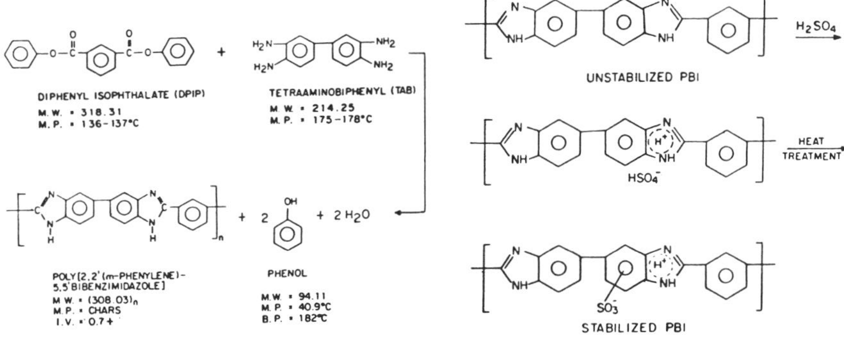

FIG. 5. Synthesis of Poly[2,2'-(m-phenylene)-5,5'-bibenzimidazole.

FIG. 6. Sulfonation of PBI fiber for flame stabilization.

PBI fiber spinning dope is prepared by dissolving dry PBI polymer and lithium chloride, in dimethylacetamide under pressure at temperatures above the boiling point of the solvent. As with polymer synthesis, solutioning is conducted under inert atmospheres. Typical polymer solids are 25% by weight, having viscosities about 1500 poise at room temperature. Lithium chloride is used to increase solution shelf life from days to several months. PBI dope is routinely filtered to remove any particulates which could interfere with fiber spinning. In fact, filterability is a critical property of polymer produced in the above process. It is characterized by minimum requirements in a Plugging Value Test which measures the weight of polymer solution that will pass a through a filter before plugging.

Fiber forming involves the extrusion of polymer solution through a spinneret containing fine holes into a heated inert atmosphere. The majority of the solvent evaporates. Residual solvent and the lithium chloride stabilizer are extracted from the fiber by washing in water. Subsequently, the fiber is dried, drawn to increase its strength, acid treated to provide stabilization to shrinkage in a flame, and finally crimped and cut to provide a material which will permit typical textile processability.

Acid treatment of PBI fiber sulfonates the polymer forming what is called Stabilized PBI. The chemistry, in FIGURE 6, indicates that a polymeric salt is formed. The polymer's spectra, thermal stability and solubility are affected [22]. The treatment reduces flame shrinkage of drawn fiber from >50% to <10%. Sulfonation of PBI by this process appears to produce some crosslinking, since sulfonated PBI is virtually insoluble.

Physical properties of PBI fiber are shown in TABLE VI. Tenacity converts to a tensile strength in the range of 50 to 100 thousand pounds per square inch (ksi) and modulus to the range of 500 to 1,000 ksi. These properties are fully adequate for textile processing but PBI is not intended to be a high modulus/high strength organic fiber for structural reinforcement. Note that PBI fiber has a high moisture regain which, undoubtedly, contributes to its comfort.

TABLE VI. Physical properties of PBI fibers.

PROPERTY	STABILIZED FIBER	UNSTABILIZED FIBER
Denier per Filament (denier)	1.5	1.5
Tenacity (g/d)[a]	2.7	3.1
Breaking Elongation (%)	30	30
Initial Modulus (g/d)	45	90
Density	1.43	1.39
Moisture Regain at 68°F, 65% RH (%)	15	13
Hot Air Shrinkage at 400°F (%)	1	2
Flame Test Shrinkage (%)	6	50
TMA Shrinkage at 400-500°C	4	10

[a] g/d denotes gram/denier, in which denier is the weight of a fiber (in grams) of 9000 m length.

Heat and flame shrinkage properties of PBI fiber are listed at the bottom of TABLE VI. Flame, fuel and smoke tests can be found in TABLE VII. TABLES VIII and IX demonstrate its resistance to acids, bases and solvents. Note that even heated solvents which dissolve PBI polymer do not affect the strength of stabilized (sulfonated) PBI fiber. TABLE X compares PBI comfort to a number of other fabrics. Significantly, PBI was perceived to be at least as comfortable as cotton and rayon.

TABLE VII. Flame test results of PBI.

TYPE OF TEST	RESULTS	
Flame Tunnel Test (ASTME-84)[a]	Smoke developed	0
	Fuel contributed	0
Vertical Burn Test (FSTM 191-5903)	After flame	0 seconds
	Char length	0.3 inches
Small Scale Flammability Test (NFPA-701)[b]	Smolder time	<1 second
	Char length	0.03 inches

[a]Asbestos = 0, Red Oak = 100; [b]Maximum accepted values are 2 seconds and 4.5 inches.

TABLE VIII. Strength retention after immersion in inorganic acids and bases.

ACID OR BASE	CONCENTRATION (%)	TEMPERATURE (°C)	TIME (HOURS)	STRENGTH RETENTION (%)
Sulfuric Acid	50	30	144	90
Sulfuric Acid	50	70	24	90
Hydrochloric Acid	35	30	144	95
Hydrochloric Acid	10	70	24	90
Nitric Acid	70	30	144	100
Nitric Acid	10	70	48	90
Sodium Hydroxide	10	30	144	95
Sodium Hydroxide	10	93	2	65
Potassium Hydroxide	10	25	24	88

TABLE IX. Strength retention of PBI after imersion in organic liquids.

LIQUID	STRENGTH RETENTION (%)
Acetic Acid	100
Methanol	100
Perchloroethylene	100
Dimethylacetamide	100
Dimethylformamide	100
Dimethylsulfoxide	100
Kerosene	100
Acetone	100
Gasoline	100

[a]All exposures were at 86°C for 168 hours.

TABLE X. Comfort ratings of various fabrics.

GARMENT	RATING AVERAGE
PBI	4.45
Rayon	4.42
Cotton	4.23
50/50 Polyester/Cotton	4.15
100% Regular Polyester	3.61
100% Polypropylene	3.27

[a]Numerical average of eight Hollies descriptors: snug, heavy, sticky, non-absorbent, damp, clingy, picky, scratchy. No attempt was made to give extra weighting to any descriptor.

PBI fiber has an impressive list of intrinsic properties [18,20,21,23] which lend to its textile as well as non-textile applications. It:

* Does not burn
* Does not melt
* Contributes no fuel to flames
* Produces little or no smoke to 1000°F
* Retains dimensional stability at high temperatures
* Retains form at high temperatures
* Forms a tough char
* Produces very comfortable fabrics
* Resists solvents.

Current PBI textile applications, include:

* Aircraft Seat Fire Blocking
* Firefighters Protective Gear
* Sock Hoods (Fire Fighters, Race Car Drivers, etc.)
* High Temperature Gloves
* Race Car Driver Uniforms
* Aluminized Crash Rescue Gear
* Flight Suits
* Coveralls (Forestry Service, Hazardous Work Areas, ect.)

All of the above applications utilize PBI's thermal and flame protection and its comfort [23,20,21,24]. Celanese has also developed many blends of PBI with other high performance fibers to optimize specific performance and minimize costs while maintaining thermal protection and comfort.

PBI fabrics' unusual properties also suit it to a number of industrial applications which utilizes PBI's chemical and thermal resistance [20,25]:

* Filtration
* Insulation (Wovens, Knits or Felts)
* Glass Handling Belts
* Expansion Joints

Applications of chopped and continuous filament PBI fiber include:

* Gasketing
* Rocket Motor Insulation
* Packings
* Resin Filler
* Filament Wound Composites
* Lacing for Reinforcement Fibers

PBI is an obvious replacement for asbestos in many applications. It doesn't melt or burn and asbestos replacements are actively being sought. Cost critical asbestos applications could not simply substitute PBI due to PBI's relatively higher cost. However, performance critical applications are being developed with PBI, in addition to those listed above. In gaskets and packings, PBI is both abrasion resistant and non-abrading providing improvements in performance rather than merely replacing asbestos [26].

Probably due to its high carbon conversion and tough char characteristics, PBI functions well as a reinforcement in rocket motor insulation [27]. PBI's relatively low modulus is beneficial as lacing or tie yarns for carbon fiber fabrics or three dimensional arrays. The PBI stabilizes the carbon fiber alignment through processing, yielding to maintain the alignments of the reinforcement filaments. PBI chopped fiber has recently been found to toughen resin matrices, which will be discussed in the next section with other new developments and applications.

NEW DEVELOPMENTS AND APPLICATIONS OF PBI

Celanese's initial interest in PBI focused on its textile applications. However, it is obvious when considering the structure in FIGURE 3, that PBI has an unusual polymeric structure that should have other applications. It is a chemically functional molecule even though its physical properties are chemically and thermally resistant. PBI is a linear thermoplastic polymer in spite of its very high Tg and melting point. Unlike other high temperature or chemically resistant materials it is hydrophilic. PBI is a polymeric amine. It will absorb and neutralize acids, forming a polycation [22]. Acids like sulfuric will react with PBI to bond an anion, providing both cationic and anionic sites on the polymer. The molecule's polarity should permit it to be wet by most liquids, whether monomeric or polymeric. The imidazole N-H is an active hydrogen which should be reactive, to covalently bond substituents to the polymer backbone. Wetting and reactivity should promote superb adhesion characteristics.

While not comprehensive, the above PBI features provide a large list of possibly useful performance benefits that could span many new markets. To provide focus, applications of readily accessible forms of PBI were considered first. FIGURE 7 provides a materials view of the Celanese fiber process that was outlined in FIGURE 4. And, while the plant was designed to produce crimped cut staple for textile applications, any of the boxed items could be accessed, some more easily than others.

Short-cut uncrimped PBI fiber was first taken into asbestos replacements which were discussed above. Multiple user feedback has verified that PBI fiber does wet-out and disperse unusually well.

A new application for PBI short cut fiber is just now being developed. Thermal stability, thermoplastic toughness, dispersibility, and chemically bonded adhesion suggested that PBI might toughen high temperature structural resins which are typically crosslinked and brittle. TABLE XI shows the first data from a study in a model system of 1/4 inch chopped sulfonated PBI fiber in a high Tg epoxy (a TGMDA type). Indeed, fracture toughness improvements are significant and the moduli are raised, too. Additional work has observed similar effects in two other epoxies, one a rubber modified RDGE resin and one a high rheology material. Table XI also shows a column for PBI Fibrid which are wisker-like particles of underivatized PBI, to be described later.

Work also has been extended into PMR-15 polyimide. With 1/4 inch chopped sulfonated PBI, the neat polyimide's K[1C] increased from 0.7 to 2.4. And, Dynamic Mechanical Analysis shows a similar modulus for the PMR-15, whether neat or containing 17% by weight dispersed PBI fiber.

It is important to note that while the above results are very promising, they are preliminary. The effect of various PBI particles, composition or level effects are still under study. Careful scientific evaluation and end-use tests, as in composites, are incomplete.

Other applications of high molecular weight PBI have focused on non-fiber forms of the polymer that are available or that can be made. Generally, prototype forms of the polymer are produced and evaluated. Specimens are sampled. And, feedback is sought from potential customers to determine if those forms can be cost effective for the customer's product needs, ...before too much is invested in developing the form. TABLE XII summarizes some of the forms and anticipated applications that are currently in development/evaluation.

PBI polymer heads the list in TABLE XII. Since PBI is thermoplastic and if care is taken to avoid oxidative cross-linking, it will fuse at some temperature and pressure. On a scouting basis, Celazole (TM), high

TABLE XI. Physical properties of a high Tg epoxy casting containing PBI fiber and fibrid fillers.

PROPERTIES	VALUES			
	FILLER TYPE AND CONTENT BY WEIGHT PERCENT			
	NONE (CONTROL)	17% FIBER	30% FIBER	17%FIBRID
Tg (°)	190	208	170	189
G' AMBIENT (KSI)	230	327	566	250
G' 120° (KSI)	152	222	425	194
K[IC] (PSI IN)	550	1684	3070	1050
G[IC] (IN-LB/SQ.IN.)	0.424	5.27	7.14	1.7
E' FLEX (DRY) (MSI)	0.62	0.66		
E' FLEX (CAL G') (MSI)	0.515	0.717	1.268	0.56
E' FLEX (WET)		0.56		
% WATER (48 hr)	5.0	2.12,2.1		

Fibers are sulfonated PBI. Fibrids are unmodified PBI.

TABLE XII. Non-textile applications of high molecular weight PBI forms.

PBI FORM	APPLICATIONS
POLYMER POWDER (CELAZOLE)	MOLDING RESIN ACID SCAVENGER
SOLUTION	SIZING ROUTE TO OTHER FORMS
FILM	THERMALLY STABLE DERIVATIZEABLE (ION EXCHANGE)
FIBRID	FILLER, PAPER BINDER
PAPER	SEE TABLE XVI
MICROPOROUS RESIN	SEPARATIONS
FOAM	THERMAL INSULATION

molecular weight PBI polymer (from the top of FIGURE 7) was pressed at 420°C, 6000 psi for 90 minutes. Properties of the fused piece on polymer were 4,900 psi tensile strength, 680,000 psi modulus and a breaking extension of 0.6% [34]. In more structurally experienced hands, tensile strengths of nearly 10,000 psi have been reported [14,15]. With development and optimization, properties approaching those of cured-in-place prepolymer, as shown in TABLE IV, should be possible.

PBI solutions can be prepared to concentrations of about 30% by weight in polar aprotic solvents like dimethylacetamide, dimethylformamide, dimethysulfoxide and N-methylpyrrolidone. Solutioning generally requires heating under pressure above the boiling point of the solvent. Viscosities, at those solids, will generally be above 2000 poise and the solution's stabilities are generally measured in days. However, PBI dissolved in dimethylsulfoxide at reflux under atmospheric pressure and a 25% solution had a shelf life of 2 weeks at room temperature and a month at 90°C. In general, the stability of solutions increases with decreasing solids content. For example, a 2% solution of PBI in N-methylpyrrolidone remains stable after three months. Additives, typically salts, stabilize PBI solutions which was discussed in the section on PBI fiber spinning dope. Being available, that solution is used to make many of the polymer forms which are discussed below, such as films, fibrids, and microporous resin.

An application for PBI solutions that is currently under evaluation is its use as a size for various reinforcement fibers as carbon, ceramic or glass. PBI's good wetting and bonding potential, together with its thermal and chemical stability suggests that it should be an ideal size. In addition, the toughening affect of PBI forms in some resins adds a tantalizing element of potential.

PBI is also soluble in dry sulfuric and phosphoric acids. Sulfuric acid solution is typically used for viscosity measurements.

Films are a convenient structure to study material properties or the effects of polymer interactions with a variety of reagents or environmental conditions. PBI films are formed by casting a polymer solution on a carrier, allowing enough of the solvent to evaporate to produce a handleable structure, washing residual solvent and stabilizing salt from the film and, finally, drying. Physical properties of PBI film are shown in TABLE XIII. Annealing PBI film raises its molecular weight and causes cross-linking as judged by an increase of inherent viscosity and the presence of insolubles after the annealing step. The reduction of moisture regain of annealed PBI films is notable. Moisture content is useful for comfort in apparel applications but generally is less desirable for most other materials applications. PBI is plasticized by allowing it to absorb phosphoric acid from a dilute aqueous solution. The physical properties in TABLE XIII are not significantly affected by the adsorbed acid but the handling characteristics of the film have a dampened character. Using a higher molecular weight polymer has a significant effect on the toughness of PBI films. Breaking extensions increase by a factor of 5 with only a 0.2 increase in the polymer's inherent viscosity.

TABLE XIII. PBI film physical property summary.

PROPERTY	CAST PBI	ANNEALED PBI	PLASTICIZED PBI	HI MW PBI	PLASTICIZED HI MW PBI
BREAK STRENGTH (PSI)	17K	27K	15K	14K	14K
BREAK ELONGATION (%)	14	24	20	97	97
MODULUS (PSI)	400K	550K	330K	460K	410K
MOISTURE CONTENT	10%	5%	12%	10%	12%
DENSITY (G/CU CM)	1.2	1.3	1.4		
LOI (PCT. OXYGEN)	>40		>70		
TMA SHRINKAGE (%)					
100°C	0.9		0.8		
200°C	2.0		1.7		
300°C	2.7		2.5		
400°C	3.0		2.7		
500°C	3.1		2.9		

Thermal Gravimentric Analyses (TGA) of annealed PBI film in air and nitrogen are shown in FIGURE 8. A few characteristics of the thermograms are different from previously published data [18,20] which were measured on sulfonated PBI. Annealed underivatized PBI film shows a lower initial weight loss and about a 50°C higher onset of weight loss in air than sulfonated fiber. And, underivatized PBI has a higher conversion to carbonaceous material than sulfonated material. If allowance is made for the moisture in PBI, the conversion of PBI polymer mass to carbon, shown in FIGURE 8, is 80%. Phosphoric acid plasticized PBI film has unusual thermo-oxidative stability which will be discussed, further, at the end of this report.

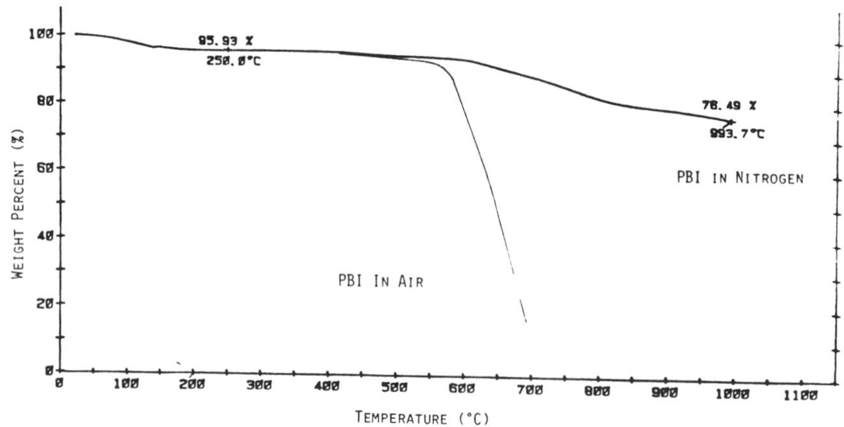

FIG. 8. TGA of annealed PBI film in air and in nitrogen, heated at 20°C per minute.

Electrical properties of PBI films are shown in TABLE XIV. As would be expected, films with higher moisture contents have lower resistivities and higher dielectric constants. PBI plasticized with phosphoric acid, which should be a polycationic polymer with an unbound phosphate anion, has a very low resistivity and a high dielectic constant which is even higher at elevated temperatures. In general, dielectric constants are higher at 100 Hz and lower at 1 MHz.

PBI's moisture affinity negatively affects its application as an electrical insulator. However, if a lack of resistivity is a drawback, moisture promoted conductivity might be advantageous. A report that PBI was proton conductive [28] supported the thought. TABLE XV lists the resistivities of several PBI film variants (typically, 1.5 mil films) in aqueous cells. PBI which had been allowed to absorb acids, making them polycationic, are much more conductive than films that are water swollen but neutral. The lowest resistivity was obtained by a sodium ion exchanged sulfonated PBI film. PBI films can be sulfonated much the same way that fiber is sulfonated, which was discussed earlier.

PBI Fibrids are fine, high aspect ratio particles of PBI that are extruded from solution. They have dimensions of 2-5 microns by 200-900 microns with surface areas of 15 to 40 square meters per gram. Figure 9 is a SEM photomicrograph of one type of PBI Fibrids. Process variations can modify their size and surface area. Applications for fibrids include binders for papers and high performance fillers for resins, see TABLE XI. As with chopped PBI fiber, user feedback indicates that fibrids disperse well.

PBI papers have been wet laid from slurries of PBI Fibrids and short cut PBI fiber on standard Fourdrinier paper making equipment [29,30]. The fibrids entangle the fiber to form a paper structure. Typical properties of PBI papers are shown in TABLE XVI. Depending on the ratio of fibrid to fiber and whether or not the paper is calendered, annealed or otherwise treated, gives rise to a wide range of densities, strengths, and chemical resistance. Possible applications for PBI papers include:

* Lightweight Heat Shields

* Transformer Wrap
* Honey Comb
* Asbestos Replacement
* Tubular Insulation
* Composites
* High Temperature / Chemical Filtration
* Battery Separators

TABLE XIV. Electrical properties of PBI films.

TEST CONDITION	RT AS RECEIVED	RT DRIED	100°C	250°C
	CAST, WASHED & DRIED FILM (~12% MOISTURE)			
VOL. RESISTIVITY (OHM-CM)	$1.6X10^{10}$	$3.1X10^{14}$	$4.0X10^{12}$	$9.4X10^{9}$
DIELECTRIC CONST. (1KHZ)	16.2	4.4	4.6	4.2
DISSIPATION FACTOR	0.57	0.024	0.030	0.079
BREAKDOWN VOLTAGE	2400		2000	4100
(KVAC, RMS)	(4.0)		(3.4)	(7.0)
	ANNEALED FILM (~5% MOISTURE REGAIN)			
VOL. RESISTIVITY (OHM-CM)	$8.5X10^{11}$	$5.1X10^{14}$	$1.4X10^{13}$	$3.0X10^{10}$
DIELECTRIC CONST. (1KHZ)	8.1	4.5	4.6	4.7
DISSIPATION FACTOR	0.19	0.013	0.015	0.089
BREAKDOWN VOLTAGE	3900		3000	4700
(KVAC, RMS)	(7.0)		(5.3)	(8.5)
	PLASTICIZED PBI FILM			
VOL. RESISTIVITY (OHM-CM)	$2.5X10^{8}$	--	$2.2X10^{7}$	$1.6X10^{7}$
DIELECTRIC CONST. (1KHZ)	19.2	--	541	343
DISSIPATION FACTOR	0.91	--	9.0	1.98
BREAKDOWN VOLTAGE	280		640	660
(KVAC, RMS)	(0.38)		(0.90)	(0.93)

TABLE XV. PBI film ion conductivities.

FILM COMPOSITION	AVERAGE RESISTIVITY (OHM.CM.)
PBI WITH NOMINAL MOISTURE	$2.85 X 10^{-7}$
WATER SWOLLEN PBI FILM	$2.67 X 10^{-6}$
WATER SWOLLEN ACIDIFIED PBI	TO $9.13 X 10^{-3}$
SODIUM SALTED SULFONATED PBI	$1.63 X 10^{-3}$

Measurements were made at Brookhaven Laboratories [16]

PBI microporous resin beads, Aurorez (TM), are also formed from PBI solution. They are spheres that range in diameter from 50 to 500 microns that are typically graded for different applications. Aurorez (TM) beads have a porous internal structure that is 85% void. Unique separation capabilities were anticipated from PBI due to its unusual chemistry, hydrophilicity, as well as thermal and chemical stability. In addition, sulfonated PBI would provide an ion exchange capability which should alter its absorption characteristics. FIGURE 10 shows a Liquid Chromatographic separation with sulfonated PBI Aurorez (TM) and FIGURE 11 shows a Gas Chromatographic separation of a series of C(5) to C(9) hydrocarbons. PBI's hydrophilic nature suggests its use as a sorbent for biological separations. Many applications are under consideration.

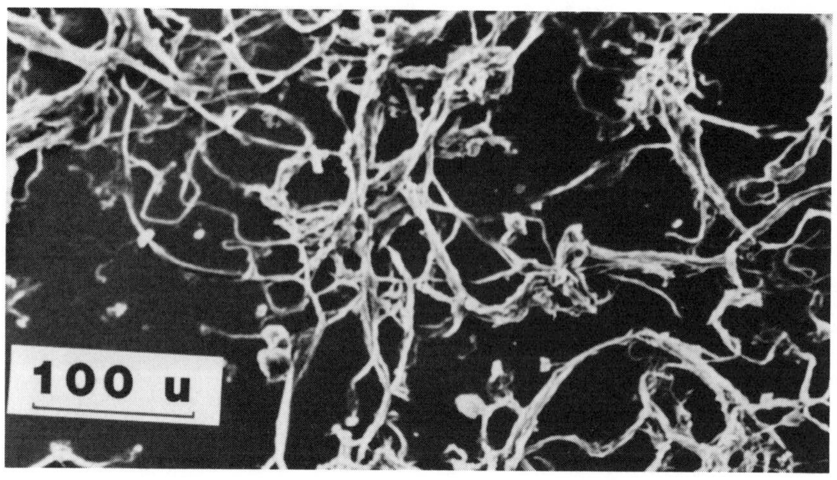

FIG. 9. SEM photomicrograph of PBI fibrids.

TABLE XVI. Typical properties of papers made from PBI fibers [29].

Color	Gold
Void volume (%)	40-80
Strip Strength (lbs/in)	>5
Breaking Elongation (%)	2-6
Thickness (mil)	4-20
Limiting Oxygen Index[a]	41
Moisture Regain (% at 20°C, 65% R.H.)[a]	14

[a](Measured on PBI fiber)

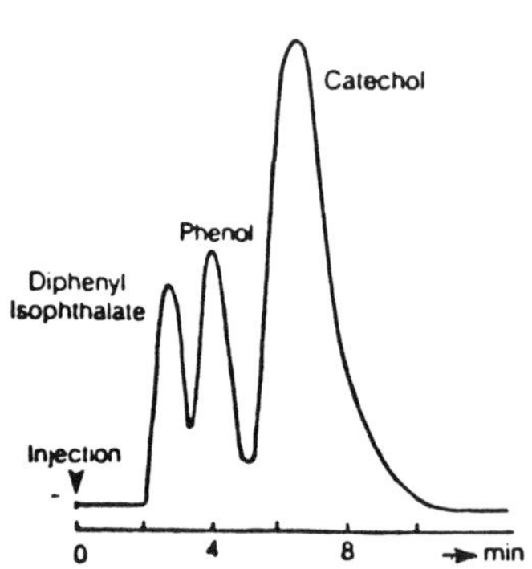

FIG. 10. Liquid-chromatographic separations by Aurorez sulfonated PBI in 50/50 acetonitrile/water at 2 ml/min [29].

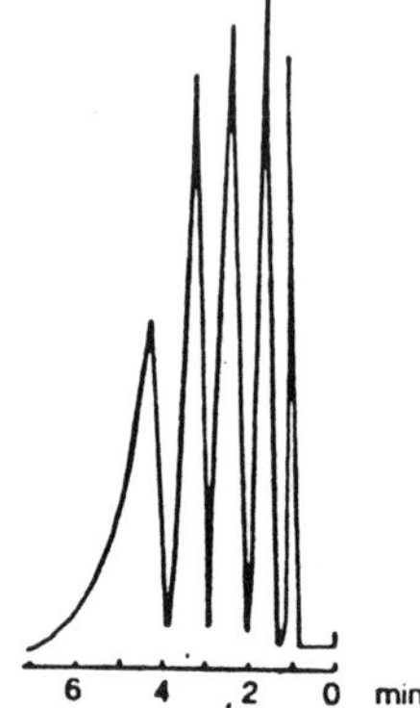

FIG. 11. Gas-chromatographic separations by Aurorez PBI in a 6 ft X 2mm (ID) column at 30 ml/min helium fow rate [29].

Phosphoric acid treatment of PBI has been reserved for the end of this report. Data was generated during development work on films and the treatment may have utility for very thermo-oxidatively stable fibers or as a chemical route to crosslink and oxidatively stabilize fused parts. In fact, the data may point the way to very stable, new polybenzimidazole polymers.

FIGURE 12, which is a Thermal Gravimetric Analysis (TGA) comparison in air of an unmodified PBI film and a phosphoric acid treated PBI film, indicates that phosphoric acid improves PBI's resistance to decomposition in air. That was verified by a TGA comparison, run isothermally. FIGURE 13 shows that PBI suffers about 50% weight loss whereas the phosphoric acid treated PBI looses less than 2% by weight when heated at 500°C in air for three hours. Also note in TABLE XIII that the LOI of phosphoric acid plasticized PBI films is significantly higher than unmodified PBI, which also indicates resistance to oxidation. (Note: differences in the initial weight loss between the as-cast and the annealed PBI films, shown in FIGURES 12 and 8, respectively, reflect differences in their moisture absorption characteristics.)

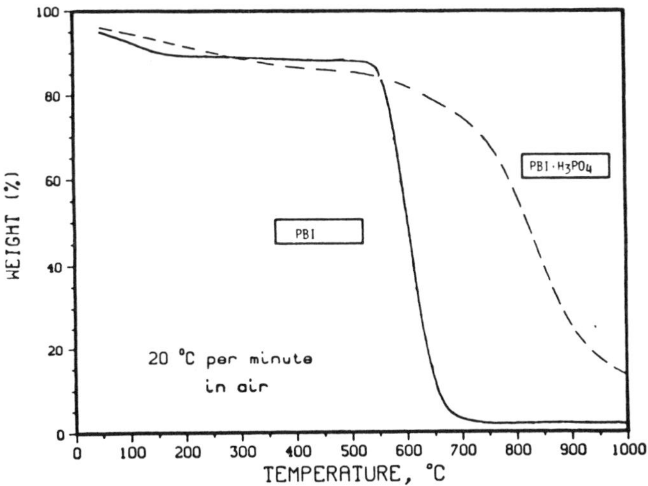

FIG. 12. TGA weight loss of PBI and phosphoric acid treated PBI in air.

The acid treatment was conducted by soaking PBI films in 2% aqueous phosphoric acid. Analyses indicated that 27 weight percent (%) acid was absorbed which was not appreciably eluted by water but was quantitatively extracted by strongly alkaline solutions. Heat treatments fixed the phosphorous to various extents depending on treatment temperatures and durations. Physical property changes, accompanying phosphorous fixation were consistent with polymer crosslinking, i.e., modulus increases, and eventually brittleness.

The improved oxidative stability of phosphated PBI may be due to formation of the benzimidazonium cation (see FIGURE 6). Early work on PBI [8] reported significantly better oxidative stability of PBI films cast from sulfuric acid compared to films cast from organic solvent. Data from other studies fail to support the observation [31]. However, discrepancies may be due to a combined effect of the thermal stability of the PBI salt and the oxidative stability of the PBI cation. PBI sulfate disassociates at temperatures above 350°C whereas PBI phosphate disassociates at temperatures above 850°C. So, a phosphate anion may be able to support a PBI cation to much higher temperatures than a sulfate anion.

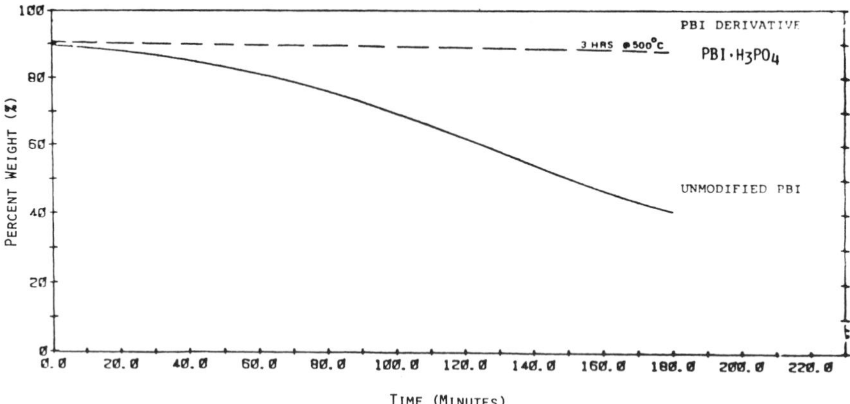

FIG. 13 Isothermal weight loss by PBI and phosphoric acid treated PBI in air at 500°C for 3 hours.

In conclusion, while PBI polymer has been known for 25 years, it is still a new polymer whose development and applications are just beginning.

REFERENCES

1. Patrick E. Cassidy, Thermally Stable Polymers (Marcel Dekker, Inc, New YorK 1980) pp. 163-173.
2. J.P. Critchley, G.J. Knight, W.W. Wright, Heat-Resistant Polymers (Plenum Press, New York 1983), pp 259-322.
3. H. Vogel and C.S. Marvel, J. Polym. Sci. 50, 511 (1961).
4. H.H. Levine, Encycl. Polym. Sci. Technol. 11, 188 (1969).
5. J.I. Jones, Macromol. Sci. C2, 303. (1968).
6. A.H. Frazer, High Temperature Resistant Polymers (Interscience, New York 1968) pp. 138-150.
7. V.V. Korshak, Heat Resistant Polymers (Izdatel'stvo "Nauka", Moscow 1969; Israli Translation: Keter Press, Jerusalem 1971, pp. 244-248.
8. R.T. Conley, J.J. Kane, S. Ghosh and S.L. Lu, Annual Technical Report to AFML/LNP (March 1972); Order: No. AD-903/6st from NTI Service.
9. R.T. Conley, J.J. Kane and S. Ghosh, Technical Report AFML-TR-71-219 (November 1971).
10. H.L. Freidman, G.A. Griffith and H.W. Goldstein, Technical Report ML-TDR-64-274 (April 1967).
11. H.H. Levine, US Patent 3,386,969, June 4, 1968.
12. Technical Bulletins, Acurex, Aerotherm Div., 485 Clyde Ave., Mt. View, CA: Polybenzimidazole Prepolymers; PBI 1850 Laminating Material; PBI 850 Adhesive; PBI SA Syntactic Foam Material; PBI PC Syntactic Foam Material.
13. J.F. Jones, J.C. Waldrop, R. Fountain, 29th National SAMPE Symposium, (April 3-5, 1984) p 777.
14. J.F. Jones and R. Fountain, Internat. Conf. of Composite Mtls. IV, AIME, Warrendale, Pa. (1985) p1591.
15. R. Fountain and J.F. Jones, Fabricating Composites Conf. SME, Hartford, Conn. (June 11 1985) MF85-505.
16. D.P. Griswald Jr., A.E. Casey, E.K. Weisberger and J.H.Weisburger, Cancer Research 28, (May 1968) pp. 924-933.
17. J. K. Gillham, CRC Critical Reviews In Macromomol. Sci. 1, 117 (1972)
18. G.M. Moulter, R. F. Tetreault, and M. J. Hefferon, Polymer News 9, 134-138 (December, 1983).

19. D.A. Koutrides, J.A. Parker, C.L. Segal, Proceedings SPE, 32nd Annual Tech. Conf., San Francisco, CA (May 13, 1974).
20. D.R. Coffin, G.A. Serad, H.L. Hicks, R.T.Montgomery, Textile Res. J. 52 (7), 466-472 (1982).
21. A. Buckley, G.A. Serad and D.E. Steutz, Encycl. Polym. Sci. and Eng. (John Wiley, New York to be published).
22. K.A. Reinhart, E.J. Powers, G.W. Calundann and C.P. Driscol, Technical Report ASD-TR-73-49 (January 1974) pp127, 179-185.
23. R.H.Jackson, Textile Res. J. 48, 314-319, (1978).
24. Technical Bulletins, Celanese Corporation, P.O. Box 32414. Charlotte, N.C.: PBI Fiber in Fire Blocking Layers for Air Craft Seat Cushions (HPF-P5 1984); PBI in Fire Service Applications (HPF-P2 1985); PBI in High Temperature Protective Gloves (HPF-P1 1983); In Protective Hoods, PBI Makes the Difference (HPF-P3 1984); In Protective Coveralls, PBI Makes the Difference (1986).
25. Keeping Hot Glass on the Move, Southern's Guardian PBI Conveyer Belts, Southern Manufacturing Inc. Charlotte, N.C. (7/84).
26. 1200-PBI Asbestos-Free Packings, Garlock Mechanical Packings Div., Colt Industries, Sodus, N.Y. (8/85).
27. Replacement of Asbestos in Flexible Insulators for Rocket Motors, Technical Report MMT 3811051 FR3 (May 1985).
28. D. Hoel, E. Grunwald, J. Phys. Chem. 81 (22), 2135 (1977).
29. Technical Bulletins, Celanese Corporation, P.O. Box 32414. Charlotte, N.C.: PBI Fibrids (SF-4 1984); Papers Made of PBI Fibers (SF-5 1984); Typical Applications for Celanese Microporous Resins (SF-6.1 1984); PBI Microporous Resins (SF-6 1984).
30. J.E. Ramirez and C.F. Dwiggins, TAPPI Non-Wovens Symposium, Myrtle Beach, S.C. (April 21, 1984) pp111-114.
31. J.R. Brown and N.Mc M. Browne, Australian Def. Scient. Serv. Mtl. Res. Lab. Report MRL-R-674, Maribyrnong Victoria (August 1976) pp4-6.
32. F.J. Salzano, Conductivity of Polybenzimidazole in Steam, Topical Report, Department of Applied Science, Brookhaven National Laboratories, June 1985.
33. R.B. Gosnell and H.H. Levine, J. Macromol. Sci. Chem. A3 (7), 1389 (1969).
34. Internal Company Communication from Dr. B. C. Ward, PBI Business Unit, Charlotte.
35. P.M. Hergenrother, Chemtech (August, 1984) p 497.

High Performance Elastomers

HIGH PERFORMANCE ELASTOMERS

Maurice Morton
Institute of Polymer Science, The University of Akron, Akron, Ohio

The term "high performance elastomers" can, of course, be rather broadly defined. This is because of the unusual situation which prevails in the field of elastomers, and which is not really paralleled in any other area of polymer science. It arises from the fact that the use of rubber in industry and commerce began over 150 years ago, and is still greatly concerned today with natural rubber. Hence the development of the synthetic rubbers, which only date back about 50 years, depended strongly on their "high performance", compared with the natural product. Of course, since those early beginnings, synthetic rubber has also filled part of the need for "general purpose" rubber. However, in selecting the particular synthetic rubbers for discussion at this symposium, I defined "high performance" as the ability of the material to withstand rather extreme ambient conditions, involving heat, chemicals, solvents, etc. Hence the fact that any given elastomer could do "as good a job as", or even slightly better than, natural rubber, or at lower cost, should not, in my opinion, put it in the class of high performance.

My choice was also influenced, to some degree, by the historical aspects, since this session was part of a larger symposium on "History of High Performance Polymers", and since the earliest synthetic rubbers developed in the United States owe their development to their unusually high performance. In this connection, it should be noted that among the principal defects of natural rubber is its susceptibility to attack by solvents, especially the hydrocarbons, such as found in petroleum and coal tar products. Hence it is not surprising that the first synthetic rubbers, at least in this country, were those with the special ability to withstand such solvent attack. Later developments led to the rise of the more sophisticated synthetic rubbers which could withstand not only solvents, but high temperatures and long oxidative attack as well.

The four types of elastomers whose history will be briefly reviewed here are (in chronological order):

Thiokol (polysulfide rubber)(1929)
Nitrile rubber (1934)
Silicone rubber (1943)
Fluorocarbon elastomers (1957)

Some of these are discussed in further detail in the following papers of this symposium. As can be seen, these elastomers represent only a small group, and there are many other synthetic rubbers available today, with special and even unusual properties, and some of these have had a powerful influence on rubber technology. The best example of this is represented by the thermoplastic elastomers, which are permanently thermoplastic and yet behave remarkably like vulcanized rubber under normal conditions. Although these have had, and are continuing to have, a profound effect on the processing and applications of rubber, they do not show much improvement in performance (in fact, their temperature resistance will, as a rule, be somewhat poorer than conventional vulcanized rubber). However, some of the newer, more unique types of these thermoplastic elastomers do show some promise of really improved temperature resistance, and have been included (vide infra).

Copyright 1986 by Elsevier Science Publishing Co., Inc.
High Performance Polymers: Their Origin and Development
R.B. Seymour and G.S. Kirshenbaum, Editors

Thiokol (polysulfide rubber)

This has the distinction of being the first synthetic rubber developed and commercially produced in the USA. It was also clearly the invention of Dr. J. C. Patrick[1], who not only discovered this polymer, but also founded the company, Thiokol Corporation, to produce and market it.

Strangely enough, its inventor, Joseph C. Patrick, earned his doctorate in the field of medicine, but never practiced, and instead returned to his first love, chemistry. His discovery was, as is often the case, quite incidental to his goal, which was the conversion of 1,2-dichloroethane into ethylene glycol. Finding that ordinary alkalies were ineffective, he turned to the sodium polysulfides, which his previous experience in the cattle products industry had shown to be a powerful depilatory of animal hides. To his surprise, instead of ethylene glycol, he obtained a sticky, gummy mass. However, instead of following the usual practice of contemporary organic chemists and discarding it, he thought it better to discover more about it.

The final result was that he developed a suitable method of carrying out this polymerization in water dispersion, shown below in modern form in Equation 1.

$$ClCH_2CH_2Cl + Na_2S_x \xrightarrow[70^oC]{Mg(OH)_2} [-CH_2-CH_2-S_x]_n \qquad (1)$$

(aqueous sol$^{\underline{n}}$) 2-6 hrs. poly(ethylene polysulfide)

In this way the polymer was obtained as a coarse dispersion, which could be separated and dried to a rubber. This proved to be a remarkable material in that it was virtually impervious to attack by solvents, especially hydrocarbons, showing very little swelling. It is still the most solvent-resistant rubber today but, because of its inferior mechanical properties (and odor problems) is not used as such. However, it does play a major role as a liquid, castable rubber, based on a process originally discovered by Patrick himself, i.e., the susceptibility of the sulfur-sulfur linkages in the chain to simple scission and coupling reactions, as shown in Equation (2).

$$\underset{\text{Rubber}}{[CH_2-CH_2-S-S]_n} \underset{\text{Oxidizer}}{\overset{\text{Na S(reductant)}}{\rightleftharpoons}} \underset{\text{Liquid polymer}}{H[S-CH_2-CH_2-S]_x} \qquad (2)$$

where n>>x

This reaction is used extensively for sealants, potting compounds, solid propellants and so on.

Nitrile Rubber

The commercial development of Neoprene (polychloroprene) by du Pont followed that of Thiokol by only a few years, and this turned out to be the second synthetic rubber to be developed in the USA, because it was far superior to natural rubber in oxidation and solvent resistance. However, it is not being included in this review, since it is "outperformed" by both Thiokol and nitrile rubber. The latter is actually the most commonly used solvent-resistant rubber today.

Nitrile rubber is a copolymer of butadiene and acrylonitrile, made by emulsion polymerization. Its structure is illustrated by Equation (3) below.

As can be seen, the acrylonitrile content varies from 10% to 40% depending on the extent of solvent resistance required, bearing in mind that higher

$$CH_2{=}CH{-}CH{=}CH_2 + CH_2{=}CHCN \xrightarrow[\text{Pzn.}]{\text{Emulsion}} [(CH_2{-}CH{=}CH{-}CH_2)_x CH_2{-}\underset{\substack{|\\CN}}{CH}]_n \qquad (3)$$

where x = 2-6

nitrile content means greater solvent resistance but also an increase in low-temperature stiffening.

The discovery of this outstanding synthetic rubber dates back to 1934, which corresponds both with the date of the basic U.S. patent of Konrad and Tschunkur[2] and with its first commercial production in Germany, as "Perbunan". This was part of the intensive effort[3] going on in the laboratories of the I. G. Farbenindustrie Company in Germany at that time to develop synthetic rubbers which were superior, both in properties and process, to the sodium-polymerized polybutadienes (Buna rubber) which was then in production. Out of this effort, of course, came the butadiene-styrene copolymers (Buna S), which were the precursors of SBR, still the dominant synthetic tire rubber today. It was most fortunate that, because of the close contacts between IGF and the Standard Oil Co. prior to World War II, all the necessary information on the emulsion copolymerization of butadiene with styrene, and other monomers, was available when this country was suddenly cut off from its main supply of natural rubber on December 7, 1941.

Silicone Rubber

The inclusion of this elastomer in the "high performance" category is well deserved, since it is outstanding in maintaining its elastic properties over the widest temperature range, and especially at low temperatures. This unusual material is discussed in more detail in one of the papers of this symposium, but it is interesting to review briefly its historical development.

The basis for silicone polymers was established by the pioneering work of Kipping (1901) on organosilicon chemistry, specifically on the synthesis and properties of chemical compounds containing the silicon-carbon bond. This later led to the development of the most popular silicone polymers, the polysiloxanes, which, at high molecular weight, exhibit rubbery properties. Such elastomers are generally obtained from a cyclosiloxane monomer by a ring-opening reaction, as shown in Equation (4). Because of the unusual ease of rotation around the silicone-oxygen bond, the polymer chain is much more flexible than carbon-based polymers, so that a semi-liquid material is obtained, even at molecular weights of several hundred thousand. Fine

$$(\underset{\substack{|\\CH_3}}{\overset{\substack{CH_3\\|}}{Si}}{-}O)_4 \xrightarrow[150^\circ C]{\text{KOH, etc.}} [\underset{\substack{|\\CH_3}}{\overset{\substack{CH_3\\|}}{Si}}{-}O]_n \quad (n \sim 7000) \qquad (4)$$

Octamethylcyclo-tetrasiloxane (liquid) — Polydimethylsiloxane (semi-liquid rubber)

silica is therefore used as a reinforcing filler to stiffen the material for suitable use.

These developments of the polymerization methods and silicone rubber compounding date back to the early 1940's, based on work done at the Dow-Corning laboratories (Warrick)[4] as well as at General Electric (Rochow)[5].

Fluorocarbon Elastomers

The foremost elastomers for maximum high temperature resistance are the fluorocarbon rubbers. They play the same role in the rubber field as polytetrafluoroethylene ("Teflon") does in plastics. Furthermore, the inert fluorine atoms also impart a high solvent resistance as well. Hence it is not surprising that these elastomers are in a "class by themselves" (also as regards cost!)

The development of these elastomers was based on the pioneering work of the inventor of du Pont's Teflon, Plunkett[6] (1941). However, in order to obtain rubbery polymers, it was necessary to modify this tough plastic, and this was first done in 1949 by Hanford and Roland[7] (du Pont), who copolymerized tetrafluoroethylene (TFE) with olefins, and by Schroeder[8] (du Pont), at the same time, who did it by copolymerizing 3,3,3-trifluoropropene with olefins.

Later work, both by workers at du Pont and the 3M Co., led to the more modern fluorocarbon elastomers[9], based on copolymers of vinylidene fluoride, as shown in Table I below together with the approximate dates of their commercial development.

Table I

Copolymers of Vinylidene Fluoride

Polymer	Co-monomer (with $CH_2=CF_2$)
Kel-F (3M Co. 1955)	$CF_2=CFCl$
Viton A (du Pont 1957)	$CF_2=CFCF_3$
Fluorel (3M Co.)	$CF_2=CFCF_3$
Viton B (du Pont 1961)	$CF_2=CFCF_3 + CF_2=CF_2$

A more detailed discussion of the fluorocarbon elastomers is provided in another paper of this symposium.

REFERENCES

1. J. C. Patrick, Rubber World, 139, 84 (1958); J. C. Patrick and S. M. Martin, Jr., Rubber Chem. Technol., 39(5), cl (1966).
2. E. Konrad and E. Tschunkur, U. S. Patent 1,973,000 (1934).
3. F. A. Howard, Buna Rubber, the Birth of an Industry, Van Nostrand, New York, 1947.
4. E. L. Warrick, Rubber Chem. Technol., 49, 909 (1976).
5. E. G. Rochow, An Introduction to the Chemistry of Silicones, John Wiley & Sons, New York, 1946.
6. R. J. Plunkett, U. S. Patent 2,230,654 (1941).
7. W. E. Hanford and J. R. Roland, U. S. Patent 2,468,664 (1949).
8. H. E. Schroeder, U. S. Patent, 2,484,530 (1949).
9. H. E. Schroeder, Chap. 14 in Rubber Technology, 3rd Ed., Van Nostrand Reinhold, New York (in press).

HISTORY OF SILICONE ELASTOMERS

F.W. GORDON FEARON
Dow Corning Corporation, 3901 S. Saginaw Road,
Midland, MI

ABSTRACT

In a little more than 40 years silicone elastomers have emerged from laboratory curiosities to a broad family of synthetic, largely inorganic, materials useful in a wide variety of applications. Key discoveries in the development of silicone elastomer technology, including monomer synthesis and polymerization, crosslinking and reinforcement of the polymers, will be reviewed. Composition and structural features responsible for the characteristic properties of these versatile elastomers will be discussed and related to their applications.

INTRODUCTION

The advances in science and technology key to silicone elastomers emerging from discovery in 1943 to a broad family of useful products can be traced back to the early 19th Century. A broad scientific base was first established over at least 100 years by a small number of academic research groups. The first industrial research began in 1933, the first patent describing a silicone elastomer was issued in 1943 [1], and since then this field has expanded very rapidly. To date over 18,000 patents describing silicone elastomers have been issued world wide. Approximately 7,000 of these have issued in the United States since 1950. Since 1945 over 6,000 publications in various scientific journals describe silicone elastomers, rubbers, caulks or sealants. In 1985, 495 U.S. patents and 203 publications describe these types of materials. Sales of silicone elastomeric products produced by at least 12 companies throughout the world approached $1 Billion in 1985.

In a rapidly moving field such as this where parallel research, development, and commercial activities are frequently pursuing similar goals in competing companies it is often difficult to learn the exact sequence of events leading to the important discoveries. Fortunately, a number of excellent sources of reference [2 - 11] are available which explore the development of silicone technology from the perspective of various authors associated with many of the companies involved. These different accounts show general agreement on the key events and discoveries which will now be reviewed with emphasis on the selection and production of the monomers, conversion of these to polymers, reinforcement and crosslinking of the polymeric materials to form elastomers, and fabrication of these elastomers to give articles able to perform under a wide variety of adverse conditions.

MONOMERS

Without doubt one of the key events of silicon science occurred in 1824 when Berzelius [12] first isolated silicon by the reduction of potassium fluorosilicate, K_2SiF_6, with potassium metal and then showed that this element reacted with chlorine to give a volatile, highly reactive material subsequently identified as silicon tetrachloride, $SiCl_4$. The next step from silicon to organosilicon science was accomplished by Friedle and Craft [13] in 1865 when they obtained tetraethylsilane by reaction of silicon tetrachloride with diethylzinc:

$$2\ Zn(C_2H_5)_2 + SiCl_4 \longrightarrow 2\ ZnCl_2 + Si(C_2H_5)_4$$

Copyright 1986 by Elsevier Science Publishing Co., Inc.
High Performance Polymers: Their Origin and Development
R.B. Seymour and G.S. Kirshenbaum, Editors

Next Ladenburg [14], who had worked with Friedle, showed that hydrolysis of diethyldiethoxysilane, $(C_2H_5)_2Si(OC_2H_5)_2$, with dilute acid gave a very viscous oil which decomposed only at "very high temperatures" and did not freeze at -15°C. The date was 1872 and already the building blocks for silicone elastomers were in hand, but the work was not pursued.

Kipping, at the University of Nottingham, methodically uncovered the basics of organosilicone chemistry. In investigations reported in a series of papers [15], of which 53 appear in a numbered series in the Journal of the Chemical Society, he established that the chemistry of silicon is only superficially similar to that of carbon and many workers have only reinforced his fundamental conclusions.

Among Kipping's major contributions to this field are the use of Grignard reagents to attach a wide variety of organic groups to silicon and the recognition that hydrolysis of a molecule, such as dimethyldichlorosilane does not give silicone, the silicon equivalent of acetone, but rather gives large molecules which he characterized at that time as "uninviting glues and oils."

$$(CH_3)_2SiCl_2 + 2\ H_2O \longrightarrow (CH_3)_2Si(OH)_2$$

$$(CH_3)_2Si(OH)_2 \not\longrightarrow (CH_3)_2Si{=}O$$

$$(CH_3)_2Si(OH)_2 \longrightarrow -(CH_3)_2SiOSi(CH_3)_2-$$

Some have suggested that Kipping should be criticized for not recognizing the importance of his discoveries. This author prefers to say that he recognized several important facts which form the basis of silicone science. In fact, he investigated these "oils and glues" quite diligently, showed that they were large molecules formed by the combination of a number of small molecules and concluded that the formation of multiple bonds between silicon and oxygen or carbon is very difficult. Subsequent workers have only confirmed his findings.

In the early 1930's the field of polymer science was beginning to open up. Sullivan, the Research Director for Corning Glass Works, became interested in the possibility of combinations of organic polymers and glasses, so he hired Hyde, an organic chemist, to investigate the field. Hyde rapidly picked up and expanded upon Kipping's work and recognized that organosilicon systems offered materials that could operate in "adverse environments." People at General Electric learned of this work and started parallel research efforts under Rochow. Further research efforts were started at Mellon Institute of Industrial Research under McGregor and in Russia under Andrianov at about the same time. World War II arrived on the scene and the pace became fast and furious, as it was shown that fluids, resins, and compounds based on silicon were able to operate under combinations of high and low temperature and extreme dielectric stress where no other polymeric materials could survive. Corning Glass Works formed a joint venture with the Dow Chemical Company in 1943, Dow Corning, a venture dedicated to development and commercialization of silicone materials. General Electric announced a dedicated facility in 1946 and in the next decade several other companies established manufacturing capabilities around the world.

Initially monomer synthesis was based on the reaction of Grignard reagents with silicon tetrachloride as taught by Kipping, a flexible but expensive route. Discovery that methylchloride reacts readily with silicon metal in the presence of certain additives opened up a route to relatively low cost methylchlorosilane building blocks, such as, Me_3SiCl, Me_2SiCl_2, and $MeSiCl_3$:

$$CH_3Cl + Si \longrightarrow \begin{array}{l} (CH_4)_4Si + (CH_3)_3SiCl + \\ (CH_3)_2SiCl_2 + CH_3HSiCl_2 + \\ (CH_3)_2SiHCl + HSiCl_3 \quad \text{etc.} \end{array}$$

This direct process has been the focus of much research, particularly in industrial laboratories during the last 40 years [8]. Published results show that it an extremely complex process with many competing reactions which is difficult to apply to systems other than methylchlorosilanes and trichlorosilane. Key monomers for production of linear polydimethylsiloxane are dimethyldichlorosilane and trimethylchlorosilane so in general the process has been optimized to produce these materials. Other monomers such as, phenylmethyldichlorosilane, methylvinyldichlorosilane, diphenyldichlorosilane and trifluoropropylmethyldichlorosilane are required to modify the properties of polydimethylsiloxane as described later. These are produced by a variety of approaches depending upon the producer.

There is no doubt that the supply of relatively low cost dimethyldichlorosilane has been key to the rapid growth of the silicone industry, but it is also apparent that the availability of monomers has focused attention on systems primarily based on polydimethylsiloxane. Many would argue that this is of no concern because this is obviously the material of choice as shown by the very rapid growth in silicone science and technology. Others, who see a materials science based on silicon that potentially parallels that based on carbon, observe the profusion of silicon monomers that have been made and wonder if there are other silicon based polymer systems of equal importance to polydimethylsiloxane which now need to be developed and commercialized to maintain rapid growth in this industry.

POLYMERS

Polymerization of siloxanes, particularly the polydimethylsiloxane system is inherently a flexible and controllable process but it has taken much effort by many research groups to understand the basic principles underlying these systems and this work still continues today [10].

Hydrolysis of most chlorosilanes is a rapid exothermic process yielding various mixtures of cyclic and linear polysiloxanes depending upon the nature of the chlorosilane used. But in general the reaction shown below for hydrolysis of dimethyldichlorosilane applies to all silicon compounds susceptible to hydrolysis and this is a wide range indeed covering halogen, alkoxy, acyloxy, and aminofunctional silicon compounds to name but a few.

$$(CH_3)_2SiCl_2 + H_2O \downarrow (CH_3)_2Si(OH)_2$$

$$(CH_3)_2Si(OH)_2 \swarrow HO[(CH_3)_2SO]_xH \qquad (CH_3)_2Si(OH)_2 \searrow [(CH_3)_2SiO]_x$$

Much of the initial work was carried out with alkoxysilanes but as chlorosilanes became available from the direct process they became the materials of choice.

The first reported "rubbery" silicones were obtained by the thermal condensation of the linear portion of these hydrolysates [1]. Iron oxide was incorporated as a filler and the whole system was crosslinked with aluminum chloride. This approach was developed using a variety of condensation catalysts, fillers and different crosslinking mechanisms [7] to give silicone elastomers which were rapidly applied in critical high temperature applications, such as gaskets for searchlights and turbochargers. However, these systems were weak and unstable, because the polymers were still of relatively low molecular weight, had uncontrolled branching, and contained residual acid catalysts.

Discovery that low molecular weight cyclic siloxanes can be polymerized by ring opening to give high molecular weight linear materials in the presence of small amounts of alkaline catalysts [16] opened up avenues to the heat cured silicone rubbers used today. This ring opening polymerization is complex [8] and apparently very different to most organic ring opening polymerizations. For instance, octamethylcyclotetrasiloxane, the work horse cyclic siloxane for production of high molecular weight polydimethylsiloxanes, polymerizes in the presence of a potassium catalyst with zero enthalphy and positive entropy of polymerization [17]. Certain cyclosiloxanes, particularly the strained ring trimers, such as hexamethylcyclotrisiloxane can give polymers with a very narrow molecular weight distribution and precise end group chemistries which in turn are used to give a wide variety of copolymer systems. In some systems, such as the trifluoropropylmethylpolysiloxane system, the strained ring trimer is an important source of all of the high polymers.

These two approaches, condensation and ring opening polymerization, have been developed by many workers through the years to the point that a wide variety of polymers and copolymers with controlled molecular weight, molecular weight distribution, and chemical composition are now available in both linear and branched forms with a variety of end group chemistries.

High molecular weight polydimethylsiloxane with 3,000 to 10,000 repeat units containing less than 1% of vinyl functionality are the basis of most heat cured high consistency elastomers. Phenyl groups are added to inhibit crystallization at low temperature and improve radiation resistances. Trifluoropropyl groups are added to improve solvent resistance. Low molecular weight polydimethylsiloxanes with 50 to 2,000 repeat units are used to make systems which cure at room temperature.

The field of siloxane polymerization is complex and fertile with many discoveries awaiting careful research. For instance, it was recently shown [24] that silanol terminated polydimethylsiloxane can be polymerized in emulsion to give high molecular weight latex systems which cure on the evaporation of water to give useful elastomeric materials. An interesting comparison, organic latexes which give useful elastomers occur naturally, whereas in silicone systems the latex materials were only discovered after much research.

Reinforcement

Crosslinked high molecular weight polydimethylsiloxane has a tensile strength of only 0.35 MPa so it must be reinforced. Warrick [18] was the first to discover that high surface area "fumed" silicas interact strongly with the siloxane chain to give useful elastomers with tensile strength in the 5 MPa range. Today these types of fillers are still used, but by careful selection and with appropriate surface treatments, tensile strengths in the 13-15 MPa range can now be obtained.

The inherent lack of strength of a crosslinked polydimethylsiloxane system is directly related to lack of interchain interactions at ambient temperatures. Several approaches have been explored to increase this interaction. Incorporation of small amounts of boron into the chain gives a material with extremely high resilience under shock loading which flows under slow deformation, the so called bouncing putty.

Copolymerization of polydimethyl siloxane with glassy or crystalline blocks, such as, organosilphenylene gives materials with tensile strengths in the 15 MPa range without use of reinforcing fillers [19].

Block copolymers of silicone, generally polydimethylsiloxane, and organic blocks such as styrenic, urethane, carbonate, imide, amide and polyester yield a wide range of materials both thermoplastic and thermosetting with tensile strengths as high as 50 MPa while maintaining many of the desirable properties of polydimethylsiloxane. Blends and so-called interpenetrating networks of polydimethylsiloxane have also been explored as an approach to higher strength materials with plastic processing characteristics [20]. Much work continues in this whole field of silicone organic copolymers and blends, but so far few products have been commercialized.

Silica reinforcement is still of prime importance and work is constantly underway to understand and improve the process. Recently new wet process hydrophobic silicas have been discovered and developed based on the hydrolysis of alkoxysilanes. These materials have controlled surface properties and narrow particle size distribution which makes them excellent for the production of optically clear elastomers [21].

CROSSLINKING

The chemistry of crosslinking, or curing of silicone elastomers is rich and complex. Aluminum chloride and heat were uses to cure the first elastomers. This was rapidly superseded by the use of organic peroxides which effectively crosslinked polydimethylsiloxane by attack on the methyl group on silicon [22]. Since this discovery many workers have reported the use of a wide variety of peroxides and other free radical sources to obtain specific properties in the crosslinked systems. Incorporation of small amounts of more reactive groups, such as vinyl on silicon, and use of so-called "vinyl specific peroxides" such as di-t-butyl peroxide, permits the formation of localized crosslinks which significantly change the dynamic response of the crosslinked polymers [10]. A valuable and specific heat cure system is based on the addition of silicon hydrogen bonds to vinyl groups attached to silicon in the presence of a platinum catalyst.

The next major advance in curing of silicones was the discovery and development of systems which cured at room temperatures, the so-called RTV, room temperature vulcanizing, systems. The first of these were two component systems based on a variety of silanol terminated polysiloxanes crosslinked with various silanes or alkoxysilanes. A major step forward was the development of one component RTV systems using silanol terminated polydimethylsiloxanes and acytoxysilanes. These systems which cure when exposed to moisture in the air [23] provided the backbone for the rapid growth of this type of material as both sealants and adhesives.

Many other systems have subsequently been discovered and patented attesting once again to wealth of silicone chemistry available to polymerize siloxane systems at relatively low temperatures. In general terms these systems fall into two classes, one component systems which crosslink on exposure to the atmosphere because the system reacts with either atmospheric moisture or oxygen and two component systems which react on mixing. Both general classes encompass addition and condensation cure chemistries. Anaerobic systems which cure in the absence of oxygen have also been described. As mentioned previously silicon rubber latex systems which cure upon evaporation of water are now being developed [26] and commercialized.

PROCESSING AND FABRICATION

The first heat cured elastomers were processed on the typical mills and mixers used by the Rubber Industry. Fabrication methods were also conventional consisting of transfer and blow molding, extrusion, calendaring, and dispersion coating. However, as the industry has developed more and more emphasis is being placed on simplification and cost reduction in the compounding and fabrication steps. Heat cured liquid silicone rubbers which cure fast enough at 200°C without formation of by-products to permit direct processing in injection molding equipment provide much flexibility for the production of large numbers of small complex parts [25] these materials were introduced to the market in the mid-1970's. Comprehensive ranges of rubber bases and modifiers which allow the user to formulate compositions in the same way that users of organic rubber have done were introduced in the same time frame [11].

Much work continues on the development and introduction of "thermoplastic silicone rubbers" and a wide variety of both silicone and silicone-organic block copolymers are being evaluated at this time. As with the liquid silicone rubbers the major fabrication advantage being sought is direct processing by injection molding techniques [20].

PROPERTIES AND APPLICATIONS

Inherent properties of silicone rubber responsible for the growth of this family of materials are generally some combination of high and low temperature stability, resistance to oxidation and radiation, good electrical properties, low activation energy for viscous flow, high gas permeability to small molecules and physiological inertness. These properties are directly related to the structural features of the polymer systems which include a strong but flexible siloxane backbone with low interchain forces, high free volume fraction at ambient temperature and a composition which is largely made up of carbon, hydrogen, oxygen, and silicon joined by single bonds. These well published properties are characteristics of the polydimethylsiloxane system or minor modifications of this versatile system which have been developed to achieve specific properties.

Silicone rubbers were first used as gaskets to seal military components, such as, searchlights and turbochargers [7] where a combination of high and low temperature flexibility was required. Not only are the silicone rubbers stable to thermal degradation at temperatures in excess of 200°C, but compared with most organic elastomers, they show much better retention of strength at high temperatures and this property is critical in a large variety of sealing applications. Recognition that the materials have good electrical insulation characteristics, particularly corona resistance and that these properties are also maintained at elevated temperature led to early use in wire and cable insulation and other applications where high performance electrical insulation was required. A unique feature of silicones compared with any other synthetic elastomers is that that after burning an insulating layer of silica still remains on the conductor so the systems remain in operation. This feature was particularly emphasized in the development and use of so-called "navy cable."

Room temperature vulcanizing (RTV) systems with outstanding resistance to weathering together with good adhesion to a wide variety of substrates form the basis of families of coatings, sealants and adhesives which find much use in exposed applications particularly in the construction industry and this is one of the most important applications of silicones today.

Elastomers based on trifluoropropylmethylsiloxane with good resistance to solvents, flexible at -60°C yet stable at temperatures in excess of 200°C find use in many mechanical sealing applications particularly in the automotive industry. RTV versions of these materials are also available.

Low viscosity materials with good dielectric properties and very low ionic contamination which can be cured by a variety of mechanisms form the basis for a broad family of electronic coatings, encapsulants, sealants and adhesives. Systems of this type which can be cured very rapidly (<1 second) [26] are important for coating optical fibers.

The observation that certain silicone polymers are compatible with living tissue has lead to a major use of these materials in medical implants and devices. This type of applications was largely developed by the medical profession assisted for many years by the "Center for Aid to Medical Research" which was established by the Dow Corning in 1959 under McGregor [10].

The high permeability of silicones, particularly polydimethylsiloxane, finds application in extended wear contact lenses where high oxygen permeability is important. The elastomers are also permeable to a wide variety of drugs. This fact together with biologically inertness leads to their use as a matrix for controlled release of drugs either through implant or transdermal devices. In this application the ability to tailor the solubility characteristics of the elastomers through a variety of copolymer systems is of growing importance [27].

This short review of the history of silicone elastomers through the last 40 years is superficial in many respects, but hopefully the leading references will permit those interested to explore the field in greater depth. One observations on closing. The silicones are a family of hybrid materials in which organic moieties are used to modify the the properties of the inorganic, siloxane backbone. It is this strong backbone which provides much of the resistance of these materials to adverse conditions, the organic moieties are generally weak links to radiation, dielectric and thermal stress. In order to respond to demands for ever increasing resistance to adverse environments, particularly high temperature and radiation resistance, it is likely that more and more emphasis will have to be placed upon reducing the organic content of these materials.

ACKNOWLEDGEMENTS

My thanks to J. C. Saam, K. E. Polmanteer, R. M. Fraleigh, J. M. Klosowski, C. L. Lee, A. S. Lee, and C. M. Woods for their help and comments and to G. L. McComb for preparing this manuscript.

REFERENCES

1. Agens, M. E., U.S. 2,448,756 (Sept. 7, 1948).
2. Rochow, E. Chemistry of the Silicones (New York: Wiley, 1946).
3. McGregor, R. R., Silicones and Their Uses (McGraw-Hill, 1954).
4. Freeman, G. G., Silicones An Introduction To Their Chemistry and Application (London, 1962).
5. Post H. W., Silicones and Other Organic Silicon Compounds (New York: Reinhold, 1948).
6. Fordham S., Silicones (London: Newes Limited, 1960).
7. Liebhafsky, H. A., Silicones Under the Monogram (New York: Wiley, 1978).
8. Noll, W., Chemistry and Technology of Silicones (New York, Academic Press, 1968).
9. Warrick, E. L., Rubber Chemistry and Technology, 1976, 48 4, pp. 909-934.
10. Warrick, E. L., Pierce, O. R., Polmanteer, K. E., and Saam, J. C., Rubber Chemistry and Technology, 1979 52 3, pp.437-525.
11. Polmanteer, K. E., in Handbook of Elastomers - New Developments and Technology. Stephens, H. L. and Bhowmick, A. K. ed. (New York: Marcel Dekker) in press.
12. Berzelius, J. J., Pogg Ann, 1824 1, pp. 169.
13. Friedle, C., Craft, J. M., Compt. Rend, 1863 56, pp. 592.
14. Ladenburg, A., Ann 1871, 159, pp. 259.
15. Kipping, F. S., and Abraham, J. T., J. Chem. Soc., 1944 81 and papers cited therein.
16. Hyde, J. F. and Johannson, O. K., U.S. 2.453,092 (Nov. 2, 1948).
17. Lee, C. L. and Johannson, O. K., J. Polymer Science A 1966 4, pp. 3013
18. Warrick E. L., U.S. 2,541,137 (Feb. 13, 1951)
19. Merker, R. L., Scott, M. J. and Haberland, G. G., J. Polymer Sci., 1964 2, pp. 31.
20. Arkles, B. C., Chemtech, 1983, pp. 542.
21. Lutz, M. A., Polmanteer, K. E., and Chapman, H. L., Rubber Chemistry and Technology, 1985 58, pp. 939-952.
22. Warrick, E. C., U.S. 2,460, 795.
23. Bruner, L., U.S. 3,077,465 (Feb. 1963).

24. Saam, J. C., Graiver, D., Baile, M., Rubber Chem. and Technology, 54 5, 1981, pp. 776.
25. Cush, R. L., and Winnan, H. W. in Developments in Rubber Technology, Whelan and Lee ed. (London: Applied Science, 1981), pp 203-230.
26. Clark, J. N., Lee, C. L., Lutz, M. A., Science and Engineering, 52, 1985, pp. 442.
27. Lee, C. L., Ulman, K. L., Larson, K. R., Drug Development and Industrial Pharmacy, 1986, in press.

ADVANCES IN FLUOROELASTOMERS

Herman Schroeder
74 Stonegates
4031 Kennett Pike
Greenville, DE

ABSTRACT

Fluoroelastomer discoveries and the prospects for future developments are reviewed. The first fluoroelastomer was poly (2-fluorobutadiene) which though an excellent rubber, had no unusual properties. It was then noted that certain copolymers of vinylidene fluoride were flexible and ethylene/trifluoromethylethylene copolymers were rubbery. Intensive research to meet defense and aviation needs for rubbers with better heat and solvent resistance led to the discovery of rubbery vinylidene fluoride copolymers with fluoroolefins. Of these the copolymer with hexafluoropropylene and the terpolymer which contained tetrafluoroethylene met the needs and became mainstays of the market for high temperature elastomers. Their properties and many improvements in these products are detailed. Advancing aerospace needs for low temperature flexibility led to the fluorosilicones which sacrificed solvent resistance and high temperature stability. Though wide ranging research to correct these deficiencies was largely unsuccessful, one commercial product of exceptional stability emerged. It is a rubbery copolymer of tetrafluoroethylene with a perfluorovinyl ether.

INTRODUCTION

This paper is concerned with the history of elastomers containing fluorine. Though devoted chiefly to polymers containing carbon chains, it will also refer to materials containing hetero atoms in the chain. The products have aroused interest because of their outstanding heat and solvent resistance.

The hydrofluorocarbon elastomers were the first entry into the field and are all offshoots of Roy Plunkett's 1938 discovery of polytetrafluoroethylene, later called TEFLON® [1]. The revelation of the inertness and resistance of the new polymer to all solvents and most chemicals stimulated intense interest in other polymerizable fluorine containing monomers. So the Du Pont chemists made a variety of them to follow up Plunkett's new lead hoping of course for new and interesting properties. With the outbreak of World War II and the frenzied search for synthetic rubber, they quickly prepared a fluorine analog of neoprene through reaction of monovinylacetylene with hydrogen fluoride to make 2-fluorobutadiene which they then polymerized (Figure 1.) [2]. Poly(fluoroprene) proved to be a good rubber but its advantages were too slight to justify the cost; it was just too much like rubber.

POLYFLUOROPRENE

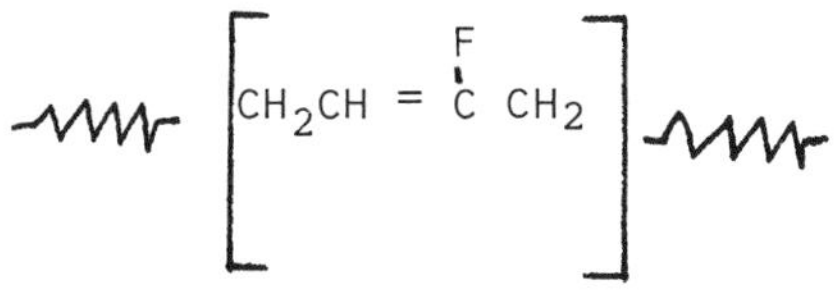

$$\sim\left[CH_2CH = \overset{F}{\overset{|}{C}}\ CH_2\right]\sim$$

FIG. 1

Copyright 1986 by Elsevier Science Publishing Co., Inc.
High Performance Polymers: Their Origin and Development
R.B. Seymour and G.S. Kirshenbaum, Editors

However, the curiosity-motivated research on fluoro-olefin polymers was well rewarded and a variety of novel products tumbled out. First came plastic polyvinyl fluoride and polyvinylidene fluoride each of which had remarkable physical properties. Then Tom Ford discovered that flexible but leathery products were produced when he copolymerized vinylidene fluoride with some unsaturated monomers[3]. Hanford and Roland discovered that a copolymer of propylene with tetrafluoroethylene was rubbery and even recommended that it be cured with radiation or peroxide (Figure 2). About the same time I found that plastic polyethylene could be changed to a limp rubbery material by attachment of as little as 5 mol percent of trifluoromethylethylene (Figure 3)[5]. However, with the urgent pressures of the war, there wasn't enough manpower to pursue these leads. Then in the press of the post World War II industrial boom these developments were put aside. There were just too many ripe apples on the tree.

PROPYLENE/TETRAFLUOROETHYLENE COPOLYMER

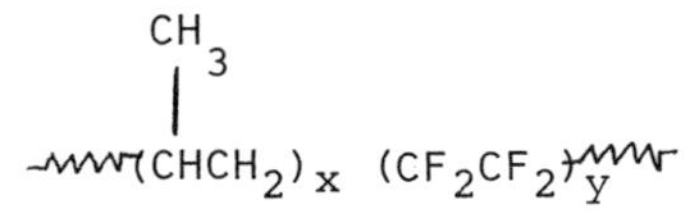

$$\sim\!\!\sim(\underset{\mathrm{CH_3}}{\mathrm{CH}}CH_2)_x\ (CF_2CF_2)_y\!\sim\!\!\sim$$

FIG. 2

ETHYLENE/TRIFLUOROPROPYLENE COPOLYMER

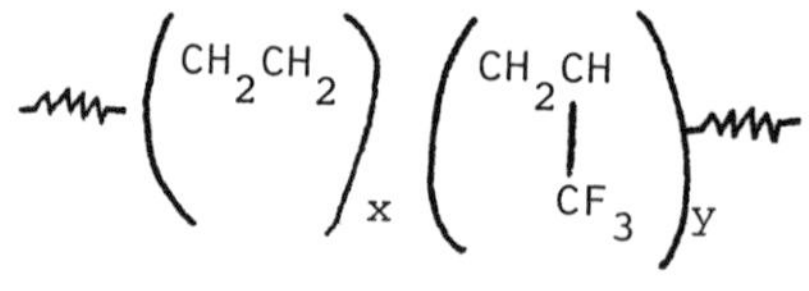

$$\sim\!\!\sim\left(CH_2CH_2\right)_x\left(CH_2\underset{\mathrm{CF_3}}{\mathrm{CH}}\right)_y\sim\!\!\sim$$

FIG. 3

In the early 1950's the emerging aerospace industry developed a critical need for an elastomer with good heat and fuel resistance for use in seals and hose in military jet engines. Available rubbers, even the silicones, were just not adequate to meet the stringent new requirements. The studies which followed, unlike the early work, were not motivated by scientific curiosity but involved deliberate use of the knowledge gained from past work on polymer structure to produce desired results. A number of attractive candidates soon emerged.

The Hydrofluoro or FKM elastomers

W. E. Hanford who had supervised some of the earlier fluoropolymer studies at Du Pont was now at M. W. Kellogg, an engineering company. Kellogg was then manufacturing a stable plastic poly-(chlorotrifluoroethylene) which had been developed in the Atomic Energy program by Bill Miller and used in certain critical applications. Hanford remembered the early work at Du Pont and turned his new group toward Ford's vinylidene fluoride (VF2) copolymers. They found that a copolymer of VF2 with chlorotrifluoroethylene (CTFE) was elastomeric and very stable[6].

This VF_2/CTFE copolymer (CTFE) was later introduced under the name KEL-F elastomer (Figure 4). Manufacturing rights to this polymer were later purchased by 3M Company.

VF_2/CTFE COPOLYMER

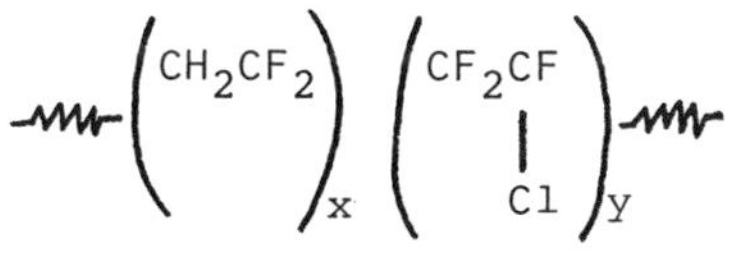

$$\sim\left(CH_2CF_2\right)_x\left(\begin{array}{c}CF_2CF\\ |\\ Cl\end{array}\right)_y\sim$$

FIG. 4

Meanwhile, the Du Pont fluorochemical group which I was now directing was looking for new opportunities to use its substantial technical and manufacturing capabilities in the fluorine field. These had been developed through long and successful experience with the FREON® refrigerants and other fluorochemicals like TEFLON®. Among other objectives we began to look at fluoroolefins again with the hope of developing new useful polymers. With our extensive knowledge of rubber gained from experience with neoprene and rubber chemicals, we thought we could make a good stable elastomer to meet the air force needs.

We chose to concentrate on copolymers of fluoroethylenes with perfluoroolefins known and potentially available to us from earlier research. One of these was hexafluoropropylene which had been synthesized by Benning, Downing and Park about 10 years earlier by pyrolysis of tetrafluoroethylene[7]. We assigned Dean Rexford to the problem and in relatively short order he found out how to run the then difficult copolymerization of vinylidene fluoride with hexafluoropropylene. He obtained a leathery elastomer which showed outstanding thermal stability and resistance to chemical and solvent attack [8]. So we optimized the monomer ratios and obtained a fine polymer, which we later called VITON® A (Figure 5). We were rather proud of our new rubber so we gave small research samples to the Airforce Materials Development group at Wright Field and to other government groups to see if it met their needs. Of course it did and pandemonium broke loose. They found our products much better than anything they had seen and had to have lots more; so we all raced the VITON® development along. We described the product at the fall 1956 ACS/Rubber Division meeting less than 3 years after starting the work. Meanwhile, in an attempt to make a yet more stable and solvent resistant product we prepared the tetrafluoroethylene terpolymer (Figure 6) [9]. These products were formally commercialized in 1958 as VITON® A and VITON® B fluoroelastomers.

VITON® A

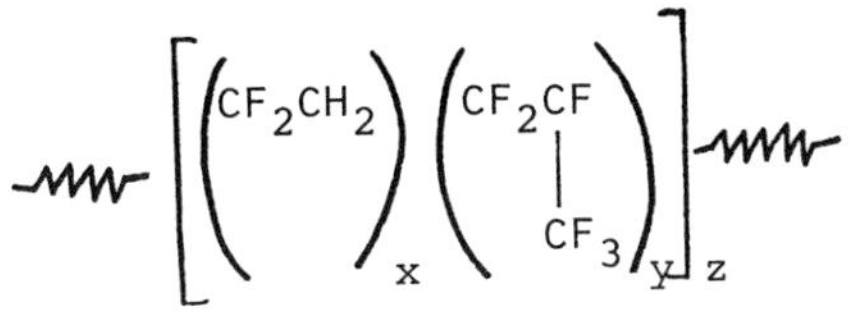

$$\sim\left[\left(CF_2CH_2\right)_x\left(\begin{array}{c}CF_2CF\\ |\\ CF_3\end{array}\right)_y\right]_z\sim$$

FIG. 5

VITON® B

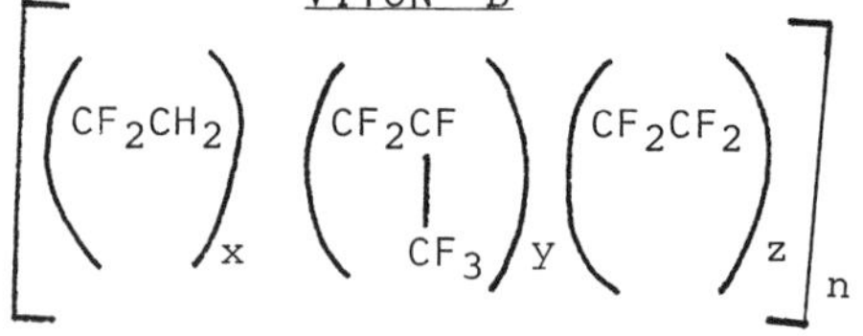

$$\left[\left(CF_2CH_2 \right)_x \left(\begin{matrix} CF_2CF \\ | \\ CF_3 \end{matrix} \right)_y \left(CF_2CF_2 \right)_z \right]_n$$

FIG. 6

The early forms of VITON® processed poorly and were hard to cure. The Kellogg chemists had discovered that their CTFE/VF_2 elastomer could be cured with hexamethylene diamine carbamate which also was effective with VITON®, but the compounds were very scorchy. Thermal stability of the cured products was good but poorer than the raw polymer because the curing reactions introduced olefinic bonds and a relatively unstable crosslink. More importantly for critical sealing applications high temperature compression set fell far short of demands. Further research delivered delayed action amine type curing agents which improved processability. Then bisphenols such as bisphenol A and also hydroquinone were found to be less scorchy and to give better cures. However, good compression set remained an elusive goal as uses of these FKM elastomers expanded rapidly in both military and civilian markets.

The Search for Exotic Elastomers

Meanwhile, developments in the military and in the aerospace industry and work on the supersonic transport (SST) imposed new performance criteria. While the hydrofluoroelastomers surpassed the silicones with their practically indefinite life at 200°C and their resistance to chemicals and solvents, their stiffness at low temperatures left them poorly equipped for arctic type applications. This led to the development of the fluorosilicones which achieve very good low temperature flexibility with some solvent resistance (Figure 7) [10]. The fluorosilicones were a good compromise, but were not as freeze resistant as the silicones nor as stable and solvent resistant as the hydrofluoroelastomers. So the need for better products persisted as the aerospace interests developed. During the 1960's these stimulated widespread academic and industrial research seeking unrealizable properties, rubbers with softness at cryogenic temperatures and stability, strength and elasticity at a red heat. While quite out of the reach of rational possibilities achievable with known elements and chemical science, these demands stimulated wide-ranging research and led to very interesting products. Some had excellent low temperature flexibility, but couldn't match the silicones. Others were remarkable in high temperature or chemical resistance, close to the limits possible for organic or organo/inorganic compounds, but in most cases they were deficient in low temperature elasticity. Out of these researches only the fluorosilicones became important commercial products.

FLUOROSILICONES

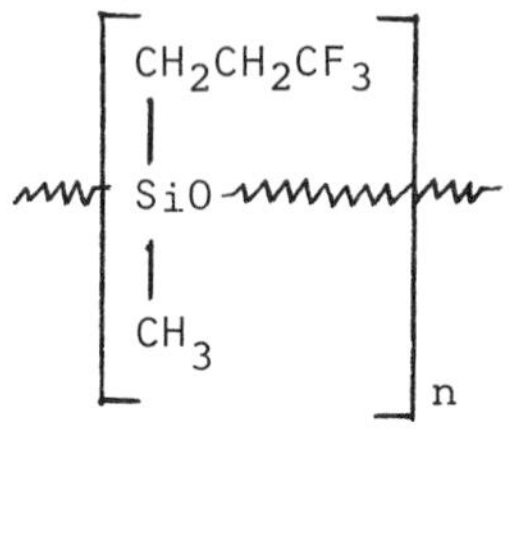

$$\sim\sim \left[\begin{matrix} CH_2CH_2CF_3 \\ | \\ SiO \sim\sim\sim \\ | \\ CH_3 \end{matrix} \right]_n \sim\sim$$

FIG. 7

The diversity of chemical types investigated is illustrated by the following examples [10]:

Nitroso Elastomers (Figure 8)

Copolymers of trifluoronitrosomethane with tetrafluoroethylene have an alternating structure wherein the easy rotation around the enchained -NO- leads to a Tg of -50°C and the high fluorine content to excellent resistance to fluids and oxidizing agents [12]. Curability is achieved through copolymerization with a nitrosoperfluorobutyric acid. Unfortunately, the polymer depolymerizes rapidly above 180°C and is very sensitive to bases.

NITROSOELASTOMERS

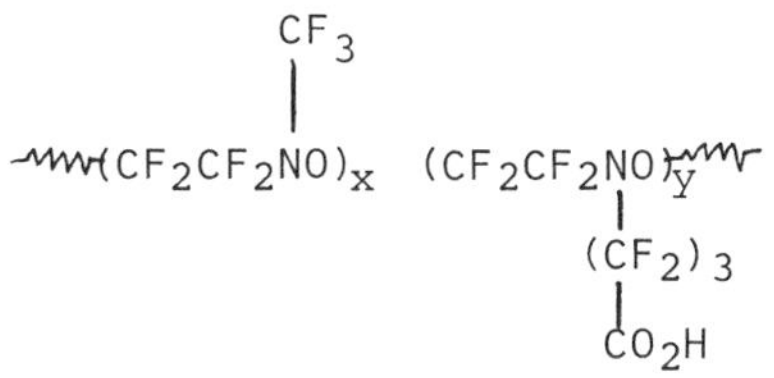

CF3
(CF2CF2NO)x
(CF2CF2NO)y
(CF2)3
CO2H

FIG. 8

Perfluorotriazines (Figure 9)

Perfluoroalkylnitriles polymerize to form triazines. When the mononitriles are copolymerized with the corresponding dinitriles, products with fair elastomeric properties and outstanding thermal and oxidative stability are formed [13]. Unfortunately, these products are subject to easy hydrolysis and have very poor low temperature flexibility.

PERFLUOROTRIAZINES

Rf
N
N
N
Rf
Rf

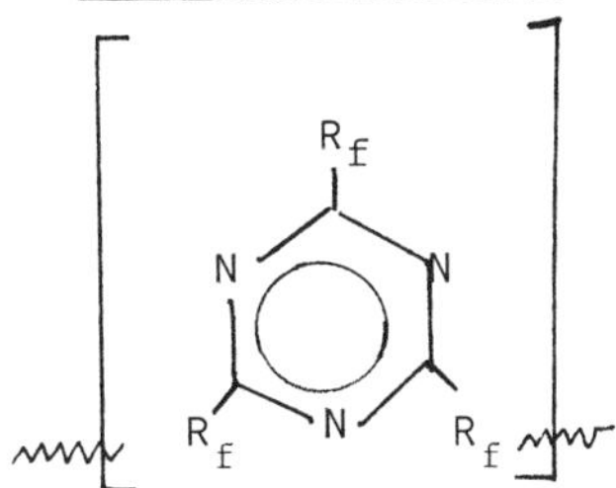

FIG. 9

Poly(perfluorothioacyl fluorides) (Figure 10)

Thiocarbonyl fluoride and certain other thioacyl fluorides can be homopolymerized or copolymerized with a variety of olefins. The homopolymer is of very high molecular weight and is in some respects the most resilient elastomer known Tg -118°C. It too depolymerizes at 160-180°C particularly in the presence of oxidants [14].

POLYTHIOCARBONYL FLUORIDE

$$\sim\!\!\sim\!\!\left(CF_2S\right)_n\!\sim\!\!\sim$$

FIG. 10

Further Hydrofluoroelastomer FKM Developments

With the realization that new exotic polymers would not easily be found, efforts to improve the FKM types were increased. Despite their high cost (over fifty times SBR) the products were very well accepted on performance, particularly for seals and O-rings in critical applications where other rubbers are inadequate. So Du Pont and 3M chemists both went to work on the urgent compression set and processability problems. A major advance in processability and compression set was soon achieved with the discovery of curing systems involving quaternary ammonium, or better, phosphonium accelerators with bisphenols, especially bisphenol AF (Figure 11) [15]. The advantages of this system are shown in the reduction of compression set (Method B) to about 15% after 70 hours at 200°C. The system was used in the new VITON® E-60 series and later in the related Fluorel 2170 products manufactured by 3M.

CURING SYSTEM

$$(C_6H_5)_3(C_6H_5CH_2)P^+Cl^-$$

BENZYL TRIPHENYL PHOSPHONIUM CHLORIDE

$$HOC_6H_4 - \underset{CF_3}{\overset{CF_3}{\underset{|}{\overset{|}{C}}}} - C_6H_4OH$$

BIS PHENOL AF

FIG. 11

An interesting mechanism for the curing reactions was postulated by Schmiegel [16]. He showed that the primary cure site was a vinylidene fluoride unit flanked by two hexafluoropropylene units. Through a series of reactions catalyzed by the phosphonium compound, a diene subject to nucleophilic attached is generated (Figure 12). The resulting cured structure, though very stable, has residual olefinic bonds and the phenol ether cross link. It is, therefore, subject to chemical and oxidative attack, and has poor resistance to steam.

CURED FLUOROPOLYMER

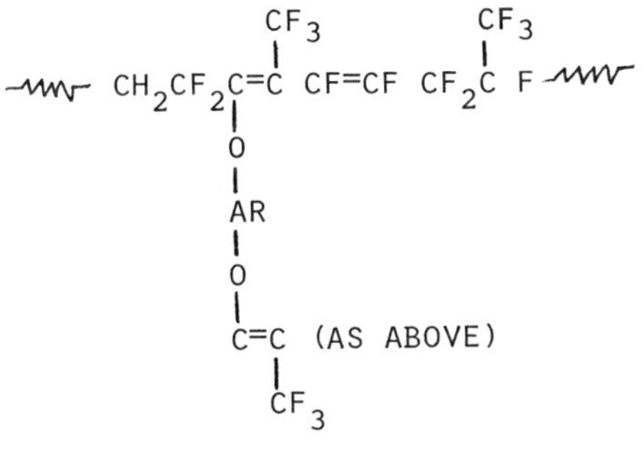

$$
\begin{array}{c}
\sim\!\!\sim CH_2CF_2\underset{\underset{\underset{\underset{\underset{\underset{\underset{\underset{CF_3}{|}}{C=C\ (AS\ ABOVE)}}{|}}{O}}{|}}{AR}}{|}}{O}}{\overset{}{C}}=\overset{CF_3}{\overset{|}{C}}\ CF=CF\ CF_2\overset{CF_3}{\overset{|}{C}}\ F\sim\!\!\sim
\end{array}
$$

FIG. 12

This problem was addressed by the development of peroxide curable polymers. While FKM elastomers do not cure easily with peroxides, good curability was achieved by introduction of peroxide sensitive substituents on the polymer chain. This was accomplished by use of a bromine containing fluoroolefin (Figure 13). Products based on this discovery were introduced under the trade name of VITON® G fluoroelastomers and included an easy processing type, a highly fluid resistant type and a specialty sol type polymer with improved low temperature flexibility[17]. With the resultant carbon-carbon cross link, the cured products are resistant to steam and hot water. This is shown in Figure 14 which compares a G type FKM elastomer with the bisphenol cured products in the Lucas Stress Relaxation test in water at 170°C. Finally, we had a product that in the cured form took advantage of the inherent chemical and thermal stability of the copolymer chain.

VITON® G TYPE

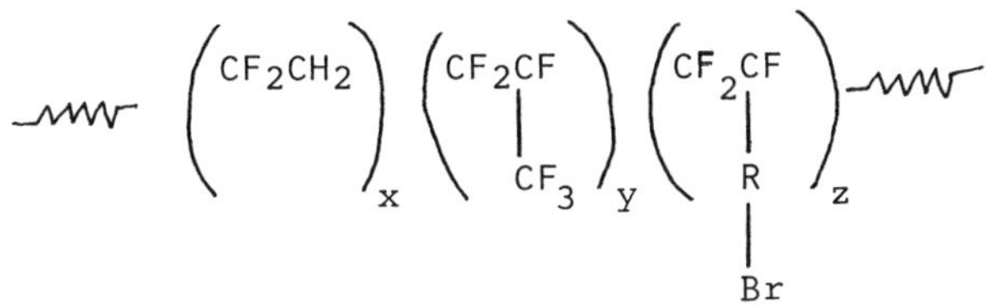

$$
\sim\!\!\sim \left(CF_2CH_2\right)_x \left(CF_2\underset{\underset{CF_3}{|}}{CF}\right)_y \left(CF_2\underset{\underset{\underset{\underset{Br}{|}}{R}}{|}}{CF}\right)_z \sim\!\!\sim
$$

FIG. 13

Other developments were the introduction by Montedison of VF2 copolymers with pentafluoropropylene-TECHNOFLON-and the commercialization by Asahi of the alternating TFE/propylene polymer under the trade name AFLAS. This old polymer had been considerably refined and improved by Brasen and Cleaver of Du Pont[18]. The product called AFLAS shows good physical properties except for its stiffness at low temperature (Tg +2°C). Though not as stable and solvent resistant as the VF2 products, its resistance to polar solvents and steam is better[19].

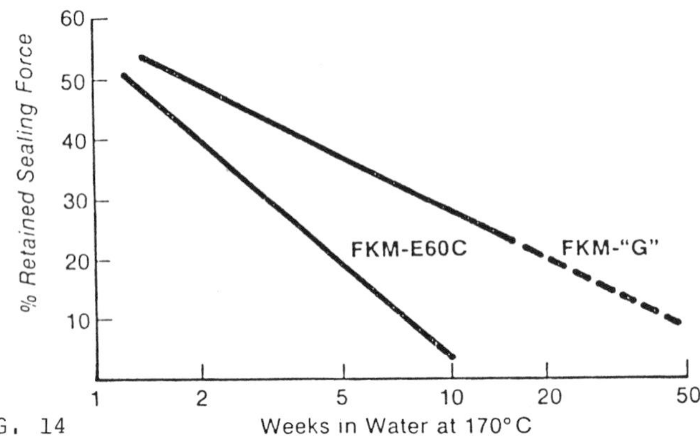

FIG. 14

Perfluorinated Elastomers

Through this period it became quite obvious that the hydrofluoroelastomers were not stable enough to meet the most rigorous requirements and that exotic chemistry was not likely to lead viable commercial products which would meet the extreme demands. Nevertheless, two new elastomer types which meet many of the needs have emerged: perfluorocarbons and perfluorophosphazenes.

The phosphazene structure represented by poly (phosphonitrile chloride) is an interesting inorganic rubber. It is well known for thermal stability and low temperature flexibility, but has poor hydrolytic stability and poor physical properties. Alcock's studies of controlled polymerization of $(PNCl_2)_3$ and subsequent reaction with alcohols led to a polymer of better, but still inadequate stability[20]. Then Rose discovered that the reaction of the polymer with fluorinated alcohols (TFE-methanol telomers) led to a rubbery polymer of great interest[21]. These structures (Figure 15) are reasonably stable hydrolytically. They also show a good balance of resistance to heat, flame, solvents and hydrolysis. Their low temperature flexibility (Tg -68°C) is equal to that of the fluorosilicones. Though they are stable only to 180°C and not equal to the FKM elastomers in mechanical properties, they are of interest in a number of applications because of their low temperature behavior. There is a reasonable possibility their properties and economics can be improved.

FLUORINATED PHOSPHAZENE

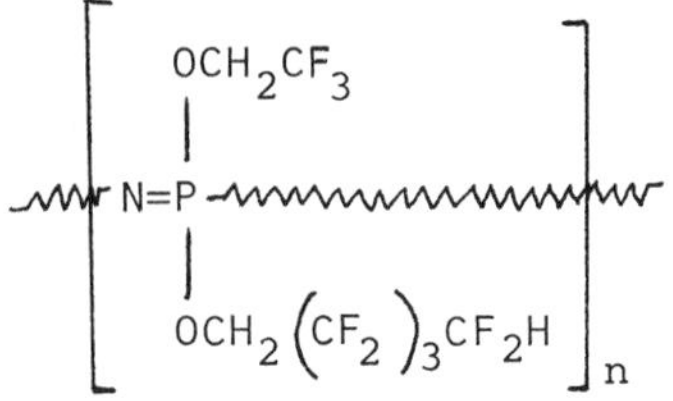

FIG. 15

Meanwhile our attention was quite naturally focused on the development of an elastomeric analog of TEFLON® poly TFE. In 1959 Du Pont chemists, H. S. Eleuterio and R. W. Meshke, had found that perfluoroolefins like hexafluoropropylene could be epoxidized[22]. Study of the chemistry of the resulting hexafluoropropylene oxide led to the synthesis of the new perfluoroalkyl perfluorovinyl ethers (Figure 16)[23]. When copolymerized in small amounts with tetrafluoroethylene, these yielded a processable plastic just about as thermally stable as poly TFE. When we heard of this, we became interested in the possibility of synthesizing a stable elastomer like TEFLON®. So we got some of the perfluorovinyl ethers and tried polymerizing massive amounts with TFE. George Gallagher was able to incorporate the ethers in amounts sufficient to make rubbery copolymers[24]. These were just as stable and solvent resistant as we had hoped, but were much too inert to cure and were impossibly low in molecular weight.

PERFLUOROETHER SYNTHESIS

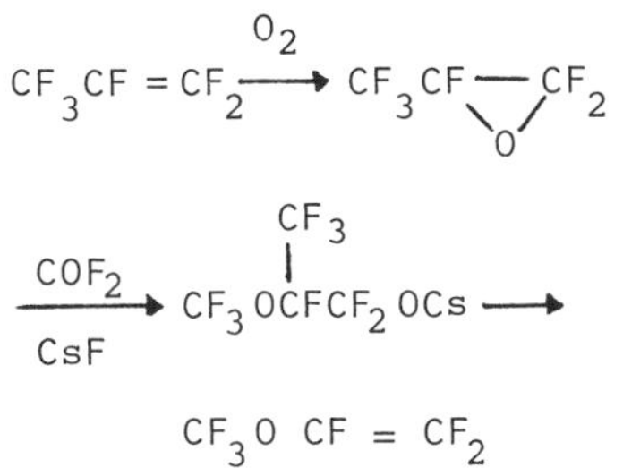

$$CF_3CF=CF_2 \xrightarrow{O_2} CF_3CF\underset{O}{—}CF_2$$

$$\xrightarrow[CsF]{COF_2} CF_3OCF(CF_3)CF_2OCs \longrightarrow$$

$$CF_3O\ CF=CF_2$$

FIG. 16

Convinced we were close to our goal, we went to work to produce a curable elastomer. It took eight years but we finally got there. First, John Keller made a good copolymer of adequate molecular weight. Then after many fruitless attempts we found some curing systems which emitted no gases during the 200°C curing reaction. These involved perfluoro monomers with pendant reactive groups like sulfonyl fluoride, cyano, and perfluorophenoxy which are hard to synthesize[25-28]. One of these discovered by Dexter Pattison turned out to be the best compromise. The new elastomer was remarkably good. Though relatively poor in low temperature flexibility (Tg -12°C; brittle point -30 to -40°C), it showed outstanding thermal and chemical stability. The preferred product is a copolymer of perfluoromethylvinyl ether with tetrafluoroethylene containing a cure site monomer (Figure 17) "PFE" and is marketed only in the form of finished products as KALREZ® parts (rods, sheets, gaskets, O-rings, etc.). It resists almost all fluids and solvents and has a continuous service temperature above 260°C. Its compression set at high temperatures is excellent and it maintains sealing force under most adverse conditions as is shown by its behavior in the Lucas Stress Relaxation test at 232°C (Figure 18).

The ever elusive goal which remains is a product which combines low temperature flexibility and good high temperature stability. There are reasonable possibilities of achieving this goal through use of enchained heteroatoms (N, O, S, P, Si) in perfluoro structures. Perfluorocarbon chains interrupted by oxygen such as in poly (tetrafluoroethylene oxide) and poly (hexafluoropropylene oxide) have the desired characteristics, but it is very difficult to develop the molecular weight needed for good physical properties[29]. Although possible in the laboratory, the polymerization is at present too difficult to be practical. A new approach to the synthesis of such structures is needed.

PERFLUOROELASTOMER (PFE)

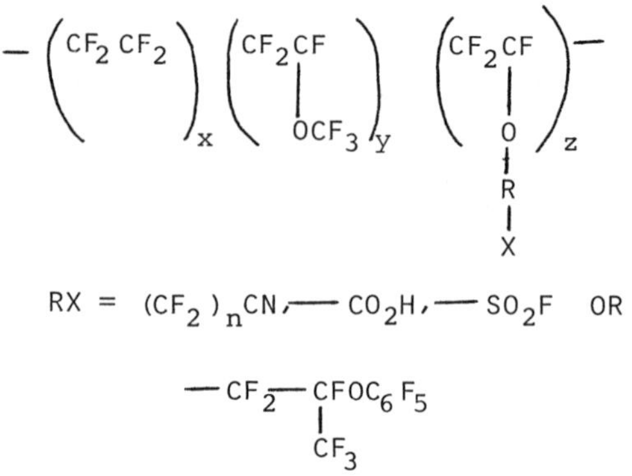

$$-\left(CF_2CF_2\right)_x\left(\begin{array}{c}CF_2CF\\ |\\ OCF_3\end{array}\right)_y\left(\begin{array}{c}CF_2CF\\ |\\ O\\ |\\ R\\ |\\ X\end{array}\right)_z-$$

$$RX = (CF_2)_nCN, -CO_2H, -SO_2F \quad OR$$

$$-CF_2-\underset{\displaystyle CF_3}{\underset{|}{CF}}OC_6F_5$$

FIG. 17

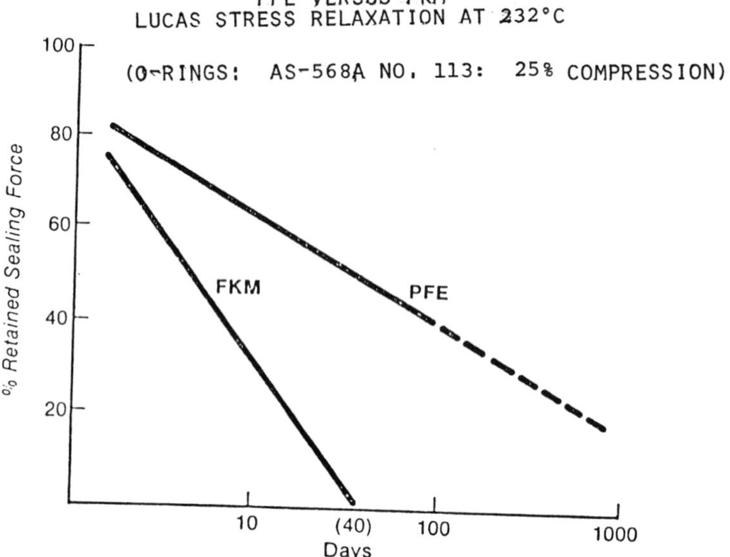

FIG. 18

REFERENCES

1. R.J. Plunkett, U.S. 2,230,654 (1941) to Du Pont.
2. W.E. Mochel et al, Ind. Eng. Chem. 40, 2285 (1948).
3. T.A. Ford, U.S. 2,458,054 (1949) to Du Pont.
4. W.E. Hanford and J.R. Roland, U.S. 2,468,664 (1949) to Du Pont.
5. H.E. Schroeder, U.S. 2,484,664 (1949) to Du Pont.
6. M.E. Conroy et al, Rubber Age 76, 543 (1955); C.B. Griffin and J.C. Montermoso, ibid. 77, 559 (1955).
7. A.F. Benning, F.B. Downing and J.D. Park, U.S. 2,406,794 (1946) to Du Pont.

8. S. Dixon, D.R. Rexford and J.S. Rugg, Ind. Eng. Chem. 49, 1687 (1957); D.R. Rexford, U.S. 3,051,677 (1962) to Du Pont.
9. J.R. Pailthorp and H.E. Schroeder, U.S. 2,968,649 (1961) to Du Pont.
10. O.K. Johannson, Can. Pat. 570,580 (1959) and U.S. 3,002,951 (1961); E.O. Brown Can. Pat. 586,871 (1959) all to Dow Corning.
11. R.G. Arnold, A.L. Barney and D.C. Thompson, Rubber Chem. and Technol. 46(3), 619 (1973).
12. D.A. Barr and R.N. Haszeldine, J.Chem.Soc. 1881 (1955); J.B. Rose, U.S. 3,065,214 (1962) to I.C.I.
13. H.C. Brown, J. Polymer Sci. 44, 9 (1960).
14. W.J. Middleton et al, J. Polymer Sci. A3, 4115 (1965); A.L. Barney et al, J. Polymer Sci. A1,4, 2617 (1966).
15. A.L. Moran and D.B. Pattison, Rubber World 103, 37 (1971).
16. W.W. Schmiegel, Kautsch. & Gummi 31, 137 (1978); Angew. Makrolmol. Chem. 76/77, 39 (1979).
17. D. Apotheker and P.J. Krusic, U.S. 4,214,060 to Du Pont; D. Apotheker et al, Rubber Chem. Technol. 55, 1004 (1982).
18. W.R. Brasen and C.S. Cleaver, U.S. 3,467,635 (1969) to Du Pont.
19. T.W. Ray and C.E. Ivy, Paper 68, International Corrosion Forum, New Orleans, LA, 1984.
20. H.R. Alcock and R.L. Kugel, J. Am. Chem. Soc. 87, 4216 (1965); H.R. Alcock et al, Inorg. Chem. 5, 1709 (1966).
21. S.H. Rose, J. Polymer Chem. Part B 6, 837 (1968).
22. H.S. Eleuterio and R.W. Meschke, U.S. 3,358,003 (1967) to Du Pont.
23. J.F. Harris and D.I. McCane, U.S. 3,291,843 (1965) to Du Pont; D.C. England et al, Proc. R.A. Welch Conf. on Chemical Research XXVI, 193 (1982).
24. G.A. Gallagher, U.S. 3,069,401 (1962) to Du Pont.
25. A.L. Barney, W.J. Keller and N.M. van Gulick, J. Polymer Sci. A1, 8, 1091 (1970).
26. A.L. Barney. G.H. Kalb and A.A. Khan, Rubber Chem. and Technol. 44, 660 (1971).
27. D.B. Pattison, U.S. 3,467,638 (1969) and U.S. 3,876,654 (1975) both to Du Pont.
28. H.E. Schroeder, Rubber Chem. Technol. 57-No.3, G86 (1984).
29. J.T. Hill, J. Makromol. Sci., Chem. A8(3), 499 (1974).

PEBAX® POLYETHER BLOCK AMIDE - A NEW FAMILY OF ENGINEERING THERMOPLASTIC ELASTOMERS

JOSEPH R. FLESHER, JR.
Atochem Inc., Polymers Division, 1112 Lincoln Road, Birdsboro, PA

ABSTRACT

This growing family of thermoplastic elastomers, known generically as polyether block amides (PEBA), is comprised of rigid polyamide blocks and flexible polyether blocks. Through the proper combination of polyamide and polyether segments, a wide range of grades that offer a variety of end-use performance characteristics are possible. In general, these polymers are characterized as having a broad range of flexibility without the use of plasticizers, extremely good low temperature flexibility characteristics, high strength and toughness. This paper covers the history of development, general chemistry and main performance characteristics of PEBA. Several particular areas of use for PEBA are also included.

INTRODUCTION

By itself, a summary of the history of polyether block amide thermoplastic elastomers would indeed be a relatively short story as these types of resins are still quite young in terms of commercial development. However, the rapid market acceptance and future growth potential for this new class of engineering resins makes the documentation of their development a worthwhile endeavor at this time. To complete the story on polyether block amides, a review of the chemistry, properties and major end-uses of these resins is also included.

HISTORY AND DEVELOPMENT

A chronological synopsis of the development of polyether block amides (PEBA) follows [1]:

1972

ATOCHEM S.A. (Paris, France) began research which was focused on development of non-plasticized flexible polymers. Being a leader in nylon 11 and nylon 12 technology, ATOCHEM was well aware of some of the drawbacks associated with plasticized resins, the most evident of which is plasticizer extraction caused by harsh chemical or high heat environments. Such extraction can hamper development of plasticized resins in certain applications, particularly in the medical and electronics industries. It was the intent of ATOCHEM to develop a line of non-plasticized polymers that would compliment its range of engineering thermoplastic resins. Since ATOCHEM was basic in nylon (polyamide) technology, including backward integration to the nylon precursors, it was decided that the research would be directed toward development of non-plasticized flexible nylons.

Copyright 1986 by Elsevier Science Publishing Co., Inc.
High Performance Polymers: Their Origin and Development
R.B. Seymour and G.S. Kirshenbaum, Editors

1974

Initial research indicated that the best route toward development of a non-plasticized flexible nylon was through the use of block copolymer technology. At the time, such technology was already being exploited. In fact, examples of several commercially available engineering resins that employed block copolymer technology could be readily cited:

Copolyether ester elastomers

Thermoplastic Polyurethanes

Styrene-Butadiene-Styrene block copolymers.

Further research showed that the most viable approach to production of an elastomeric nylon via block copolymer technology was through the combination of a polyamide and a polyether. Such reactions, using various chemical means to join the polyether and polyamide, had been documented. However, apparently none of these methods resulted in the production of a commercially viable thermoplastic elastomer.

ATOCHEM focused on the use of a dicarboxylic polyamide and a polyether diol as the building blocks for the polyether block amide. Due to the incompatability of these two segments, however, an efficient polymerization could not be achieved initially. It was not until the discovery of an appropriate catalyst that the melt polycondensation reaction could occur in a homogeneous phase. Following this discovery, Deleens [2] studied the kinetics of the reaction and the first patents were granted in 1974 [3]. Several additional patents covering the chemistry, process and certain applications have been granted since.

1980

During the period of 1974 to 1980, development of the polyether block amide technology continued. In this phase of the development, many different copolymers were synthesized using combinations of various polyethers and polyamides. A viable manufacturing process was also under development. Following successful evaluation of several potential formulations, a pilot plant was brought on-stream in 1980.

1981

By this time it was evident that a commercial market for PEBA did exist. Several applications targeted during the early development of the product appeared to be feasible and a need for commercial quantities of the resin was imminent. In 1981 a commercial production line was brought on-stream in Europe, and a unique new family of polyether block amide thermoplastic elastomers under the tradename of PEBAX® polyether block amide was formally introduced at Interplas in Birmingham, England.

1985

Following the U.S. introduction of PEBAX in 1981, a market for the product rapidly developed. Based on the apparent future demand for this new family of resins, a commercial production facility for polyether block amide thermoplastic elastomers was brought on-stream in the U.S.

CHEMISTRY

The PEBA are produced by a molten state polycondensation reaction of a dicarboxylic polyamide and a polyether diol. The reaction of the rigid polyamide (hard) segments and amorphous polyether (soft) segments in the presence of heat, vacuum and catalyst yields the polyether block amide with the general formula shown below.

$$HO \left[\overset{O}{\overset{\|}{C}} - PA - \overset{O}{\overset{\|}{C}} - O - PE - O \right]_n H$$

In this formula, PA represents the polyamide segment and PE represents the polyether segment. As shown in the general formula, the polyether block amide consists of a linear chain of polyamide and polyether segments (blocks) in an ordered pattern.

Polyamide segment

The dicarboxylic polyamide, or diacid, is obtained from the reaction of a polyamide monomer with a chain limiter. The molecular weight of the diacid is regulated by the amount of chain limiter present during the reaction. Several types of polyamides can be selected for production of the PEBA; among these are:

nylon 6	nylon 66
nylon 11	nylon 6/11
nylon 12	nylon 6/12

Polyether segment

As with the diacid, several types of polyether diols are available for the production of PEBA. For each particular type of polyether listed below, a range of molecular weights is available:

PEG - Polyethylene glycol

PPG - Polypropylene glycol

PTMEG - Polytetramethylene ether glycol

The selection and combination of the appropriate types and amounts of the polyamide and polyether blocks makes it possible to modify the physical and chemical characteristics of the polyether block amide. Due to the wide variety of polyamide and polyether blocks available, a broad range of polyether block amides can be synthesized.

Polyether block amide copolymer

The combination of the hard polyamide segment and the soft polyether segment yields a block copolymer that exhibits a two-phase structure: a crystalline phase due to the polyamide and an amorphous phase due to the polyether. The crystalline phase allows the copolymer to behave as a thermoplastic and the amorphous phase imparts elastomeric qualities.

In addition, each segment lends its own particular set of characteristics to the copolymer. To describe the characteristics donated by each segment, four factors that affect the properties of the polyether block amide have been identified:

The type of polyamide affects the melting point of the copolymer. For example, PEBA produced with nylon 6 would exhibit a relatively high melting point when compared to those produced with a nylon 6/12 copolymer. The type of nylon present also affects the specific gravity and chemical resistance of the copolymer.

The type of polyether affects the glass transition temperature of the copolymer. It also plays an important role in the hydrophilic and antistatic properties of the polyether block amide.

The molecular weight of the polyamide block also has an effect on the melting point of the copolymer.

The relative amounts of polyamide and polyether determine the hardness and flexibility of the copolymer. This allows for a range of polymers with different flexibilities to be produced using the same type of polyamide and polyether segments.

The ability to control these four basic factors allows the polyether block amide to be custom-designed to meet specific performance properties.

PROPERTIES

Due to the diversity of polyether block amide thermoplastic elastomers that can be produced, a discussion of their specific properties would be a rather lengthy exercise. Even to list the properties of all the possible combinations of a particular set of polyamide and polyether segments could prove cumbersome as the amount of either segment in the copolymer may vary between ten and ninety percent. Therefore, a discussion of the properties of PEBA can be best accomplished by citing overall performance characteristics and ranges of values. Several properties worth noting include:

Tensile properties

Current PEBA are formulated to achieve tensile strengths at break of between 29 MPa and 57 MPa with corresponding elongations at break of 715% and 430%. At low strain, the corresponding stress of the PEBA is quite high when compared to other elastomers; this high load-bearing capacity of the PEBA often allow their use in thinner wall sections, which can result in lower weight and potential overall part cost savings. In addition, the polyether block amides tend to retain their load-bearing capability at both low and high temperatures.

Flexural Properties

Currently available PEBA show room temperature flexural moduli of between 16 MPa and 370 MPa. However, this does not represent the total range of flexibility achievable with such technology; both softer (more flexible) and harder (more rigid) PEBA can be formulated. For the softer PEBA in particular, the flexural modulus is relatively constant over the range of -40°C to +80°C.

TABLE I. Physical properties of Pebax® polyether block amide.

Property[a]	ASTM Test Method	Units	Extrusion/Molding Grades 6333	5533	4033	3533	2533	Molding Grades 6312	5512	5562	4011
Density	D792		1.01	1.01	1.01	1.01	1.01	1.11	1.10	1.06	1.14
Melting Point	D2117	°C	173	168	168	152	148	205	195	120	195
Water Absorption	D570	%	1.2	1.2	1.2	1.2	1.2	6.4	6.1	3.5	119
Hardness	D2240										
Shore D			63	55	40	35	25	63	55	55	40
Shore A			--	--	--	83	75	--	--	--	--
Tensile Strength At Break	D638	MPa	51	44	36	34	29	57	48	51	32
Elongation At Break	D638	%	380	455	485	710	715	430	550	510	530
Flexural Modulus	D790	MPa	370	165	75	20	16	320	175	210	110
Impact Strength	D256-A	J/m									
+20°C Notched			NB[b]	NB	NB	NB	NB	NB	NB	NB	NB
-40°C Notched			100	NB	NB	NB	NB	35	90	90	90

a) Samples were injection molded and conditioned 14 days at 23°C and 50% relative humidity.

b) NB = No Break.

Impact Properties

PEBA show very good resistance to impact at both high and low temperatures. This is due, in part, to the low glass transition temperature of the copolymers.

Dynamic Properties

Common to most PEBA is an excellent resistance to fatigue during flexing. Due, in part, to the high resilience and low hysteresis of the polyether block amide, these properties allow the use of PEBA in applications that may undergo rapid and quite possibly major deformation. The resistance to fatigue during flexing can lead to an overall longer continual service life of a part in a given end-use.

Chemical Resistance

Generally, the chemical resistance of a particular polyether block amide is dependent on the type and amount of nylon present in the copolymer; harder grades tend to show better chemical resistance than softer grades. In certain solvents, PEBA may exhibit a high degree of swelling; this is particularly true of the softer grades. In many cases, however, the swelling is not accompanied by polymer degradation.

Thermal Properties

In their pure forms, both polyamides and polyethers exhibit their own unique melting points and glass transition temperatures. When combined into the polyether block amide, however, each component gives up one of these thermal traits to produce a copolymer with its own unique set of thermal properties. Fortunately, the PEBA retain the high melting point of the polyamide and the low glass transition temperature (Tg) of the polyether. The effects of the Tg of the polyamide, which is relatively high compared to the polyether, and the melting point of the polyether, which is relatively low compared to the polyamide, are, for the most part, not seen in the copolymer. This allows the polyether block amide to be formulated with a high melting point and a low Tg. Typical PEBA are available with melting points in the range of 120°C to 210°C. The glass transition temperature of most PEBA is in the range of -60°C to -70°C.

Other Properties

The specific gravity of a particular polyether block amide is dependent on the type of nylon present in the copolymer; typically, the specific gravity is in the range of 1.01 to 1.14.

The moisture absorption characteristics and, to some extent, the antistatic properties of the PEBA are dependent on the type of polyether found in the copolymer. Due to the diverse types of polyethers available for production of PEBA, a moisture absorption (ASTM D570) of as low as 1.2% and as high as 120% is achievable.

PROCESSING

Polyether block amide thermoplastic elastomers are processed on conventional thermoplastic processing equipment. Typical processing temperatures are in the range of 195°C to 260°C. Being a condensation polymer, the resin should contain no more than 0.1% moisture when processed to avoid hydrolysis. Most commercial grades of PEBA can be processed by a variety of conversion techniques, such as:

Injection Molding

Extrusion

Blow Molding

In some cases, however, the low melt viscosity of particular PEBA may preclude their use in profile extrusion.

APPLICATIONS

Due to the wide range of performance and processing characteristics available with PEBA, many application areas have been developed for these polymers; a few specific applications are detailed below.

High Performance Athletic Footwear

In these applications, PEBA offer the processing advantage of overmolding; using this process allows sections of different flexibilities to be joined during molding without the use of adhesives. This allows for production of a unique sole unit that is tailored for specific performance needs. The excellent abrasion resistance, flex-fatigue resistance and stable flexibility of the PEBA over a wide temperature range are also exploited.

Hose and Tubing

PEBA are used in these applications because of their high strength, chemical resistance and elastic recovery attributes. Due to the high melt strength achievable with these polymers, good dimensional control is also maintained during processing.

Molded Goods

Another application is in flexible keypads, where PEBA replace typical metal and spring composites. The use of such resins allows the keypad to be injection molded in one piece. The tactile nature of these components is maintained by varying the thickness of the resin within each key on the pad. By varying the thickness, the force required to depress the key can be altered. In this application, PEBA offer flex-fatigue resistance, excellent melt flow characteristics during molding and dimensional stability.

Powder Coatings

For these applications, the PEBA are cryogenically ground and applied to a substrate by an electrostatic powder spraying or fluid bed dipping technique. When fused, the flexible coating protects the substrate from a potentially harsh environment. The use of a low melting point polyether block amide can allow tempered metals to be coated; the temper of such metals would most likely be destroyed during exposure to the high oven temperatures required to cure other flexible coatings.

Compounding

PEBA are typically very compatible with other resin and filler systems. Some examples of compatible resins include PVC, SBS and polyolefins. In addition, PEBA can be highly filled to impart properties to the resin that are not inherent in the chemistry. In one application, a semi-conductive polyether block amide is injection molded into a car antenna. This flexible antenna, which contains no internal metal support, allows for clear AM-FM reception.

These are but a few of the many applications in which polyether block amides have found commercial use. Many others exist in such industries as automotive, wire and cable, sporting goods, hose and tubing, electronics and compounding.

SUMMARY

Although a relatively new family of engineering thermoplastic elastomers, polyether block amides have found a place in the growing family of engineering resins. The versatility of the polyether block amide chemistry and the wide variety of achievable performance characteristics has resulted in very rapid growth of this class of resins. ATOCHEM expects this growth to continue for many years to come as further research and development unveils new formulations, stabilization systems, polymerization processes and applications.

ACKNOWLEDGEMENTS

The author would like to thank G. Deleens and J.M. Huet for their assistance in the preparation of this paper.

REFERENCES

1. Discussions with G. Deleens.
2. G. Deleens Ph.D Thesis Rouen (1975).
   G. Deleens, P. Foy, E. Marechal, Eur Pol Journal Vol. 13, pp. 337-342, 343-351, 351-360 (1977).
3. G. Deleens, P. Foy, C. Jungblut French Patent 2,273,021 (1974); U.S. Patent 4,230,838 (1980); U.S. Patent 4,331,786 (1982); U.S. Patent 4,332,920 (1982).

ENGINEERING POLYESTER ELASTOMERS AND THE FUTURE FOR TPE'S

Louis F. Savelli, Donald F. Brizzolara, Du Pont Company

In the ideal world, new technologies arise in response to the definable needs of society and the marketplace. This is just another way of saying that necessity is the mother of invention.

While this may be so, it's also true that those seeking novel technological solutions are often confronted with problems that couldn't be defined or foreseen even with the best available information. And although we know that it takes quite a bit of extra vision to be a leader in any field, experience teaches us that pioneering advances frequently arise from unexpected places.

These are some of the elements that make up the story of the creation of high performance polyester elastomers at Du Pont and their acceptance in the marketplace.

Early Efforts in Urethane Chemistry

In the early 1950's, the discovery of polyurethane elastomers engendered a flurry of research activity. Within a few years we had a product, called ADIPRENE synthetic rubber, that we believed would make swift inroads into the automotive tire market. Unfortunately, although the polymer showed terrific potential for long tread wear, a field friction test - in which tires were spun against a curb - led to severe softening and decomposition of the tire. That test graphically demonstrated to us, once again, that any new flexible polymer trying to make significant inroads into traditional rubber markets would have to face the crucial issue of performance integrity at high temperatures.

The test result provided a basis for a research program, the ramifications of which are still being felt today. When we analyzed the cause of the failures, we discovered that some chain linkages in the polymer structure were unstable and unzipped to yield isocyanate and macromonomers:

RNHCO-Polymer $\longrightarrow$ RNCO + H-Polymer

This revelation spurred a search for a new kind of material, and eventually led to a new polymer based on polyurethane, but which did not have the sensitive hydrogen linkages which caused the experimental tires to fail.

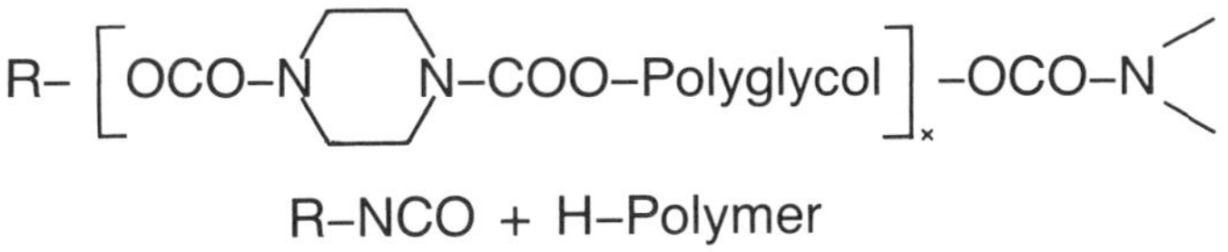

$$\text{R-}\left[\text{OCO-N}\bigcirc\text{N-COO-Polyglycol}\right]_x\text{-OCO-N}\langle$$

R–NCO + H–Polymer

Copyright 1986 by Elsevier Science Publishing Co., Inc.
High Performance Polymers: Their Origin and Development
R.B. Seymour and G.S. Kirshenbaum, Editors

Although the new polymer softened at high temperatures, it didn't decompose. This led us to consider it as a candidate for another search we were engaged in, the search for a family of materials which behaved like rubbers but could be fabricated using efficient thermoplastic processing. Although the urethane product we were testing had excellent properties and was a thermoplastic, it was prohibitively expensive as a replacement for rubbers.

Dr. Rudy Pariser, a Du Pont researcher, suggested that we try using bifunctional monomers, such as those used in polyesters and polyamides, to create a chemistry that was economically viable and might be functionally similar to the urethane. At the same time, similar work was being done by our Textile Fibers research group in search for a new elastic fiber.

This effort bore fruit in 1968, with the production (under the direction of Bill Witisiepe) of the material known today as HYTREL® – a polyether-ester copolymer which has excellent thermal and melt stability, and combines desirable elastomeric properties with thermoplastic processability. Because this completely new resin was a block copolymer of randomized hard and soft segments, it offered a combination of resilience and strength that had never been seen in any thermoplastic. And it is this combination that today positions polyester elastomers perfectly for economic replacement of parts currently made from other flexible materials.

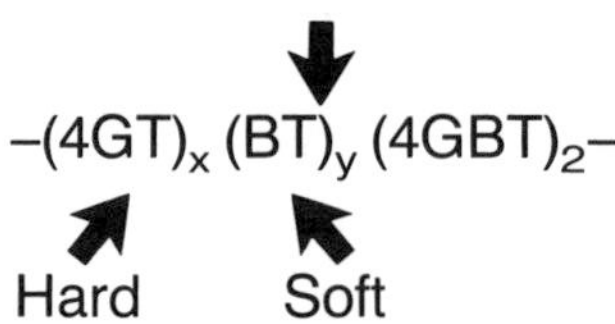

Ratio x/y/z Controls Modulus

Early Market Experience with Polyester Elastomers

In our early marketing experience with HYTREL, we first positioned it as a competitive replacement for medium to high performance specialty rubbers. Then, as now, rubber products were positioned on the basis of selected properties to fit particular applications. As a result, HYTREL was defined primarily as a rubber, albeit one with thermoplastic processing characteristics, and it was marketed on the basis of its elastomeric properties and heat and chemical resistance.

The earliest successful applications with this approach were in a major market segment of the rubber industry – hydraulic hose and tubing. We also replaced thermoset elastomers in applications such as molded flexible couplings, air bladders for sports balls, shoe parts and other traditional rubber applications.

During this early period, one of the greatest challenges we faced came from designers, engineers and manufacturers in the rubber industry, who were skeptical about the ability of a thermoplastic to perform in hitherto untried areas. They asked a number of difficult questions: Don't plastics melt? Aren't they too stiff for truly flexible applications? Will they creep under sustained load? Are they resilient enough to resist failure during flexing?

As it turned out, we were able to demonstrate that polyester elastomers did have the elastomeric "guts" to perform as an elastomer under even the most demanding conditions.

Polyester elastomers proved themselves on their own merits, persuading a reluctant market that these high performance engineering plastics could, and should, be used in products that heretofore had relied on conventional rubbers.

Heat and Oil Resistance of Automotive Elastomers
(Based on SAE J200 Specification System)

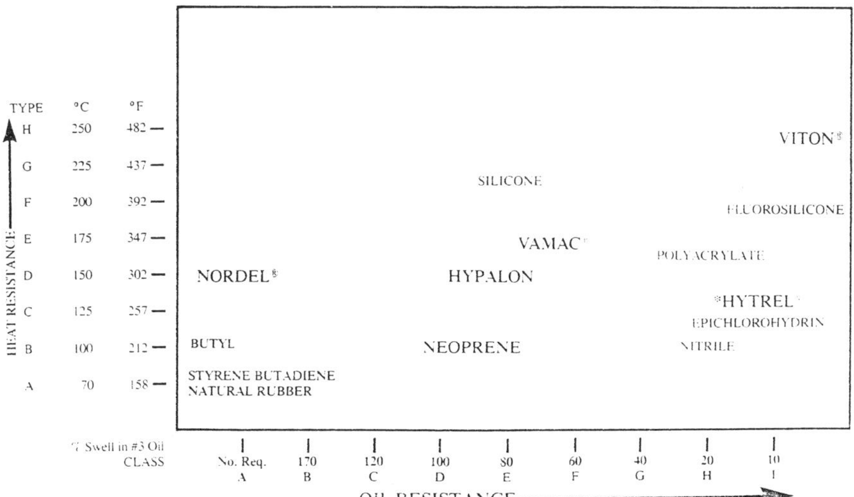

EXAMPLE: The first two steps in any elastomer specification are to designate the *Type* (heat resistance) with a single letter, such as "B" for 100°C (212°F) and to designate *Class* (oil resistance) with a second letter, such as "C" for 120% (maximum percent volume swell in ASTM No. 3 oil at 100°C).

Thus, an elastomeric application with these requirements would be designated as "BC". *HYTREL® is a thermoplastic elastomer and, as such, is not included in the thermoset elastomer SAE J200 specification system. It has been positioned in this chart to demonstrate its resistance characteristics to oil and heat.

With the benefit now of 15 years of experience, it's easy to view that marketing approach as conservative. At the time we couldn't envision the manifold opportunities to use polyester elastomers in applications that were well beyond the reach of rubbers. But, as was cited earlier, some important advances come out of unexpected places.

During the product development phase of TPE hose, it became apparent that a new kind of seal had to be fashioned to join the hose to its metal coupling. Simply put, in the process of commercializing the thermoplastic hoses, we were made very much aware that TPE's and rubbers were not conceptually the same. This experience led to an expanded program to assess the properties of high-performance thermoplastic elastomers. Subsequent research efforts were devoted to finding ways to redesign applications so to make best use of the material characteristics. This led us to work more closely with plastics designers and processors on unique ways to apply polyester elastomers.

Important New Applications for Polyester Elastomers

The principle of engineering redesign for improved performance and economics has led to a wealth of opportunities for polyester elastomers, with remarkable commercial successes in some areas in which rubber dominated, and other areas in which using rubber was simply not feasible.

For example, HYTREL is now the standard material for CVJ boots in American automobiles. This represents an instance of dramatic redesign to take advantage of strength, resilience, economics and durability of the block copolymer.

It is produced at one half the weight and with four times the tear strength of rubber boots, and can be processed on plastics blow molding equipment in one twelfth the time. Car makers are now saying that the boot of HYTREL lasts six times longer than the older rubber designs. What makes this story so powerful is that a polyester elastomer has succeeded, far beyond the capability of any rubber, in an application where many "experts" insisted that no thermoplastic could survive.

Because polyester elastomers combine strength and flexibility, they are ideally suited for applications requiring spring action, recovery from deformation, resistance to high impact and the ability to withstand repeated flexing. They are now being used as heavy duty shock absorbers, actually replacing hydraulic systems, with a completely new concept in shock absorber design. A 170-ton earth mover, designed to transport materials at mine sites, uses shock absorbing cylinders which are filled with discs of HYTREL. Field use demonstrates that shock absorber performance failure rates have been significantly reduced and maintenance nearly eliminated. Another similar application is the Daimler-Benz bumper mount where HYTREL acts as a shock absorbing member.

The Yamaha portable headphones were designed with the unique properties of HYTREL in mind, in conjunction with the advantages of other thermoplastics such as DELRIN® and ZYTEL®. The polyester elastomer spring side pieces, replacing metal and built into the headband, cause the assembly to coil when not in use, with the springs rolled protectively around the earpieces. Only a high performance elastomer could meet the requirements of this design.

Ordinary rubbers would not provide the necessary high strength and tear resistance. Perhaps more significant, conventional rubbers simply do not meet the demands of this product for aesthetic consumer appeal. It is probably fair to say that, had thermoplastic elastomers never been discovered, these headphones would have never been produced.

These examples illustrate some key features about the historical development of polyester elastomers. Based on a rather limited view of what high performance polymeric materials could do, they were originally thought of as a new class of rubber. However, once a broad program of research and development was underway based on a new understanding of the market, it was discovered that the material could be used in unique ways, unanticipated either by us, developers, or our customers and end-users.

The Future

While a process of iterative redesign is often the only way to proceed when considering the use of high performance polyester elastomers, completely new thermoplastic elastomers are being invented with a range of properties and cost savings which permit them to be substituted directly for conventional thermoset rubber.

Examples of these products are dynamically cured EPDM-polypropylene and chlorinated ethylene copolymer TPE's recently introduced by Monsanto and Du Pont under the tradenames SANTOPRENE and ALCRYN®. These new products are opening up new application areas for TPE's by providing low cost, direct substitutes for medium performance thermoset rubbers. Work in this product area is expanding rapidly and many new products are expected to be introduced in the near future.

It's clear that all this activity has important consequences for the rubber and plastics industries. It has been a long time since there has been a comparable explosion of possibilities in the field. If the future of thermoplastic elastomers can be extrapolated from its history, we may expect to see a continuing cycle of growth, discovery, development and design that will lead to the replacement of existing materials like thermoset elastomers, but also to totally new applications never before possible. The growth of TPE's has been phenomenal in the past 15 years, but with the introduction of new products and the development of deeper understanding of how to use the new technologies, the future looks even brighter than the past.

High Barrier Packaging Materials

PET -- A GLOBAL PERSPECTIVE

Nathaniel C. Wyeth
Retired Senior Engineering Fellow & Consultant
E. I. Dupont De Nemours and Company

I would like to begin my remarks with a disclaimer. The title of my speech -- "PET: A Global Perspective" -- was provided by our public relations people and frankly covers more ground than I am prepared for today or, for that matter, ever. I will, however, try to cover some of the ideas implied by that ambitious title, including:

o A brief update on where PET technology stands today, and where it is headed;

o An account of the events leading to the invention of the PET bottle which spanned nearly two decades and, I am afraid, are far less glamorous than I care to admit;

o And finally, I'd like to offer a few concluding thoughts about creativity and engineering -- two subjects I am still learning about.

Since the PET bottle was plucked off the lab bench and went into production in the late seventies, PET's explosive growth has continued to outstrip everyone's expectations -- including my own. Some analysts have gone so far as to say that the PET bottle is the most significant invention in plastic packaging during the postwar era.

The PET bottle market grew rapidly to 600 million pounds per year following the introduction of the 2-liter soft drink bottle in 1978. Today, there are more than 6 billion PET bottles produced worldwide each year, and we fully expect that number to double within the next five years as PET edges its way into liquor, wine, food and even beer markets.

For several years, these other growth areas have been essentially stagnant, awaiting improvements in barrier technology, hot-fill technology and bottle-making equipment. Those technology improvements are now occurring and the PET bottle industry appears poised for another period of rapid growth, spurred initially by the 1/2-liter soft drink bottle.

Indeed, one marketing research firm, Predicasts, forecasts a steady 14 percent per year growth rate through the late eighties -- with a conservative estimate of 27.6 billion PET containers produced a year by 1995.

Demand for PET resin will grow similarly to about 1.2 billion pounds per year by 1990 (a 15 percent per year growth rate).

What I find most gratifying is that PET technology has achieved truly global acceptance. Du Pont now licenses the technology to almost 70 PET container manufacturers in 12 countries, and the numbers are still growing.

Published 1986 by Elsevier Science Publishing Co., Inc.
High Performance Polymers: Their Origin and Development
R.B. Seymour and G.S. Kirshenbaum, Editors

At the risk of overlooking some important developments, I'd like to briefly highlight some of the new directions for PET containers.

First, let's look at markets.

As I indicated earlier, the 1/2-liter soft drink bottle is the driving force for the rapid growth beginning in 1986. The 1/2-liter bottle got off to a shaky start because of repetitive problems with the initial bottle-making equipment and because of concerns that an 8 to 10-week shelf life was insufficient. Recently, however, the equipment problems have been resolved, and market tests have shown the 8 to 10-week shelf life to be satisfactory. Current expectations call for very rapid growth -- from the current level of 1 billion units per year to about 8 billion units per year in 1990.

The 1-liter bottle is expected to see steady, sustained growth at about 15 percent per year and will reach about 1.1 billion units per year by 1990.

The 2-liter bottle market is essentially saturated and growth is expected at a 2-3 percent per year rate. The 3-liter bottle, introduced last year, is getting mixed reviews, and growth is likely to be at the expense of the 2-liter bottle.

Liquor is still a largely untapped market. The 1.75-liter liquor bottle is beginning to enjoy a fair degree of success, but may face substantial competition from PVC which was recently approved for use.

The 50-milliliter PET liquor bottle has met with great success and currently holds about half of this market. The major factor is a weight reduction of about 80 percent versus glass. The estimated fuel savings alone have been set at $25,000 a year or more for the typical passenger airliner. Penetration of this market is expected to continue, with volume reaching about 180 million units per year by 1990.

In wine, as in liquor, there remains some perception of consumer resistance to plastic, but the 1.5-liter, 3-liter and 4-liter wine bottles are showing steady growth. While I seriously doubt we will be toasting one another with Pommerol, Chateau Petrus or some of the expensive California chardonays from PET bottles, the opportunity for large PET bottles for the wine market remains substantial.

Beer, of course, is the most tempting target of all. About 50 billion containers per year are currently used in the beer market -- for an ultimate potential of about 2 billion pounds of PET resin per year! Even a 10 percent share of this market is a worthwhile target. The obstacles -- including oxygen barrier, CO_2 retention, hot-fill capability and consumer acceptance -- are significant, but are slowly being overcome. Beer is already being packaged in PET containers in Europe, but European brewers pasteurize their beer in a flash pasteurization process outside the container and have a relatively short turnover time from brewing to consumption. Apparently Europeans drink more beer and drink it faster than Americans! U.S. brewers (with the exception of Coors) pasteurize beer in the individual container. Thus a PET bottle must be able to withstand the heat of pasteurization -- about 66°C.

A recent European development, known as the Petainer plastic can, reportedly is able to withstand this temperature. A joint venture called Petainer Development Company, comprised of Metal Box of England, PLM of Sweden and Sewell Plastics of the U.S., is building a pilot plant in Atlanta. This plant is scheduled to start up in May or June and will produce about 100 million cans annually.

Coca-Cola has exclusive rights for 5 years to market carbonated soft drinks in the Petainer can. Soft drinks accounted for 30 billion cans last year, which represents an opportunity for PET of over 1 billion pounds of resin per year. My friend, Jack Pollock, of Sewell summed it up very nicely: "In our view, all things being equal, a consumer would prefer to see the contents of a beverage being consumed."

Perhaps the most exciting opportunity of all for PET containers is food. PET bottles have already been introduced for mustard, pickles, honey, edible oils, salad dressing, peanut butter and juices.

Driving the aggressive entry of PET into food markets are the truly innovative new technology developments that are leading to improved barrier properties and improved processing properties -- in both the food and packaging industries.

I would be remiss if I neglected to mention that Du Pont has research under way on barrier resins to enhance the oxygen and CO_2 barrier properties of PET containers. In fact, Du Pont has developed a family of resins under the trademark "Selar" for a wide range of barrier needs in both food and non-food packaging.

As barrier technology improves, we expect a growth rate of 10 percent per year for this market -- with an upside potential of as much as 20 percent per year.

Any effort to describe PET markets and technology would fall far short of even the public relations people's rosiest forecasts if I didn't mention recycling.

I personally believe recycling is a vital issue -- whether we are talking about ultimately reclaiming the polymer in the containers, or simply burning the containers after they are used to obtain the fuel value of the feedstocks.

Last year, 100 million pounds of post-consumer plastic soft drink bottles were recycled in the U.S. -- about 20 percent of all the PET in the bottle market. According to the Plastic Bottle Institute, by 1988, more than 230 million pounds of soft drink bottles will be recycled annually into new products.

Most of the development work is being carried out by the Plastics Recycling Foundation, a non-profit group founded in 1984 and funded by the plastics industry, packaging and soft drink companies.

In short: we are very bullish on the outlook for PET recycling, PET markets, and PET technology.

If anything, I hope I have communicated at least the flavor of our enthusiasm for the future of PET or, as a recent edition of the London Economist put it, "Wyeth's PET's." The enormous sense of optimism I have about the future of PET is matched with an equal dose of nostalgia about the effort that led up to our discovery.

My fascination with the problem began in the early 1950's when I was a young engineer at Du Pont. It occurred to me that if you could orient molecules in yarn or film -- just as Wallace Carothers did when he invented nylon at Du Pont in the 1930's -- why couldn't you apply the same techniques to plastic bottles?

With this in mind, I ran some very crude "cold" blown bottle tests in the 1950's. The results were disastrous. They convinced my laboratory supervisors that this would be a very difficult undertaking at best -- and at the same time gave me a great deal of respect for the "problem."

Nevertheless, I persisted in spending a great deal of personal time on the notion that if I could only produce a container which had the self-equalizing properties of a rubber balloon -- I would have the answer. Meanwhile, I kept trying to convince others at Du Pont that we should keep working on the bottle because of the potential for plastics to displace glass in the immense glass soda bottle market.

Finally, in mid-1961 Du Pont's plastics department agreed to finance my research and assigned two very bright young engineers to work with me.

Our first trial approach was to "cold" extrude a thick walled tube of linear polyethylene over a bottle-shaped mandrel. The results were encouraging. We had a number of quite successful runs, but failed to deal with one minor problem which -- frankly -- detracted from our effort in management's view. We had yet to produce a bottle with a bottom!

The know-how we developed through this approach, however, was patented and used later successfully in manufacturing operations at Remington Arms to produce plastic shotgun shells.

From 1962 to 1968 I worked on a variety of products and processes ranging from textile fibers to films, and even to electronics -- all the time looking for an opportunity to get back to my "pet" project: making the bottle -- although this time I was going to go for a bottle with a bottom.

By late 1968, we were back in business again.

My first approach was to "cold" blow the bottle to produce an oriented small neck container with a bottom.

The early results were primitive at best. We were clamping a sample of high density polyethylene sheet about 1/16th inch thick in a single cavity mold, then heating the entire assembly to just below the crystalline melt point. Finally we would remove it from the oven and blow it with compressed air. Invariably, the plastic sample would rupture.

We then raised the oven temperature to the melt point of the sample, and finally got a fully blown shape without a rupture.

We still failed to get enough orientation, however, because we were working at temperatures which were too high.

Eventually, we were able to successfully blow two sheets completely, which were at a 90 degree angle from the extrusion axis, below the melt point, and get orientation without a rupture.

The results of this twin sheet cold blow molding test led the way to what later became known as "SOM", or the slug oriented molding process. We had a successful combination of "cold" extrusion and blow-forming.

We now had a process which would produce a biaxially oriented bottle out of either high density polyethylene or polypropylene. But two significant and seemingly insurmountable obstacles remained: neither material had adequate strength or creep resistance to handle the pressure of carbonation or sufficient impermeability to contain CO_2.

So we started experimenting with new materials. Finally, another breakthrough came. Frank Gay, a research chemist at Du Pont's Experimental Station in Wilmington, Delaware, suggested that we use a material we hadn't considered: polyethylene teraphthalate or PET resin. PET hadn't been tried before because we did not believe it could be oriented in the form of thick-walled shaped articles.

The first time we ran PET resin, our little machine groaned and grunted like it never had before. Our fear was that we not only were going to fail to produce an acceptable bottle again, but that the experiment would damage our test equipment. When we opened the mold, all that remained was a small pile of granular PET resin.

We then tried a fully amorphous preform.

I remember the mixture of expectation and depression I felt when we opened our molding machine. It was late in the evening, and there was little cross light hitting the mold.

After months of frustration during which we had produced nearly 10,000 bottles by hand, we had grown used to seeing blobs of resin caked on the mold, or crude shapes that looked nothing like a bottle.

This time, at first glance, it looked as if the mold was empty.

A closer look revealed something else: a crystal clear bottle.

Since then I have seen countless truly beautiful PET bottles. But none of them will ever be as memorable as the first.

The invention of the PET bottle also drove home an invaluable lesson which my father spent years trying to teach me while I was a young man. Father used to say, "An engineer is just as much an artist as a painter."

Father, needless to say, was an artist. Indeed, almost everyone in my family is an artist. My brother Andy is. So are two of my sisters, my late wife's mother, brother and sister, my brothers-in-law, one of my sons, and of course, my nephew Jamie.

Perhaps the difference between creativity in engineering and art is that our "paintings" also have to work.

Today, we face no shortage of challenges, from formulating new resins -- as Arnold Collins did when he and Carothers invented neoprene -- to designing needed machinery that will provide four, five or even tenfold improvements in productivity. It might interest you to know that Du Pont announced the invention of neoprene at the 1931 ACS meeting in Akron.

I remain convinced that the future of plastic packaging will belong to those individuals and those companies who are the most creative.

So, with this in mind, I would like to share three notions about creativity which I learned while working on the PET bottle.

First, one of the most fundamental elements of the creativity process is recognizing the true nature of the problem. As Einstein said, "The formulation of a problem is often more essential than its solution, which is mathematical skill." Often we waste energy, to say nothing of time and money, by trying to solve the wrong problems. In retrospect, this was certainly the case with PET.

Second, another notion which I believe is essential to creativity -- in art or engineering -- is the role of the subconscious. I'm sure all of you have experienced that feeling that your best ideas surface while taking a shower, or waking from a nap. I once had a brilliant engineer working with me who insisted his best ideas came to him while he was napping.

I mention this not because I am suggesting that every engineering lab should be equipped with beds, but rather that we ask ourselves whether we offer our colleagues or employees an environment which nurtures creativity.

Another dimension of the environment question is the ability to concentrate. I suspect some people inherit this; others no doubt develop it.

It may have been an accident, but one of the most creative designers I ever worked with at Du Pont had very poor hearing and kept his hearing aid turned off most of the time. You can't help wondering whether Thomas Edison's deafness was a factor in his great inventiveness. Did you know that he was granted over 1,000 patents?

Third, and finally, I'm convinced that the best solutions are often the ones which are counterintuitive -- that challenge conventional thinking -- and end in "breakthroughs."

Nylon, the silicon chip, the transistor, the helicopter, and polio vaccine -- all in their fashion -- were discoveries which challenged conventional wisdom, but seem to us today to be as obvious as a natural phenomenon like gravity.

Creativity -- in a phrase -- demands daring.

I offer these concluding remarks about creativity not because I am urging you all to start sleeping on the job, wearing earplugs, and challenging authority -- but because I believe we all need to ask ourselves how to nurture creativity in engineering and in our businesses.

I believe that the future of plastics in packaging is assured and that the question is not "if" plastics will displace metal and glass in global packaging markets but "when".

The winners in this packaging revolution will be those companies and individuals who can harness creativity in pursuit of untapped market opportunities.

For, while PET occupies an ever-growing foothold in worldwide packaging markets, the truly exciting inventions are yet to come.

ETHYLENE VINYL ALCOHOL COPOLYMERS

A. L. BLACKWELL

Eval Company of America, P.O. Box 3565, Omaha, Nebraska

INTRODUCTION

Ethylene vinyl alcohol copolymers generically are known as EVOH. EVAL® is a trademark for EVOH. The mark is owned by Kuraray Co., Ltd., and licensed to EVAL Company of America (EVALCA) for manufacture and sale of EVAL® resins. EVALCA is a 50/50 joint venture company between Norchem, Inc. of Omaha, Nebraska, and Kuraray Co., Ltd. of Osaka, Japan. EVALCA will market EVAL® resins in North and South America, including Central America, Canada and the United States.

EVALCA is currently building a 22-million pound EVAL® resin manufacturing facility in Houston, Texas. Completion date is July 1986. This facility will be capable of producing all commercial grades of EVAL® resins presently manufactured by Kuraray in Japan. EVALCA is currently importing commercial quantities of EVAL® resins from Japan. All commercial grades are readily available for immediate shipment in commercial quantities from U. S. warehoused stocks. Many development grades are also available for immediate shipment in less-than-truckload quantities, also from U. S. warehoused stocks.

History

Early U. S. patents referencing EVOH resins appeared in 1945. Process and application patents continued to issue in the 50's, 60's and 70's.

Kuraray patents first appeared in the 60's and have continued to issue periodically ever since. Commercialization of EVAL® resins began in Japan by Kuraray in the early 1970's.

Introduction of EVAL® resins into the U. S. began in the mid-1970's. EVALCA was formed in October 1983, and the EVALCA plant was announced in July of 1984 for completion in July of 1986.

Chemistry

EVAL® resins are copolymers of ethylene and vinyl alcohol. Since vinyl alcohol does not exist in the free state, the raw materials for EVAL® resins are ethylene and vinyl acetate. The basic chemical equations for the syntheses of EVAL® resins are shown as follows:

Copyright 1986 by Elsevier Science Publishing Co., Inc.
High Performance Polymers: Their Origin and Development
R.B. Seymour and G.S. Kirshenbaum, Editors

$$CH_2\text{-}CH_2 + CH_3\text{-}\overset{\overset{\displaystyle O}{\|}}{C}\text{-}O\text{-}CH{=}CH_2 \longrightarrow$$

Ethylene Vinyl Acetate

$$\left(CH_2\text{-}CH_2 \right)_m \left(CH_2\text{-}\underset{\underset{\displaystyle O\text{-}\underset{\underset{\displaystyle O}{\|}}{C}\text{-}CH_3}{|}}{CH} \right)_n \xrightarrow[HOH]{NaOH}$$

Poly (Ethylene Vinyl Acetate)

$$\left(CH_2\text{-}CH_2 \right)_m \left(CH_2\text{-}\underset{\underset{\displaystyle OH}{|}}{CH} \right)_n + CH_3\text{-}\overset{\overset{\displaystyle O}{\|}}{C}\text{-}OH$$

Poly (Ethylene Vinyl Alcohol) Acetic Acid

The process flow diagram for manufacture of EVAL® resins, published by H. Iwasaki, K. Sato, K. Akao and K. Watanabe, is shown as Figure 1.

FIGURE 1

Process Flow Sheet For Production Of Ethylene Vinyl Alcohol Resin

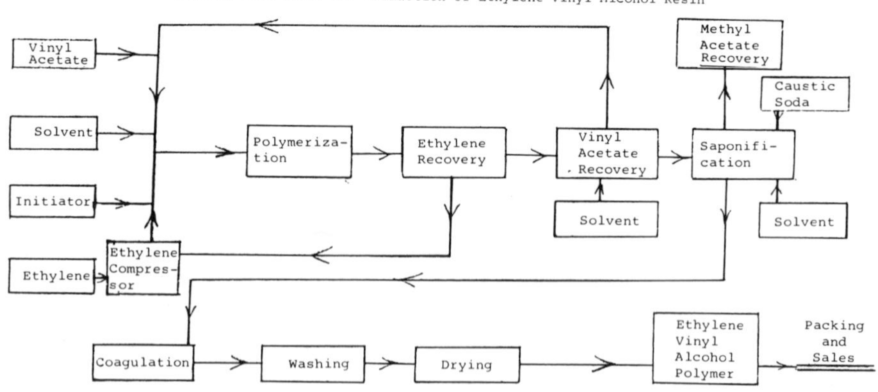

TABLE 1

RATIO OF OXYGEN TRANSMISSIONS RATE OF SELECTED POLYMERS COMPARED TO EVOH

	5°C	23°C	35°C	50°C
EVAL® EP-F	1	1	1	1
Saran 5253 PVDC	8	15	17	23.5
Barex 210 PAN	100	80	80	75.7
Oriented Nylon 6	327	178	132	--
Non-oriented Nylon 6	960	508	400	--
Oriented PET	440	230	204	207
Oriented Polypropylene	--	16,300	8,120	--
HD Polyethylene	--	15,000	11,480	--
LD Polyethylene	--	55,400	29,300	--
Polystyrene	--	26,000	--	--

Properties of EVAL® Resins

Gas barrier is, without question, the single most important property of EVAL® resins. EVOH resins are the most effective oxygen barrier polymer known. Today, it is this property of oxygen barrier which accounts for the rapid penetration of EVAL® resins in the packaging industry. As shown in Table 1, it is 16,000 times better than oriented polypropylene (PP), 200 times better than oriented nylon, 80 times better than polyacrylonitrile. EVAL® resin provides up to 15 times the oxygen barrier property as does polyvinylidene chloride (PVDC) which, for many years, has been the standard of the industry.

Oil and solvent resistance of EVAL® resins is excellent. Thus, EVAL® resins are suitable for packaging oily foods, edible oils, agricultural pesticides and organic solvents.

High gloss and low haze result in excellent optics for package enhancement.

Antistatic properties may provide applications in electronic areas.

Weather resistance, light stability and printability are excellent. They are FDA and USDA approved for direct food contact. They are also acceptable for direct food contact in Canada. EVAL® resins are non-toxic and can be recycled for use in multi-walled structures. EVAL® resins can be used in retortable food containers. They are also used to produce hollow fibers for kidney dialysis machines.

EVAL® resins are moisture sensitive and show reduced barrier properties at high humidities as shown in Figure 2. The barrier properties of EVAL® resins at 85% R.H. are equivalent to those of high barrier Saran. For this reason, most EVAL® resins are used in multi-layer coextruded structures with polyolefin materials. These are usually five-layer structures consisting of polyolefin/adhesive/EVAL® resin/adhesive/polyolefin.

FIGURE 2

OXYGEN TRANSMISSION RATE VS. RELATIVE HUMIDITY

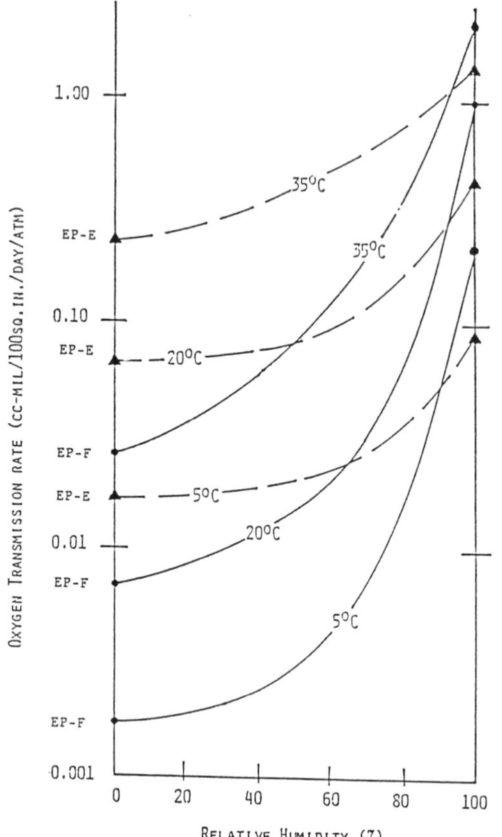

Figure 3 compares the oxygen transmission rate versus relative humidity at 77°F with EVAL® EP-F, EVAL® EP-E and Saran 5253 PVDC.

FIGURE 3

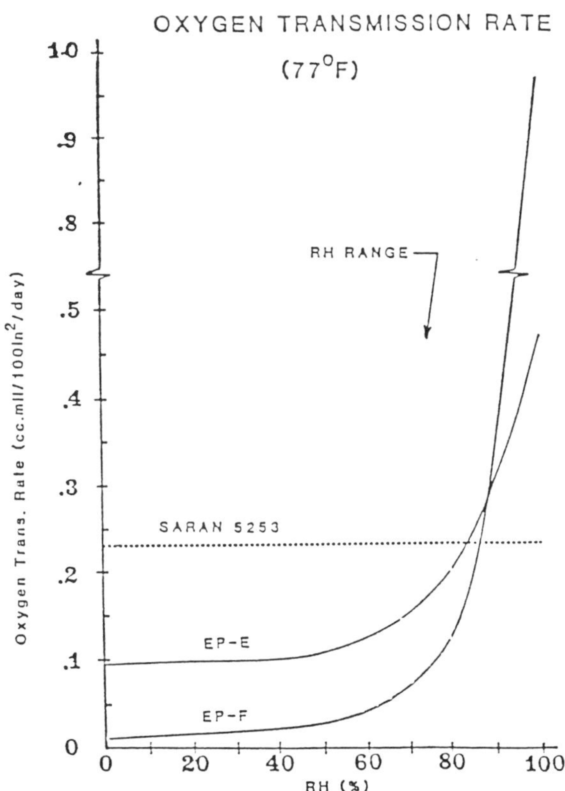

Figure 4 is an enlargement of that area of confluence. It dramatically demonstrates that EVAL® EP-F has the highest oxygen barrier properties until R.H. in excess of 85% is reached. At 77°F, the oxygen barrier properties of EVAL® EP-E and EVAL® EP-F become equal at 90% R.H. or above.

FIGURE 4

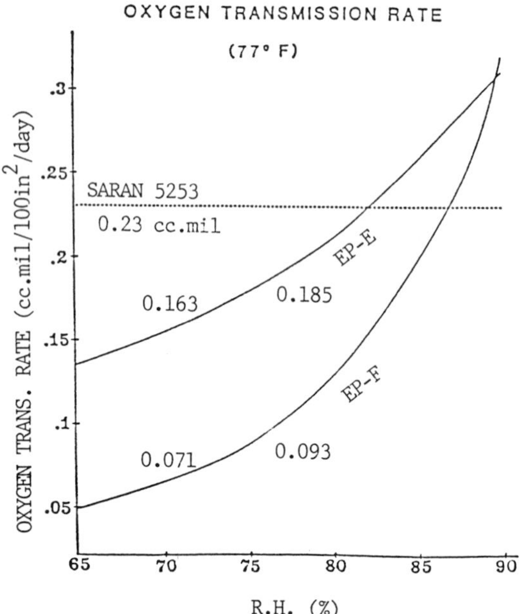

Although PVDC resins have been around since the 1930's, a period of 50 years, it has been just since the availability of EVOH resins over the past ten years that high barrier plastics packaging has grown so rapidly.

The technology of EVAL® resins can be summarized as shown in Figure 5. They are copolymers containing only polyvinyl alcohol (PVOH) and polyethylene (PE).

As the mol ratio moves in the direction of PVOH, the properties become more like those of PVOH. Moisture vapor transmission rate (MVTR) and moisture sensitivity increase, specific gravity increases, melting point increases, and processing sensitivity increases. Oxygen barrier increases, gas transmission decreases and stiffness goes up.

FIGURE 5

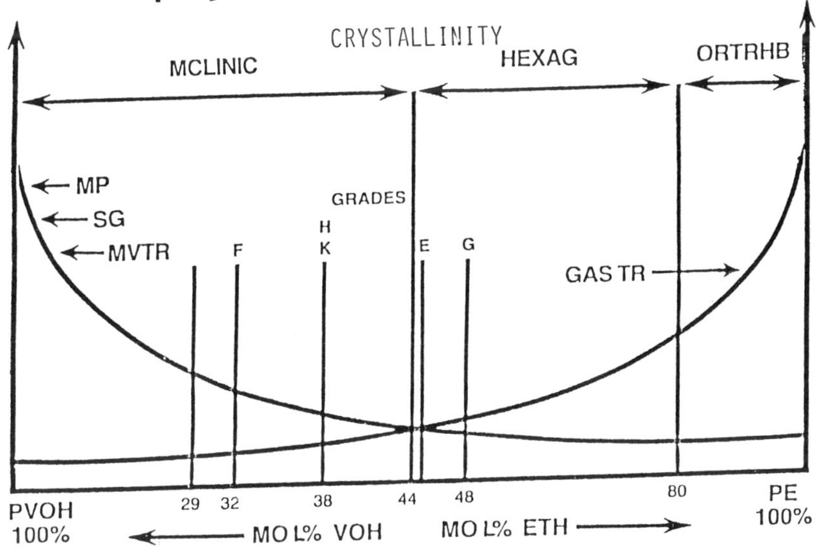

Crystallinity, form and degree have a great deal to do with the barrier level achieved. It also dictates temperatures to be utilized in the thermoforming process. The monoclinic type crystals on the left are most like those of polyvinyl alcohol (PVOH). That is, they are small, densely packed, high gas barrier producing entities. As you may realize, PVOH is an excellent gas barrier material, but its water solubility properties make it unsuitable for most food packaging applications. At about 41 - 42 mol percent ethylene, the crystal form changes and becomes hexagonal in nature. These are larger crystals, less densely packed entities, and consequently, have lower gas barrier properties. Above 80 mol percent ethylene, the crystal type changes again. Here it becomes orthorhombic in nature and exhibits properties more akin to those of polyethylene. It has lower gas barrier properties and better moisture-resistant properties when compared to lower ethylene content EVOH materials.

In this chart, the critical area for thermoforming is at the 44 mol percent ethylene level. At 32 mol percent ethylene, EVOH crystallizes most rapidly at 161°C. This is the ideal temperature for solid-phase pressure forming (SPPF) of polypropylene. Polypropylene is one of the more popular structural materials for EVOH resin barrier containers. EVOH resins above 38 mol percent ethylene can be formed by SPPF methods. At levels below 38 mol percent ethylene, melt-phase forming at temperatures above 161°C is employed. EVAL® resins at 38 - 44 mol percent ethylene are specially formulated for SPPF processing.

EVAL® resin grades presently available are as shown between 29 - 48 mol percent ethylene.

Typical properties of EVAL® resins are shown in Tables 2, 3 and 4.

TABLE 2

TYPICAL PROPERTIES OF EVAL® RESINS

ITEM	UNITS	GRADE EP-F101A	EP-H101A	EP-E105A
ETHYLENE CONTENT	MOL %	32	38	44
MELTING POINT	°C	181	175	164
MELT INDEX	g/10 MIN	1.3	1.6	5.5
DENSITY	g/cc	1.19	1.17	1.14
O_2TR	(1) cc. $15\mu/m^2$ 24 hrs atm	0.4~0.6	1~2	3~5
WVTR	(2) g. $30\mu/m^2$ 24 hrs	40~80	20~40	15~30
	APPLICATION	BOTTLES SHEETS FILMS	BOTTLES SHEETS FILMS	FILMS SHEETS

TABLE 3

MECHANICAL PROPERTIES OF EVAL® RESINS

ITEM	UNITS	GRADE EP-F101A	EP-H101A	EP-E105A
TENSILE STRENGTH (YIELD)	kg/cm^2	790	680	600
TENSILE STRENGTH (BREAKING)	kg/cm^2	730	600	520
ELONGATION (YIELD)	%	8	8	7
ELONGATION (BREAKING)	%	230	250	280
YOUNG'S MODULUS	kg/cm^2	$2.7x10^4$	$2.4x10^4$	$2.1x10^4$
IZOD IMPACT STRENGTH	kg.cm/cm	1.7	1.5	1.0
ROCKWELL SURFACE HARDNESS	M	100	93	88
TABOR ABRASION	mg	1.2	2.0	2.2
STIFFNESS	kg/cm^2	$3.7x10^4$	$3.4x10^4$	$3.1x10^4$
BENDING STRENGTH	kg/cm^2	1220	1100	--
BENDING MODULUS	kg/cm^2	$3.6x10^4$	$3.2x10^4$	--

TABOR ABRASION AND STIFFNESS MEASURED WITH HOT PRESS MOLDED SPECIMENS. OTHER ITEMS MEASURED WITH INJECTION-MOLDED SPECIMENS. ALL ITEMS MEASURED AT 20°C, 65% R.H.

TABLE 4

THERMAL PROPERTIES OF EVAL® RESINS

ITEM	UNITS		GRADE EP-F101A	EP-H101A	EP-E105A
MAX. EXTRUSION TEMPERATURE	°C		240	240	250
MELTING POINT	°C		181	175	164
CRYSTALLIZATION TEMPERATURE	°C		161	151	142
GLASS TRANSITION POINT	°C		69	62	55
MELT FLOW INDEX	g/10 min.	190°C, 2160g	1.3	1.6	5.5
	g/10 min.	210°C, 2160g	3.1	3.8	13
COEFFICIENT OF LINEAR EXPANSION	1/°C	above glass trans. point	$11x10^{-5}$	$12x10^{-5}$	$13x10^{-5}$
		below glass trans. point	$5x10^{-5}$	$7x10^{-5}$	$8x10^{-5}$

In Table 2, one should note the high melting points for EVAL® polymers (for example, 181°C for EP-F and 164°C for EP-E). The specific gravity of 1.19 and 1.14 for EP-E is noteworthy.

In Table 3, the high stiffness values, especially for EVAL® EP-F, are about twice that for a polypropylene homopolymer.

Table 4 reports typical thermal properties of EVAL® resins. Crystallization temperatures of 161°C - 142°C become important in thermoforming by both melt and SPPF methods in multi-layer structures. For example, EVAL® EP-F can be used for melt forming, but EP-K or EP-E should be used for SPPF. This is because the maximum crystallization temperature of 161°C for EP-E is the same as the ideal SPPF temperature for polypropylene.

Over 95% of EVOH resins are processed in multi-layer coextruded systems. A vast majority of these systems are five-layer or greater and utilize polyolefin layers in a major portion of the structures. Nylon, PET and polycarbonate structure are also used. Frequently, nylon polycarbonate and polystyrene are used in combination with polyethylene or polypropylene in the same structure with EVOH resins.

EVAL® resins have little or no adhesion characteristics to most other polymers except nylon. Therefore, adhesive resins, usually modified polyolefins, have been developed for use in coextrusion applications. A typical coextruded structure is shown in Figure 6. While this is shown as five different materials, a minimum of three can be utilized. For example, A and E can be the same or different. B and D also can be the same or different. In addition, one or two regrind layers can be added between the outer layers and the adhesive layer. Thus, 6- and 7-layer structures are realized. With extrusion coating, coextrusion coating and laminating procedures, a larger number of layers are achieved.

FIGURE 6

TYPICAL MULTILAYER BARRIER STRUCTURE

(BALANCED STRUCTURE)

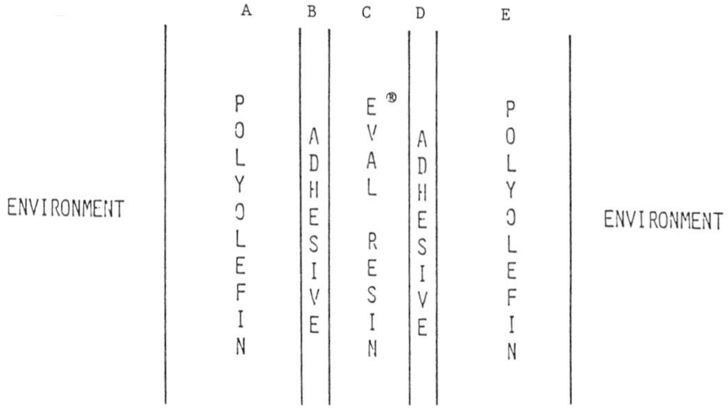

	A	B	C	D	E	
ENVIRONMENT	POLYOLEFIN	ADHESIVE	EVAL® RESIN	ADHESIVE	POLYOLEFIN	ENVIRONMENT

Figure 7 shows how different MVTR's can be utilized with EVOH resins to package either high or low moisture content foods. The higher MVTR material is placed on the higher humidity side. Conversely, the lower MVTR resin is placed on the lower humidity side. In these cases, combinations of polyolefins and polycarbonates or combinations of polyolefins and nylons can be used to control moisture content of the EVAL® resins and subsequently the oxygen transmission rate of the barrier layer. In this example, we have used the same material inside and outside but unbalanced the structure so that 80% of moisture barrier is toward the high moisture side. This results in an average R.H. of the EVOH at something less than the arithmetic mean of the inside and outside R.H. and thus develops higher oxygen barrier levels.

FIGURE 7

Barrier Moister Content Control

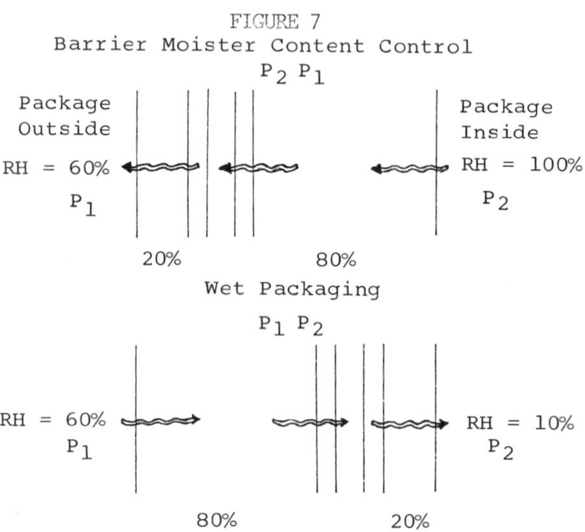

Examples of this procedure are shown in Table 5 wherein selected multi-layer film structures are listed and their respective oxygen barrier properties are reported under different inside and outside R.H. conditions.

TABLE 5

Oxygen Transmission Rate of EVAL® Composite Films (1)

Film Structure			Inside 100% RH				Inside 10% RH			
			Outside 65% RH		Outside 80% RH		Outside 65% RH		Outside 80% RH	
Outside mils	Middle mils	Inside mils	Middle % RH	O_2TR	Middle % RH	O_2TR	Middle % RH	O_2TR	Middle % RH	O_2TR
OPP (0.8)	EF-F (0.6)	LDPE (2.0)	84	0.11	91	0.24	34	0.01	40	0.02
OPP (0.8)	EF-F (0.6)	Ct'd PP (2.0)	80	0.08	89	0.19	41	0.02	49	0.02
PET (0.5)	EF-F (0.6)	LDPE (2.0)	70	0.04	83	0.10	57	0.02	69	0.04
ON (0.6)	EF-F (0.6)	LDPE (2.0)	66	0.04	81	0.08	63	0.03	77	0.06
Ct'd Nylon (0.8)	EF-E (0.8)	LDPE (2.0)	65	0.10	80	0.16	64	0.10	78	0.15
EF-XL (0.6)	---- ----	LDPE (2.0)	--	0.02	--	0.04	--	0.02	--	0.04

(1) cc/100 in^2/24 hrs./atm.

This information is utilized today in barrier packaging encompassing EVAL® resins as the premier barrier resin for applications including retort, hot-fill and aseptic packaging in film, pouches, bottles, thermoformed sheet and molded applications. Foods include tomato products, juices, sauces, barbeque sauces, salad dressings, apple products, cranberry products, meat, cheese flavorings and fish products. Other applications include pesticides, medical and pharmaceutical, and chemical products. These applications are expanding daily.

AUTHOR INDEX

COMPANY INDEX

SUBJECT INDEX

A

B

C

D

I

O

P

R

S

T

U

V

W

X